Lecture Notes in Mathematics

Edited by A. Dold and B. Eckmann

1248

 W9-ADT-870

Nonlinear Semigroups, Partial Differential Equations and Attractors

Proceedings of a Symposium held in
Washington, D.C., August 5–8, 1985

Edited by T. L. Gill and W. W. Zachary

Springer-Verlag
Berlin Heidelberg New York London Paris Tokyo

Editors

Tepper L. Gill
Department of Electrical Engineering
Howard University
Washington, D.C. 20059, USA

Woodford W. Zachary
Naval Research Laboratory
Washington, D.C. 20375, USA

Mathematics Subject Classification (1980): 34C35, 35A07, 35B30, 35K55, 35K57, 35K60, 35Q20, 47D05, 47H20

ISBN 3-540-17741-8 Springer-Verlag Berlin Heidelberg New York
ISBN 0-387-17741-8 Springer-Verlag New York Berlin Heidelberg

Library of Congress Cataloging-in-Publication Data. Nonlinear semigroups, partial differential equations, and attractors. (Lecture notes in mathematics; 1248) "Proceedings of the Symposium on Nonlinear Semigroups, Partial Differential Equations, and Attractors held at Howard University in Washington, D.C."– Pref. Bibliography: p. 1. Semigroups of operators–Congresses. 2. Differential equations, Partial–Congresses. 3. Nonlinear operators–Congresses. 4. Differentiable dynamical systems–Congresses. I. Gill, T.L. (Tepper L.), 1941-. II. Zachary, W.W., 1935-. III. Symposium on Nonlinear Semigroups, Partial Differential Equations, and Attractors (1985: Howard University) IV. Series: Lecture Notes in mathematics (Springer-Verlag); 1248.
QA3.L28 no. 1248 [QA329.8] 510 s [515.7'248] 87-9448
ISBN 0-387-17741-8 (U.S.)

© Springer-Verlag Berlin Heidelberg 1987
Printed in Germany

Printing and binding: Druckhaus Beltz, Hemsbach/Bergstr.
2146/3140-543210

PREFACE

This volume constitutes the proceedings of the Symposium on Nonlinear Semigroups, Partial Differential Equations, and Attractors held at Howard University in Washington, D.C. on August 5-8, 1985. This symposium was sponsored by a special grant from the President of Howard University, Dr. James E. Cheek. The original intention of the symposium organizers was to include a fairly narrow range of featured topics by a group of experts on some recent techniques developed for the investigation of nonlinear partial differential equations. However, it soon became clear that the dynamical systems approach to nonlinear problems interfaced significantly with many important branches of applied mathematics. As a consequence, the scope was increased to allow for a broader spectrum of research fields.

The local support committee for Washington, D.C. area universities consisted of Avron Douglis (University of Maryland), James Sandefur (Georgetown University), and Isom Herron (Howard University). We also thank Alicia Taylor for her extensive help with the organizational details of the conference and express our sincere gratitude to Mary McCalop for her typing and correcting services.

Tepper L. Gill
W. W. Zachary
Washington, D.C.
October 1986

TABLE OF CONTENTS

The following papers were presented at the conference, but could not be included in these proceedings:

Shui-Nee Chow and Reiner Lauterbach, On Bifurcation for
              Variational Problems

Milan Miklavčič, Stability for Semilinear Parabolic Equations in
              the Critical Case

S. Rankin, Existence for Semilinear Parabolic Equations in L^P when
              the Nonlinear Term Contains Derivatives.

Michael I. Weinstein, Remarks on the Dynamics of Singularity
              Formation For the Nonlinear Schrodinger Equation

PARTICIPANTS LIST

Symposium on Nonlinear Semigroups
Partial Differential Equations, and Attractors

HOWARD UNIVERSITY
Washington, D.C.

August 5-8, 1985

Adams, Charles
Department of Mathematics
Howard University
Washington, D.C. 20059

Asano, Chris
Department of Mathematics
Brown University
Providence, RI

Belbas, Stavros A
Department of Mathematics
University of Alabama
University, AL 35486

Bogdam, Victor M.
Department of Mathematics
Catholic University
Washington, D.C. 20064

Douglis, Avron
Department of Mathematics
University of Maryland
College Park, MD 20742

Evans, Lawrence E.
Department of Mathematics
University of Maryland
College Park, MD 20742

Goldstein, Jerome A.
Department of Mathematics
Tulane University
New Orleans, LA 70118

Anile, A.M.
Department of Physics
Virginia Polytechnic Institute
and State University
Blacksburg, VA 24061

Avrin, Joel
Department of Mathematics
and Computer Science
University of North Carolina
at Charlotte
Charlotte, NC 28223

Berger, Melvyn S.
Department of Applied Math.
University of Massachusetts
Amherst, MA 01003

Chow, Shui-Nee
Department of Mathematics
Michigan State University
East Lansing, MI 48824

Engler, Hans
Department of Mathematics
Georgetown University
Washington, D.C. 20057

Gill, Tepper L.
Department of Electrical Engineering
Howard University
Washington, D.C. 20059

Hale, Jack K.
Lefschetz Center for Dynamical
Systems, Division of Applied Math.
Brown University
Providence, RI 02912

Handy, Carlos R.
Department of Physics
University of Atlanta
Atlanta, GA 30314

Hughes, Rhonda J.
Department of Mathematics
Bryn Mawr College
Bryn Mawr, PA 19010

Kim, Young S.
Dept. of Physics & Astronomy
University of Maryland
College Park, MD 20742

Mallet-Paret, John
Div. of Applied Mathematics
Brown University
Providence, RI 02912

Massey, William A.
AT & T Bell Laboratories
600 Mountain Avenue
Murray Hill, NJ 07974

Rankin, Samuel
AFOSR/NM
Bolling Air Force Base
Washington, DC 20332

Reed, Michael
Department of Mathematics
Duke University
Durham, NC 27701

Sadosky, Cora S.
Department of Mathematics
Howard University
Washington, DC 20059

Sandefur, James
Department of Mathematics
Georgetown University
Washington, DC 20057

Herron, Isom H.
Department of Mathematics
Howard University
Washington, DC 20059

Johnson, Raymond
Department of Mathematics
University of Maryland
College Park, MD 20742

Lauterbach, Reiner
Institute für Mathematik
Universität Augsburg
Memmingen Str. 6
D-89 Augsburg
Federal Republic of Germany

Martin, Jr. Robert H.
Department of Mathematics
North Carolina State University
Raleigh, NC 27607

Miklavčič, Milan
Department of Mathematics
Michigan State University
East Lansing, MI 48824

Raphael, Louise A.
Department of Mathematics
Howard University
Washington, DC 20059

Russo, A.
Department of Physics
Virginia Polytechnic Institute
and State University
Blacksburg, VA 24061

Schechter, Eric
Department of Mathematics
Vanderbilt University
Box 21, Station B.
Nashville, TN 37235

Schmeelk, John
Department of Mathematics
Virginia Commonwealth Univ.
Richmond, VA 23211

Seidman, Thomas I.
Department of Mathematics
University of Maryland
Baltimore County
Catonsville, MD 21228

Srivastav, Ram P
Mathematical Research
U.S. Army Research Office
Triangle Park, NC 27709

Svendsen, E.C.
Dept. of Math. Sciences
George Mason University
Fairfax, VA 22030

Tippett, Jessie
Department of Mathematics
Howard University
Washington, DC 20059

Weinstein, Michael
Department of Mathematics
Princeton University
Princeton, NJ 08544

Wolfe, Peter
Department of Mathematics
University of Maryland
College Park, MD 20742

Schwartz, Ira
Naval Research Laboratory
Code 4520
Washington, DC 20375

Sesay, Mohamed
Department of Mathematics
Univ. of the District of Columbia
4200 Connecticut Avenue, NW
Washington, DC 20008

Steadman, Vernise
Department of Mathematics
Howard University
Washington, DC 20059

Szu, Harold
Naval Research Laboratory
Code 5709
Washington, DC 20375

Vogt, Andrew
Department of Mathematics
Georgetown University
Washington, DC 20057

Williams, Daniel
Department of Mathematics
Howard University
Washington, DC 20059

Zachary, Woodford W.
Naval Research Laboratory
Code 4603-S
Washington, DC 20375

CONVERGENCE PROPERTIES OF STRONGLY-DAMPED

SEMILINEAR WAVE EQUATIONS

Joel D. Avrin

Department of Mathematics
University of North Carolina at Charlotte
Charlotte, North Carolina 28223

ABSTRACT. We consider the strongly-damped nonlinear Klein-Gordon equation

$$u_{tt} + \alpha(-\Delta + \gamma)u_t + (-\Delta + m^2)u + \lambda|u|^{p-1}u = 0$$

over a domain Ω in $\mathbb{R}^3$. Let u^α be a solution of this equation with $\alpha > 0$. Aviles and Sandefur show that such solutions are unique, strong, and exist globally for any $p \geq 1$ and arbitrary initial data $u(0)$, $u_t(0) \in D(\Delta)$. We establish here, in the case of a bounded Ω, the existence of a weak global solution with $\alpha = 0$ and a subsequence α_k such that $\alpha_k \downarrow 0$ and $\lim_{k\to\infty} u^{\alpha_k} = v$ in $C([0,T]; L^2(\Omega))$ for any $T > 0$. We conclude with a few remarks concerning the difficulty of extending this result to the case $\Omega = \mathbb{R}^3$.

Consider the strongly-damped nonlinear Klein-Gordon equation

$$u_{tt} + \alpha(-\Delta + \gamma)u_t + (-\Delta + m^2)u + \lambda|u|^{p-1}u = 0, \quad u = u(x,t),$$

where x ranges over a domain Ω in $\mathbb{R}^3$. Here $\alpha,\gamma,m,\lambda \in \mathbb{R}$ with α, $\lambda > 0$ and γ, $m \geq 0$, while $\Omega = \mathbb{R}^3$ or a bounded domain in $\mathbb{R}^3$ with smooth boundary. In the latter case $\Delta = \sum_{j=1}^{3} \partial^2/(\partial x_j)^2$ is equipped with zero Dirichlet boundary conditions.

For arbitrary initial data $u(0)$, $u_t(0) \in D(\Delta)$, Aviles and Sandefur ([2]) established existence and uniqueness of global strong solutions of (1) for all integers $p \geq 1$. This contrasts sharply with the undamped case $\alpha = 0$, where such a result is only known for $p \leq 3$; global existence results for $p > 3$ and $\alpha = 0$ are only known for weak solutions or for small initial data. A convenient summary of the undamped case can be found in [3].

Here we examine the following question posed by Aviles and Sandefur: let $p > 3$ and, for each $\alpha > 0$, let u^α be the strong global solution of (1) described above. Let v be a global weak solution of (1) with $\alpha = 0$. Does $\lim_{\alpha \to 0} u^\alpha$ exist in some strong sense? What is the relation between u^α and v ?

The purpose of this paper is to study this question in the case of bounded Ω; we will prove the following result:

Theorem. Consider equation (1) in the bounded domain case outlined above, and for each $\alpha > 0$ let u^α be as above. Then there exists a global weak solution v of (1) with $\alpha = 0$, and a sequence of positive numbers α_k such that $\alpha_k \downarrow 0$ as $k \to +\infty$, and $\lim_{k \to \infty} u^{\alpha_k} = v$ in $C([0,T]\,;\,L^2(\Omega))$ for all $T > 0$.

Proof. For each $\alpha > 0$ set

$$E^\alpha(t) = \frac{1}{2}\left[||u_t^\alpha||_2^2 + ||Bu^\alpha||_2^2 + \frac{2\lambda}{p+1}\int_\Omega |u^\alpha(x,t)|^{p+1}dx\right], \tag{2}$$

where $B^2 = -\Delta + m^2$. A standard energy argument (see e.g. [2]) shows that

$$\frac{d}{dt}\,E^\alpha(t) = -\alpha\left[||\nabla u_t^\alpha||_2^2 + \gamma||u_t^\alpha||_2^2\right]. \tag{3}$$

Consequently, $E^\alpha(t) \leqslant E^\alpha(0) \equiv E_0$ so that, in particular,

$$||u_t||_2^2 + ||Bu^\alpha||_2^2 \leqslant 2E_0. \tag{4}$$

Thus, $||u^\alpha_t||_2$ is uniformly bounded on each interval $[0,T]$ for any $T > 0$; meanwhile, the embedding $W_0^{1,2}(\Omega) \to L^2(\Omega)$ is compact ([1]), so by the Arzela-Ascoli theorem, for any sequence $\{\alpha_k\}_{k=1}^\infty$ such that $\alpha_k \downarrow 0$, there exists a subsequence (which we also denote by $\{a_k\}$) and a function $v \in C([0,T]\,;\,L^2(\Omega))$ such that

$$\lim_{\alpha_k \to 0} ||u^{\alpha_k}(t) - v(t)||_2 = 0 \tag{5}$$

uniformly on $[0,T]$. By a standard diagonal sequence argument, we can choose $\{\alpha_k\}$ such that (5) holds for any $T > 0$.

By the standard reduction-of-order procedure (see e.g. [3]), we rewrite (1) as the system

$$\emptyset_t^\alpha(t) = -iA\emptyset^\alpha(t) + F_1(t) + F_2(t) \tag{6}$$

on $D(B) + L^2(\Omega)$, where $\emptyset = \langle u, u_t\rangle$, $F_1(t) = \langle 0, \lambda|u^\alpha|^{p-1}u\rangle$, $F_2(t) = \langle 0, -\alpha(-\Delta+\gamma)u_t\rangle$, and

$$A = i\begin{pmatrix} 0 & I \\ -B^2 & 0 \end{pmatrix}. \tag{7}$$

As noted in [3], A is self-adjoint on $D(B) + L^2(\Omega)$ and generates the unitary group

$$e^{-itA} = W(t) = \begin{pmatrix} \cos(tB) & B^{-1}\sin(tB) \\ -B\sin(tB) & \cos(tB) \end{pmatrix}. \tag{8}$$

From (6) and (8) we see that $\emptyset^\alpha$ satisfies the variation-of-constants formula

$$\emptyset^\alpha(t) = W(t)\emptyset^\alpha(0) + \int_0^t W(t-s)F_1(s)\,ds$$
$$+ \int_0^t W(t-s)F_2(s)\,ds\,,\qquad(9)$$

hence the first component u^α of $\emptyset^\alpha$ satisfies

$$u^\alpha(t) = \cos(tB)f + B^{-1}\sin(tB)g$$
$$+ \int_0^t B^{-1}\sin[(t-s)B][-\lambda|u^\alpha(s)|^{p-1}u^\alpha(s)]\,ds$$
$$+ \int_0^t B^{-1}\sin[(t-s)B][-\alpha(-\Delta+\gamma)u_t^\alpha(s)]\,ds\qquad(10)$$

where $f = u^\alpha(0)$ and $g = u_t^\alpha(0)$. Let $w \in C_o^\infty(\Omega)$, then

$$(u^\alpha(t)\,,\,w) = (\cos(tB)f\,,\,w) + (B^{-1}\sin(tB)g\,,\,w)$$
$$+ \int_0^t (-\lambda|u^\alpha(s)|^{p-1}u^\alpha(s)\,,\,B^{-1}\sin[(t-s)B]\,w)\,ds$$
$$+ \int_0^t (-\alpha)(u_t(s)\,,\,(-\Delta+\gamma)B^{-1}\sin[(t-s)B]\,w)\,ds\,.\qquad(11)$$

By (2), the last term on the right-hand side of (11) is bounded in absolute value by $\alpha T(2E_o)^{1/2}||(-\Delta+\gamma)w||_2$, meanwhile $-\lambda|u^{\alpha k}|^{p-1}u^{\alpha k} \to -\lambda|v|^{p-1}v$ in $L^1(\Omega\times[0,T])$ by arguments similar to those found in [3] and [4]. Hence, if we replace α by α_k in (11) and let $\alpha_k \downarrow 0$, we have

$$(v(t)\,,\,w) = (\cos(tB)f\,,\,w) + (B^{-1}\sin(tB)g\,,\,w)$$
$$+ \int_0^t (-\lambda|v(s)|^{p-1}v(s)\,,\,B^{-1}\sin[(t-s)B]w)\,ds\,.\qquad(12)$$

We can now differentiate both sides of (12) as in [3] or [4] to obtain

$$\frac{d^2}{dt^2}(v(t)\,,\,w) + (v(t)\,,\,B^2w) = (-\lambda|v(t)|^{p-1}v(t)\,,\,w)$$

$$(v(0)\,,\,w) = (f\,,\,w),$$

$$\frac{d}{dt}(v(t)\,,\,w)\Big|_{t=0} = (g\,,\,w),\qquad(13)$$

for all $w \in C_o^\infty(\Omega)$ on $[0,T]$. As this holds for all $T > 0$, we conclude that v is a global weak solution of (1) with $\alpha = 0$; this completes the proof of the theorem.

The difficulty of extending this theorem to the case $\Omega = \mathbb{R}^3$ lies in extracting a convergent subsequence $u^{\alpha k}$. The way this problem is handled in [3], [4] is to exploit the finite propagation speed of regularized solutions u^n of (1) with $\alpha = 0$ (where $f,g \in$

$C_o^\infty(\Omega)$). This allows the application of the usual Sobolev compact embedding theory for bounded domains. But here the approximating solutions u^α do not have finite propagation speed. One can see this by taking the Fourier transform of both sides of (1) with $\alpha > 0$ and $\lambda \equiv 0$, and then applying the Paley-Wiener theory.

BIBLIOGRAPHY

1. Adams, R.A., Sobolev Spaces, Academic Press, New York, 1975.

2. Aviles, P., and Sandefur, J., "Nonlinear second order equations with applications to partial differential equations", J. Diff. Equations, to appear.

3. Reed, M., Abstract Non-Linear Wave Equations, Springer-Verlag, Berlin/Heidelberg/New York, 1976.

4. Strauss, W., "On weak solutions of semilinear hyperbolic equations", Anais Acad. Brazil Ciencias, 42 (1970), pp. 645-651.

NUMERICAL SOLUTION OF CERTAIN NONLINEAR PARABOLIC

PARTIAL DIFFERENTIAL EQUATIONS

S. A. Belbas
Department of Mathematics
University of Alabama
University, AL 35486

ABSTRACT

This paper contains the extension to the parabolic case of methods developed in [1] and [2] (for the elliptic problem) for the numerical solution of nonlinear partial differential equations arising in stochastic optimal control.

1. <u>General statement of the problems and general results on convex-diagonally dominant equations.</u>

In this section, we review briefly results associated with discrete Bellman equations, variational and quasi-variational inequalities, involving diagonally dominant matrices. All these problems arise as discretized versions of nonlinear partial differential equations, which in turn can be obtained as the optimality conditions (dynamic programming conditions) associated with the optimal control of diffusion processes. These problems can be found in [3,4,6] and will not be repeated here. For applications to other problems in Mechanics and Physics, cf. [5].

The discretized versions of the nonlinear elliptic partial differential equations of stochastic optimal control have been studied in [1,2]. Here, we present the corresponding results for the associated parabolic problems. For simplicity, we will present only results for the Bellman equation; the necessary modification for variational and quasi-variational inequalities are easy to infer from [1,2].

The general forms of discrete Bellman equations, variational inequalities, or quasi-variational inequalities, are as follows:

Bellman equation:

$$\max_{a}\{A_{ij}^{a}x_{j} - f_{i}^{a}\} = 0 , \tag{1.1}$$

Variational inequality with two-sided obstacle:

$$\max\{\min\{A_{ij}x_{j} - f_{i}, x_{i} - \phi_{i}\}, x_{i} - \psi_{i}\} = 0, \tag{1.2}$$

Quasi-variational inequality of switching type:

$$\max\{A^a_{ij}\, x^a_j - f^a_i,\ x^a_i - (M^a_x)_i\} = 0, \tag{1.3}$$

where

$$(M^a x)_i = \min_{b:b\neq a}\{x^b_i + k(a,b)\}. \tag{1.4}$$

Quasi-variational inequality of impulse-control type:

$$\max\{A_{ij}\, x_j - f_i,\ x_i - (Mx)_i\} = 0, \tag{1.5}$$

where

$$(Mx)_i = \inf_{j\varepsilon K_i}\{x_j + k(i,j)\};$$

K_i is a cone of indices j, depending on the index i. $\tag{1.6}$

We make one of the following assumptions about the matrices A^a (or A):

(A1). (Strong diagonal dominance).

$$A^a_{ii} > 0,\quad \sum_{j:j\neq i} A^a_{ij} > -A^a_{ii},$$

$$A^a_{ij} \le 0 \text{ for } i \neq j.$$

(A2). (Weak diagonal dominance).

$$A^a_{ii} > 0,\quad \sum_{j:j\neq i} A^a_{ij} \ge -A^a_{ii},$$

$$A^a_{ij} \le 0 \text{ for } i \neq j.$$

Together with conditions (A1), (A2), we consider the corresponding normalized conditions:

(A1$'$). (A1) holds, and in addition $A^a_{ii} = 1$.

(A2$'$). (A2) holds, and in addition $A^a_{ii} = 1$.

Let us define the matrices B^a by

$$B^a_{ij} = \delta_{ij} - A^a_{ij} \tag{1.7}$$

where δ_{ij} denotes Kronecker's delta.

If the matrices A^a have properties (A1$'$) or (A2$'$), then the matrices B^a have the following corresponding properties:

(B1).

$$B^a_{ii} = 0,$$

$$\sum_{j:j\neq i} B^a_{ij} < 1,$$

$$B^a_{ij} \ge 0 \text{ for } i \neq j.$$

(B2).
$$B_{ii}^a = 0,$$

$$\sum_{j:j\neq i} B_{ij}^a \leq 1,$$

$$B_{ij}^a \geq 0 \text{ for } i \neq j.$$

These conditions are related to the concept of an M-matrix, cf. [7].

Under condition (B1), it is easy to see that the affine mapping

$$x \longmapsto B^a x + f^a, \ \mathbb{R}^N \longmapsto \mathbb{R}^N \qquad (1.8)$$

is a contraction.

In order to obtain a similar result for the mapping (8) in case B^a satisfies condition (B2), we need the following non-degeneracy condition.

(Cr). It is possible to partition the coordinates of $\mathbb{R}^N$ into r disjoint subsets $C_0, C_1, C_2, \ldots, C_r$, with $C_0 = \emptyset$, with the following property: if $z \in \mathbb{R}^N$, $\|z\| = 1$, $\|z_i\| < 1$ for $i \in C_0 \cup C_1 \cup \ldots \cup C_k$, then $|(B^a z)_i| < 1$ for $i \in C_{k+1}$. This property must hold for all $k = 0, 1, \ldots, r - 1$. ▱

Under conditions (B2),(Cr), the rth power of the mapping (1.8) is a contraction.

In a similar way, under conditions (B1) or (B2,Cr) we can show that the operators associated with variational and quasi-variational inequalities and Bellman equations also have the property that either the operator itself or a power of that operator is a contraction. These operators are defined as follows:

Bellman equation:

$$(Tx)_i = \min_a \{B_{ij}^a x_j + f_i^a\}. \qquad (1.9)$$

Two-sided variational inequality:

$$(Tx)_i = \min\{\max\{B_{ij} x_j + f_i, \phi_i\}, \psi_i\}. \qquad (1.10)$$

Quasi-variational inequality of switching type:

$$(T^a x)_i = \min\{B_{ij}^a x_j^a + f_i^a, x_i^a - (M_x^a)_i; \qquad (1.11)$$

where $(M^a x)_i$ is given by (1.4).

Quasi-variational inequality of impulse-control type:

$$(Tx)_i = \min\{B_{ij} x_j + f_i, x_i - (Mx)_i\} \qquad (1.12)$$

where $(Mx)_i$ is given by (1.6).

In all cases, the original problem is reduced to a fixed-point problem of the form

$$(Tx)_i = x_i \qquad (1.13)$$

for problems (1.1,1.2,1.5), and

$$(T^a x)_i = x_i \qquad (1.14)$$

for problem (1.3).

For the quasi-variational inequality of switching control, we make the following assumption:

(L). The set of constants $\{k(a,b): 1 \leq a,b \leq m\}$ contains no loop of zero cost, i.e., no family $a_1,a_2,\ldots,a_n$ such that
$$k(a_1,a_2) = k(a_2,a_3) = \ldots = k(a_{n-1},a_n) = k(a_n,a_1) = o. \quad \boxtimes$$

For the quasi-variational inequality of impulsive control, we make the assumption:

(K). $k(x,\xi)$ satisfies $k(x,\xi) \geq k_0$ for some constant $k_0 > 0$, and $\sum\limits_j A_{ij} k_0 + f_i \geq 0.$
$$\boxtimes$$

Then, from the results of [1,2], we have that, under conditions (B1) or (B2,C_r), and with the additional assumptions (L),(K) for the quasi-variational inequalities of switching and impulsive type, respectively, the iterates $x_{(k)} = T^k x_0$ or $x_{(k)}^a = T^a(x_{(k-1)})$, for (1.3) or (1.4) respectively, converge to the solutions of (1.1),(1.2),(1.3),(1.5), with geometric rate of convergence. (For the quasi-variational inequality of impulsive type, we must take $x_0 \geq -k_0$; for the other problems, x_0 is arbitrary.)

2. Iteration scheme for the parabolic Bellman equation.

We consider the following discrete "parabolic" Bellman equation:

$$\frac{dx_i}{dt} + \max_{a \in \tilde{A}}\{A_{ij}^a x_j - f_i^a\} = 0; \quad x_i(0) = x_i^0. \tag{2.1}$$

We shall make one of the following assumptions about the matrices A^a: A1,A2,A1' or A2' of section 1.

We note that the assumption $A_{ii}^a = 1$ is essentially equivalent to $A_{ii}^a > 0$; indeed, if $A_{ii}^a > 0$, we can rescale the coefficients in the system (2.1) so that the resulting new system is equivalent to the original system, and for the new system we have $A_{ii}^a = 1.$

We rewrite the system (2.1) in the form

$$\frac{dx_i}{dt} + x_i - \min_{a \in \tilde{A}}\{B_{ij}^a x_j + f_i^a\} = 0; \quad x_i(0) = x_i^0, \tag{2.2}$$

where $B_{ij}^a = \delta_{ij} - A_{ij}^a.$

We consider the following iteration scheme for the solution of (2.2):

$x_i^{(0)}(t)$ is an arbitrary continuous function on $[0,T]$, satisfying $x_i^{(0)}(0) = x_i^0$; for $k = 0,1,2, \ldots,$ $x^{(k+1)}(t)$ is the solution of

$$\frac{dx_i^{(k+1)}}{dt} + x_i^{(k+1)} = \min\{B_{ij}^a x_j^{(k)} + f_i^a\}; \quad x_i^{(k+1)}(0) = x_i^0. \tag{2.3}$$

Let S be a mapping from $\mathbb{R}^N$ into itself, defined by

$$(Sx)_i = \min_{a \in \tilde{A}}\{B^a_{ij} x_j + f_i\}. \tag{2.4}$$

For any $t_1, t_2 \in [0,T]$, $t_1 < t_2$, and for any $x^0 \in \mathbb{R}^N$, let $\mathcal{G}^{x^0}$ denote the operator from $C^0(t_1,t_2;\mathbb{R}^N)$ into itself, defined by

$$(\mathcal{G}^{x^0} x)(t) = x^0 + \int_{t_1}^{t} S(x(t)) dt . \tag{2.5}$$

We have the following:

Lemma 2.1. Consider the following properties of the operator S:

(B). There exists a constant K such that S maps the ball $\{x \in \mathbb{R}^N : \|x\| \le K\}$ into itself.

(C). The operator S is Lipschitz, i.e., $\|Sx - Sy\| \le c\|x - y\|$ for $x,y \in \mathbb{R}^N$, for some constant $C > 0$.

Then, for the operator $\mathcal{G}^{x^0}$ we have the following:

Under condition (B), the operator $\mathcal{G}^{x^0}$ maps the ball

$$\{x \in C^0(t_1,t_2;\mathbb{R}^n): \|x - x^0\|_\infty \le K\}$$

into the ball

$$\{x \in C^0(t_1,t_2;\mathbb{R}^n): \|x - x^0\|_\infty \le K(t_2 - t_1)\}.$$

Under condition (C), the operator $\mathcal{G}^{x^0}$ satisfies the inequality

$$\|\mathcal{G}^{x^0} x_1 - \mathcal{G}^{x^0} x_2\|_\infty \le C(t_2 - t_1)\|x_1 - x_2\|_\infty.$$

The proof is quite straightforward. As a corollary, we have that, when $(t_2 - t_1)$ is sufficiently small, then the operator $\mathcal{G}^{x^0}$ is a contraction on the ball

$$\{x \in C^0(t_1,t_2;\mathbb{R}^N): \|x - x^0\|_\infty \le K\}.$$

Thus, under conditions (B) and (C) for the operator S, we can obtain a solution of the equation (2.1) on a small time interval by the iteration scheme

$$x^{(k)} = \mathcal{G}^{x^0}(x^{(k-1)});$$
$$x^{(0)} \in C^0(t_1,t_2;\mathbb{R}^N),$$
$$x^{(0)}(0) = x^0, \quad \|x^{(0)} - x^0\|_\infty \le K. \tag{2.6}$$

Now, we turn to the discretized problem that involves discretization in time, as well as in space.

We consider the discretization scheme:

$$\frac{x_{i,k} - x_{i,k-1}}{\tau} + \max\{A^a_{ij} x_{j,k-1} - f^a_{i,k-1}\} = 0, \tag{2.7}$$

where the index k corresponds to the time variable, and the indices i,j correspond to the space variables.

The system (2.7) can be written in the form

$$\max\{D^a_{ik,j\ell} x_{j\ell} - \tilde{f}^a_{ik}\} = 0, \qquad (2.8)$$

where

$$D^a_{ik,ik} = \frac{1}{\tau},$$

$$D^a_{ik,i(k-1)} = -\frac{1}{\tau} + A^a_{ii},$$

$$D^a_{ik,j(k-1)} = A^a_{ij} \quad \text{when } j \neq i,$$

$$D^a_{ik,j\ell} = 0 \quad \text{in all other cases,}$$

$$\tilde{f}^a_{ik} = f^a_{i(k-1)}. \qquad (2.9)$$

It is readily seen from (2.9) that, when

$$\tau < \frac{1}{\max\limits_a A^a_{ii}}, \qquad (2.10)$$

then the matrices $[D^a_{ik,j\ell}]$ have the property of strict diagonal dominance.

Dividing the coefficients $D^a_{ik,j\ell}$ and the constants $\tilde{f}^a_{ik}$ by $D^a_{ik,ik}$, we obtain from (2.8) a new system:

$$\max\{E^a_{ik,j\ell} x_{j\ell} - g^a_{ik}\} = 0, \qquad (2.11)$$

$$E^a_{ik,j\ell} = \frac{D^a_{ik,j\ell}}{D^a_{ik,ik}}, \; g^a_{ik} = \frac{\tilde{f}^a_{ik}}{D^a}. \qquad (2.12)$$

We have the following:

<u>Lemma 2.2.</u> Under condition (2.10), the matrices $[E^a_{ik,j\ell}]$ satisfy conditions (N1) or (N2) if the matrices $[A^a_{ij}]$ satisfy conditions (M1) or (M2). ▨

Consequently, Proposition (1.1) applies to system (2.11).

Next, we consider the following discretization scheme:

$$\frac{x_{i,k} - x_{i,k-1}}{\tau} + \max\{A^a_{ij} x^a_{j,k} - f^a_{i,k}\} = 0. \qquad (2.13)$$

The system (2.13) can be written as

$$\max\{F^a_{ik,j\ell} x_{j\ell} - f^a_{ik}\} = 0 \qquad (2.14)$$

with

$$F^a_{ik,ik} = \frac{1}{\tau};$$

$$F^a_{ik,j(k-1)} = \frac{\tau}{1+\tau} A^a_{ij}, \quad \text{when } j \neq i;$$

$$F^a_{ik,i(k-1)} = \frac{1}{1+\tau} + \frac{\tau}{1+\tau} A^a_{ii},$$

$$F^a_{ik,j\ell} = 0 \quad \text{in all other cases.} \qquad (2.15)$$

Again, it is easy to see that:

Proposition 2.3. If the matrices $[A_{ij}^a]$ satisfy condition (N1) or (N2), then the matrices $[F_{ik,j\ell}^a]$ satisfy condition (M1) or (M2), respectively, for any value of τ.

▨

Consequently, for the discretization scheme (2.13), the results of section 1.1 apply.

3. <u>Stability</u> <u>and</u> <u>periodicity</u> <u>for</u> <u>the</u> <u>discrete</u> <u>Bellman</u> <u>equation</u>.

We observe that we can always assume that

$$\max\{-f_i^a\} = 0. \tag{3.1}$$

Indeed, let $\tilde{x}_i$ be the (unique) solution of the stationary Bellman equation

$$\max_{a\varepsilon\tilde{A}}\{A_{ij}^a \tilde{x}_j - f_i^a\} = 0 \tag{3.2}$$

so that, by defining $\hat{f}_i^a = -A_{ij}^a \tilde{x}_j + f_i^a$ we get (3.1). Then the initial value problem (2.1) is equivalent to

$$\frac{d\hat{x}_i}{dt} + \max_{a\varepsilon\tilde{A}}\{A_{ij}^a \hat{x}_j - \hat{f}_i^a\} = 0; \quad \hat{x}_i(0) = \hat{x}_i^0, \tag{3.3}$$

where $\hat{x}_j = x_j - \tilde{x}_j$, $\hat{x}_i^0 = x_i^0 - \tilde{x}_i$, and $\hat{f}_i^a$ is given above.

We have the following:

Proposition 3.1. Under condition (A1), and assuming $\hat{x}_i^0 \geq 0$, the solution $\hat{x}(t)$ of problem (3.3) converges to zero.

<u>Proof</u>. For every i, let a_i be the index in $\tilde{A}$ such that $\hat{f}_i^a = 0$. Then, we have

$$\frac{d\hat{x}_i}{dt} \leq -A_{ij}^{a_i} \hat{x}_j, \tag{3.4}$$

from which

$$|\hat{x}_i(t)| \leq |\hat{x}_i^0|e^{-\lambda t}, \tag{3.5}$$

where $\lambda = \min_{a\varepsilon\tilde{A}} \lambda^a$, and λ^a is the smallest eigenvalue of A^a. This shows the asymptotic stability property.

▨

Now we consider the case in which f_i^a depends on time, and is a smooth periodic function of t,

$$f_i^a(t+T) = f_i^a(t), \quad \text{for all } t > 0, \tag{3.6}$$

while the coefficients of the matrices A^a remain constant. Consider the problem:

$$\frac{dx_i}{dt} + \max_{a\varepsilon\tilde{A}}\{A_{ij}^a x_j - f_i^a\} = 0; \quad x_i(0) = x_i(T). \tag{3.7}$$

We assume condition (A1) throughout.

Let S denote the Poincaré map associated with (3.7), i.e., for $x \in \mathbb{R}^N$, $Sx = y(T)$ where $y(t)$ is the solution of the initial value problem

$$\frac{dy_i}{dt} + \max_{a \in \tilde{A}}\{A_{ij}^a y_j - f_i^a\} = 0; \quad y_i(0) = x_i. \tag{3.8}$$

The problem of finding a solution to (3.7) is equivalent to finding a fixed point of the mapping S.

We note that we can assume, without loss of generality, that $f_i^a(t) \geq 0$ for $t \in [0,T]$, for all $a \in \tilde{A}$ and $1 \leq i \leq N$ (cf. also [1,2]).

For fixed $a \in \tilde{A}$, let $y^a(t)$ be a periodic solution of

$$\frac{dy_i^a}{dt} + A_{ij}^a y_j^a - f_i^a = 0; \quad y_i^a(0) = y_i^a(T). \tag{3.9}$$

Such a solution exists, since, under condition (A1), the only solution of the corresponding homogeneous system $\dfrac{dy_i^a}{dt} + A_{ij}^a\, y_j^a = 0$ with periodic conditions $y_i^a(0) = y_i^a(T)$ is the zero solution, so that we can invoke standard results concerning the existence of periodic solutions of linear systems.

We have:

__Lemma 3.1.__ The solution $y_i^a(t)$ defined above satisfies $y_i^a(t) \geq 0$.

__Proof.__ Suppose that $i, t_1 \in [0,T]$ are such that

$$y_i^a(t_1) = \min_{\substack{t \in [0,T] \\ 1 \leq j \leq N}} y_j^a(t),$$

and assume $y_i^a(t_i) < 0$.

Then

$$\frac{d}{dt} y_i^a(t_1) = 0 \tag{3.10}$$

and

$$A_{ij}^a y_j^a = y_i^a - B_{ij}^a y_j^a \leq y_i^a(1 - \sum_j B_{ij}^a) < 0. \tag{3.11}$$

This inequality is true because $B_{ij}^a \geq 0$ and $\sum_j B_{ij}^a < 1$. (3.10) and (3.11) imply that $f_i^a < 0$, which contradicts our earlier assumption that $f_i^a \geq 0$. ▨

__Lemma 3.2.__ The mapping S maps the set

$$C = \{x \in \mathbb{R}^N: 0 \leq x_i \leq y_i^a(0)\}$$

into itself.

__Proof.__ Using an argument based on the discrete maximum principle, as before, we can show that the solution of the initial value problem (3.8) is an increasing function of the initial data. Consequently, when $0 \leq x_i \leq y^a(0)$, we obtain $0 \leq y(t) \leq y^a(t)$,

13

thus

$$0 \leq y(T) = S x \leq y^a(T) = y^a(0).$$

Proposition 3.2. Under condition (Al), the system (3.7) has a unique solution.

Proof. Existence of solution follows from lemma 3.2 and Brouwer's fixed point theorem.

For the uniqueness, assume that x^1 and x^2 are two distinct solutions of (3.7), and let i, t_1 be such that

$$0 = x_i^1(t_1) - x_i^2(t_1) = \sup_{\substack{0 \leq t \leq T \\ 1 \leq j \leq N}} |x_j^1(t) - x_j^2(t)|. \tag{3.12}$$

Then

$$\frac{dx_i^1(t_1)}{dt} = \frac{dx_i^2(t_1)}{dt},$$

and

$$\frac{dx_i^2(t_1)}{dt} + A_{ij}^{a_2} x_j^2 - f_i^{a_2} = 0, \text{ for some } a_2,$$

while

$$\frac{dx_i^2(t_2)}{dt} + A_{ij}^{a_2} x_j^2 - f_i^{a_2} \leq 0.$$

Thus, by subtracting the two relations above,

$$A_{ij}^{a_2}(x_j^1 - x_j^2) \leq 0$$

or

$$0 \geq (x_i^1 - x_i^2) - B_{ij}^{a_2}(x_j^1 - x_j^2) \geq (x_i^1 - x_i^2)(1 - \sum_j B_{ij}^{a_2}) > 0,$$

which is a contradiction; thus $x_i^1(t) \leq x_i^2(t)$; the symmetric inequality yields uniqueness.

Acknowledgement: This work was supported by a summer research grant from the University of Alabama.

REFERENCES

[1] BELBAS, S.A., I.D. MAYERGOYZ, Applications des méthodes du point fixe aux equations de Bellman discrètes et a des inéquations quasi-variationnelles discrètes, C. R. Acad. Sci. Paris, 299 (1984), 233-236.

[2] BELBAS, S.A., I.D. MAYERGOYZ, detailed paper, to appear.

[3] BENSOUSSAN, A., J.L. LIONS, Applications des inéquations variationnelles en contrôle stochastique, Dunod, Paris, 1978.

[4] BENSOUSSAN, A., J.L. LIONS, Contrôle impulsionnel et inéquations quasi-variationnelles, Dunod, Paris, 1982.

[5] DUVAUT, G., J.L., LIONS, Inequalities in mechanics and physics, Springer-Verlag, Berlin, 1976.

[6] LENHART, S.M., S.A. BELBAS, A system of nonlinear partial differential equations arising in stochastic control with switching costs, SIAM J. Appl. Math., 43, 3, (1983), 465-475.

[7] VARGA, R.S., Matrix iterative analysis, Prentice-Hall, Englewood Cliffs, N.J., 1962.

The Explicit Solution of Nonlinear Ordinary
and Partial Differential Equations
I. Conceptual Ideas

by

M.S. Berger*
University of Massachusetts
Amherst, MA 01003

For many years, solutions of non-linear equations both ordinary and partial, have been attempted via integration by quadrature. This means that by making a succession of changes of variables, the problem at hand can be explicitly solved by a simple integration. Sometimes these changes of variables are quite elaborate as, for example, in the inverse scattering method for treating the initial value problem for the Korteweg-de Vries equation:

$$u_t + uu_x + u_{xxx} = 0.$$

However, in all cases integration by quadrature and its modern developments can be reduced to linearization procedures.

In this article I would like to discuss some important new developments in the explicit solution of nonlinear differential equations recently developed by me that are independent of such linearization ideas. The equations treated relate to applied mathematics in a special way since one might say that the nonlinear equations have a special simplicity not possessed by more general nonlinear systems. In fact, in this article I will point out the exact nature of this simplicity in the Riccati equation which cannot be integrated by quadrature except in special cases (c.f., G.N.

*Research partially supported by the N.S.F.

Watson, Theory of Bessel Functions, Cambridge University Press, Chapter 3).

Our idea is to attempt to extend the notion of eigenfunction expansion for linear equations to a nonlinear context. However, our idea is based on rigorous mathematical ideas and so to date has been limited to the study of a few classic cases, as will be described below. This paper is a combination of conceptual and computational ideas. I treat these two directions at the same time, via explicit examples instead of a general theory which might seem too obscure.

1. The Basic Ideas

The fundamental idea in our work is to regard an equation of the form:

$$Au = f, \quad f \text{ prescribed}, \tag{1}$$

as an operator equation acting between the domain of A, which we assume will be a Banach Space X, and its range, another Banach Space denoted by Y. Fundamental in our work is the case in which A has no inverse, so that the solution of the equation (1); 1) may or not exist and 2) may not possess unique solutions.

We attempt to study Eq. (1) by reducing it to a canonical form, by determining a "global normal form" for the operator. We determine this global normal form by attempting to determine smooth, global changes of coordinates h_1, h_2, acting separately on the domain X and the range Y. Pictorially we describe our situation as this:

$$\tag{2}$$

C = canonical non-linear map

The canonical map C in question is the global normal form desired and in the simplest case the question reduces to the problem, can we diagonalize an operator A by global changes of coordinates h_1, and h_2? In case A is an N × N matrix acting between N-dimensional vector spaces, the notion of conjugacy of linear mappings provides linear matrices h_1, and h_2 (different in general), such that every matrix can be diagonalized. The canonical map C can thus always be chosen to be a diagonal map with zeros and ones on the diagonal. The number of ones being called the rank of the matrix.

In the case when A is a non-linear differential operator, the situation is of course, more complex, but the mathematics is available to address the issues involved. The spaces X and Y will generally be chosen to be Sobolev spaces of functions that generally incorporate the desired boundary conditions in their definition. In the diagonal maps the associated canonical map D will be a diagonal map acting between orthonormal bases for these functional spaces. In this way, for example in ordinary differential equations, Fourier series play a crucial role.

The diagram (2) does occur in a non-linear context in terms of the geometry of singularities. We are going to find our new methods of explicit solution by extending the notions of singularity theory in two ways. First we shall choose the mappings to be infinite dimensional, and secondly, we shall extend the theory of singularities to a global context. Thus we regard A as being defined globally as an operator between function spaces. The effect of this approach is that bifurcation phenomena are explicitly brought into the "integrability process". Bifurcation is of course a nonlinear phenomenon and thus our ideas will be able to incorporate general nonlinear phenomena with explicit solutions of nonlinear problems.

From another point of view our global study of Eq. (1) is a study of a nonlinear Fredholm alternative for the operator equation in question. Such an alternative is extremely hard to find but has arisen in numerous examples in differential geometry, the periodic solutions of Hamiltonian systems, etc. . . Before proceeding further it is important to write down a number of explicit cases that we shall discuss in the sequel. One case will be the equation:

$$\frac{dy}{dx} + P_k(y) = f(x). \tag{3}$$

Here, $P_k(y)$ is a given polynomial of degree k, and f(x) is a given inhomogeneous term in the equation. We generally regard f(x) as a T-periodic function of x and search for T-periodic solutions of (3). Another case will be the equation:

$$\Delta u + G(x,u) = f(x) \qquad \text{in} \qquad \Omega \tag{4}$$

$$u = 0 \qquad \text{on} \quad \partial\Omega.$$

Here Ω is a bounded domain in $\mathbb{R}^N$, $\partial\Omega$ is its boundary, f(x) is a given inhomogeneous term, and G(x,u) is the given nonlinearity.

In this article I discuss the equation (3). Equation (4) is treated elsewhere.

2. Invariant Properties for A.

In order to discuss our ideas on explicit solutions with some degree of generality it is of great interest to discuss the invariative properties of the operator A. These properties will be so chosen that they are independent of changes of coordinates in the Banach space context. These properties will have the additional virtue of "stability". This means that when the operator A changes to $\tilde{A}$ these properties do not change. This is one of the key ideas of the operator approach to nonlinear analysis. Here is an outline of some of the key invariant properties we can use.

1. <u>Proper mapping property</u>. A mapping is called proper if the inverse image of a compact set under the mapping A is also compact. This property will be of fundamental importance in determining what happens in passing from the local to the global aspects of the mapping A.

2. <u>Singular points of A and their geometry</u>. (Bifurcation points). A very important infinite-dimensional manifold connected with A is a set of points at which the inverse function theorem breaks down. Such points are generally called bifurcation points of the mapping A. Under our equivalence diagram we note that these points, denoted here by S, are mapped homeomorphically by h_1 onto the singular points of the canonical map C. Since one can compute the set S for A and for C explicitly, we will be able by this means to construct at least partially the mapping h_1. This study differs from previous approaches in that we need to study the <u>global</u> geometry of the set S. For example, is S connected, is S a manifold, is S homeomorphic to a hyperplane of co-dimension one?

3. <u>Singular values of A, A(S), and their geometry</u>. The set of images of S under the mapping A is called the set of singular values of A and denoted by A(S). For a broad class of mappings A it is known that the number of solutions of Eq. (1), Au = f, for f fixed, is constant and so is each component of Y = A(S) provided the mapping A is proper. This result gives us a clue about how to construct explicit solutions based on the mapping A, namely, we must compute the infinite-dimensional manifold A(S) and determine the basic geometry of this set. For this set determines precisely how the solutions of Eq. (1) change. Moreover, the topology of the set A(S) is invariant under the change of coordinates h_2, so that the singular points of the canonical map C, C(S), and A(S) must be homeomorphic. This property

in turn determines the global coordinate transformtion h_2.

4. <u>Fredholm mapping property</u>. An operator A has the Fredholm mapping property if its Frechet derivative is a linear Fredholm operator between the spaces X and Y. This means that the linear mapping has a finite dimensional kernel, a finite dimensional co-kernel, and a closed range. Again this property is invariant under the change of coordinates h_1, and h_2, so that if A is a Fredholm operator so is C. Moreover, we generally require that A will be a Fredholm operator of index zero. That is, that dimker A'(x) = dimcoker A'(x). This property is crucial for the linear Fredholm theorems to be valid. What we shall achieve by using geometry is to generalize these results to a global context.

5. <u>Local classification of singular points</u>. H. Whitney and R. Thom studied the local classificaation of singular points under changes of coordinates in finite dimensions. In so doing, they obtained local normal forms for the mapping A near a singular point. Our goal is to extend this work in two ways to an infinite-dimensional context of Banach spaces and secondly, to a global context. The possibility of doing this work is simplified because we shall study some classic operators A of applied mathematics. Although these operators have been studied for some time, the methods we suggest are <u>intrinsic non-linear approaches</u> instead of the usual linearizations. For example, given a singular point $\underline{x}$ of a nonlinear (index 0) Fredholm operator A acting between real Hilbert spaces X and Y, when is x a (Whitney) fold? To answer this question we need to consider the second term in the Taylor series of A expanded about x and to determine its nondegeneracy. We call a Whitney fold of a C^2 nonlinear Fredholm operator of index zero A, a singular point x of A, such that after a local coordinate change, A can be written (near x) as $(t,v) \to (t^2,v)$, where $t \in \mathbb{R}$ and v is the associated orthogonal complement. For example, I proved (together with P. Church)

21

Theorem 1. Let A be a C^2 Fredholm map of index zero between two Hilbert spaces H_1 and H_2 with a singular point at x. **Suppose**

   (i) dim Ker A'(x) = 1 and

   (ii) (A"(x)(e_0,e_0),h*) ≠ 0, where e_0 ε Ker A'(x) and
        h* ε Ker[A'(x)]* are both not identically zero. Then A is a
        local Whitney fold near x.

The Combination of Math Techniques Utilized

The work of Riccati dates back to the early seventeenth century, and the subsequent study of integrability of Euler and Liouville showed that the Riccati equation could not be integrated by quadrature in any conventional setting. This has been born out by modern studies of RITT and many others. This means that for even the simpler equations of the form (3) stated above with k = 2, our ideas have to be extended. When k = 3 the resulting differential equation is called the Abel equation after the famous Norwegian mathematician of the 19th century. Our idea is to use the methods of Fourier series on this problem, together with the functional analysis setting of Hilbert space, in particular Sobolev spaces, and to analyze the resulting problem by the methods of nonlinear analysis. What we shall achieve by our point of view is a simple (nonlinear) Eigenfunction expansion.

Our idea is to attempt to find a global normal form for the differential operator A involved, to expand A and the inhomogeneous term g and appropriate Eigenfunction expansions, and to solve the equation coordinate-wise, after preliminary coordinate transformations have been made.

In this way, we concentrate on the constructive aspects of nonlinear analysis, (since explicit solutions will be obtained) and avoid the preoccupation of regularity theory and existence questions that have preoccupied much theoretical work up to now.

Our point of view is that nonlinear studies of differential equations have a long history and, instead of focusing on the most difficult existing problems, we wish to return to the roots of the subject by studying the most basic problems that have resisted analysis for centuries.

A unique aspect of our ideas on integrability is its stability under smooth perturbations. For the equations we study of the form $Au = g$, its perturbed version $\tilde{A}u = g$ will also be integrable if the original equation was. The reason for this fact is that, although $\tilde{A}$ is distinct from A, a small perturbation of A perturbs the changes of variables h_1 and h_2, but in our cases preserves the form of the "canonical map" C [see Eq. (2)]. The details of this work are rather lengthy (see a joint paper of Berger and Church, Indiana University Journal of Math., 1980). We shall return to this aspect at the end of this paper.

3. The Notion of Global Normal Form

As a simple first step, we suppose that the spaces X and Y are separable Hilbert spaces that have a mutual orthonormal basis that we denote $[e_1, e_2, e_3, e_4, \ldots]$. Then a general point x, a member of X, can be written:

$$x = \sum_{i=1}^{\infty} x_i e_i \quad \text{or} \quad x = (x_1, x_2, x_3, \ldots).$$

Then a mapping D is diagonal if

$$D(x_1, x_2, x_3, \ldots) = (d_1(x_1), d_2(x_2), \ldots),$$

where the continuous functions $d_i(x_i)$ are functions of <u>one variable</u>. A mapping T is called "triangular" in this notation if:

$$T(x_i, x_2, x_3, \ldots) = (f_1(x_1), f_2(x_1, x_2), f_3(x_1, x_2, x_3) \ldots).$$

The significance of these results is that the equations $Dx = y$ and $T(x) = y$ can be solved explicitly, in terms of the inverses of the functions d_i and f_i.

More generally, if a nonlinear system $Au = g$ can be reduced to these forms by changes of variables defined in the picture (2). The equation $Ax = g$ can be solved explicitly as above by supplementing it with the inverse transformations h_1 and h_2.

The general case of integration by quadrature means that a system of the form $Au = g$ can be linearized globally. Our idea is to go beyond this picture by assuming this also for the canonical map C (by assuming that after change of coordinates a given mapping can be made into a diagonal map or more generally, an upper triangular one).

A simple example of a diagonal map is a Whitney $W(x)$ Fold. It has the form:

$$W(x_1, x_2, x_3, \ldots) = (x_1^2, x_2, x_3, \ldots).$$

The significance of this result is that the Whitney fold is a non-linear diagonal map that exhibits bifurcation phenomena. A simple case of our ideas is then easily stated. Consider the Riccati operator

$$A(x) = \frac{dx}{dt} + x^2$$

acting between the Sobolev spaces $W_{1,2}(0,T) \to L_2(0,T)$ with T-periodic boundary conditions.

THEOREM 2: The Riccati operator so defined is globally equivalent by a C^1 change of variables to a Whitney fold map.

Outline of Proof. The proof can be divided into 2 parts. **Part 1**.

Local Coordinate Changes. We first show the result near an arbitrary singular point x of **A**. First we observe that A (as defined above) is a nonlinear Fredholm operator of index 0. To this end, we compute A'(x)y near a singular point $A'(x)y = \frac{dy}{dt} + 2xy$ with the same T-periodic boundary conditions for y. Thus, via the Sobolev imbedding theorems, A'(x)y can be represented as the sum of an invertible linear map plus a compact map. This fact establishes our first goal. Then we classify each singular point of A, using the result of Section 2 on Whitney folds. Now, by Taylor series expansion,

$$A(x + h) = A(x) + [h' + 2xh] + h^2.$$

This shows that the second derivative of A at **x** is "nondegenerate", so every singular point of A is a fold and so has the desired local normal form.

Indeed, by Theorem 1 of Section II,

(i) Ker $A'(x)y = \text{Ker}(\frac{dy}{dt} + 2xy)$ with T-periodic boundary conditions can be at most one-dimensional.

(ii) $x \in S(A)$ if and only if $\int_0^T x(t)dt = 0$.

Proof. Fox $x \in S(A)$, we require Ker A'(x) to be nontrivial. Thus, A'(x)y = 0 has a nontrivial solution. Now, $A'(x)y = \frac{dy}{dt} + 2xy = 0$ has a nontrivial nonzero solution y(t) (as is clear by ordinary calculus). Thus $(\frac{1}{y}) \frac{dy}{dt} + 2x(t) = 0$, i.e., by integrating over a period T, $\int_0^T \frac{d}{dt} [\log (t)]dt + 2 \int_0^T x(t)dt = 0$.

(iii) Thus, to verify $x \in S(A)$ is a fold we note that

$$(A''(x) (e_0,e_0),h^*) = \int_0^T e_0^2(t)h^*(t)dt$$

and this integral is necessarily nonzero since h* is a nonzero
solution of the equation $\frac{dy}{dt} - 2xy = 0$ which necessarily is of one
sign.

Part 2. <u>Global Coordinate Changes</u>. Referring to Eq. (2), we
discuss here the general construction of the coordinate
transformations h_1 and h_2. This part consists of four steps:

<u>Step 1</u>. "Layering" the mapping A in accord with Part I so that the
layering reduces A to a one-dimensional problem. Let the
layering be denoted by the diffeomorphism α_1.

<u>Step 2</u>. "Translation" of the singular points of A to those of C by a
diffeomorphism α_2.

<u>Step 3</u>. "Translation" of the singular points of A to those of C by a
diffeomorphism α_3.

<u>Step 4</u>. Construction of the final change of coordinates α_4.

Indeed, after step 3 we find

(*) $$\alpha_3 A \alpha_1 \alpha_2 = (a(t,\omega),\omega).$$

We represent the right-hand side of (*) as the composition
$\gamma = C\alpha_4$.

Then we find $\alpha_3 A \alpha_1 \alpha_2 = C\alpha_4$, so $A = \alpha_3^{-1} C \alpha_4 \alpha_2^{-1} \alpha_1^{-1}$. Thus $h_1 = \alpha_3^{-1}$
and $h_2 = \alpha_4 \alpha_2^{-1} \alpha_1^{-1}$.

In addition to the Riccati operator, a famous example that is not
integration by quadrature is called the Abel equation. It has the
form:

* $\frac{dy}{dx} + P_3(y) = f(x)$, where $P_3(y)$ is a polynomial of degree three.

Again we shall require that the inhomogeneous term f(x) is a
T-periodic function, and we seek T-periodic solutions of *. The
explicit solutions of this equation cannot be written down in general,

and so we attempt to utilize our ideas on global changes of variable
in both the range and domain.

The amazing fact about the Abel equation is that it has a simple
global normal form. To define this normal form we first define a
Whitney cusp map as follows: A global Whitney cusp map W between
appropriate Hilbert spaces is defined to be a mapping of the form:

$$W(x_1, x_2, x_3, \ldots) = (x_1, x_2 x_1 - x_1^3, x_3, \ldots).$$

Now we shall state a theorem about the appearance of Whitney cusp maps
in the types of equations we are studying. To this end we consider
the operator:

** $$Ay = \frac{dy}{dx} - \lambda y + y^3.$$

We consider this operator to be the canonical form of the Abel
operator (that is, the right-hand side of *) with λ regarded as a real
parameter. Now we state,

THEOREM 3: The Abel operator Ay defined by ** above is a global
Whitney cusp. That is, when A is regarded as a mapping between the
appropriate Sobolev spaces of periodic functions, after change of
coordinates both on the range and on the domain, it can be given the
form of the Whitney cusp relative to appropriate orthonormal bases for
the Sobolev space.

We end this paper by considering a C^2 perturbation of the
periodic Riccati equation,

$$\dot{x} + f(x) = g(t), \quad g(t) \text{ T-periodic.}$$

Once again we can show that if $f(x)$ is sufficiently close to x^2 in the
C^2 topology, then the operator $Ax = \dot{x} + f(x)$ is equivalent to a global

Whitney fold. This is the type of stability result referred to at the
end of Section 2, and is the unique feature of our approach to
integrability in this new global setting.

General Reference:

M.S. Berger, Nonlinearity and Functional Analysis, Academic
Press, New York - San Francisco-London, 1977.

UNIFORM BOUNDNESS AND GENRALIZED INVERSES IN LIAPUNOV-SCHMIDT METHOD FOR SUBHARMONICS

Whei-Ching C. Chan

and

Shui-Nee Chow[*]
Mathematics Department
Michigan State University
East Lansing, Michigan 48824

Consider the equation

$$(1) \qquad \ddot{x} + g(x) = -\lambda\dot{x} + \mu f(t)$$

where λ, μ are real parameters, $g(x)$ is piecewise continuous and $f(t)$ is periodic with period 1. For $\lambda = \mu = 0$, assume the system

$$(2) \qquad \ddot{x} + g(x) = 0$$

has a nontrivial periodic solution $p(t)$ with least period k, where k is an integer. Let $\Gamma = \{(p(t), \dot{p}(t): 0 \leqslant t < k\}$. The problem to find periodic solutions of (1) with least period k in a sufficiently small neighborhood of Γ for small λ, μ is of great interest. If such solutions exist, they are called subharmonic solutions of order k since their period is k times the period of $f(t)$.

For this discussion, it is convenient to use a different coordinate system near Γ. In fact, there exists a diffeomorphism from a neighborhood of Γ onto $[o,k) \times \{a : |a| < a_0\}$ where $\alpha \in [o,k)$ shows the position on Γ and a indicates the distance between the orbit Γ and the point (x,y). More precisely, we have

$$x = p(\alpha) + a\ddot{p}(\alpha)$$
$$y = \dot{p}(\alpha) - a\dot{p}(\alpha) \cdot$$

If $x(t)$ is a k-periodic solution of (1) in a small neighborhood of Γ, then there exists a unique (α, a) such that

[*]Partially supported by DARPA and NSF grant DMS8401719

$$x(0) = p(\alpha) + a\ddot{p}(\alpha)$$
$$\dot{x}(0) = \dot{p}(\alpha) - a\dot{p}(\alpha) \ .$$

Therefore we can write $x(t)$ in the form

(3)
$$x(t) = p(t + \alpha) + z(t + \alpha)$$

where $z(t + \alpha)$ has small magnitude, $(z(\alpha),\dot{z}(\alpha))$ is orthogonal to $(\dot{p}(\alpha),\ddot{p}(\alpha))$, and α is determined by the initial condition.

Let (3) be applied to (1). We get

$$\ddot{z}(t + \alpha) + g'(p(t + \alpha))z(t + \alpha)$$
$$= -\lambda\dot{z}(t + \alpha) - \lambda\dot{p}(t + \alpha) + \mu f(t) + G(t + \alpha,z)$$

where $G(t + \alpha,z) = -g(p(t + \alpha) + z(t + \alpha)) + g'(p(t + \alpha))z(t + \alpha) + g(p(t + \alpha))$. Note that $G(\cdot,z) = 0(|z|^2)$, and "," denotes the derivative with respect to x. Replacing $t + \alpha$ by t, we obtain the following equation

(4)
$$\ddot{z} + g'(p)z = -\lambda\dot{z} - \lambda\dot{p} + \mu f_\alpha(t) + G(t,z)$$

where $f_\alpha(t) = f(t - \alpha)$. Hence the problem now is to find k-periodic solutions of (4) with $(z(\alpha),\dot{z}(\alpha))$ orthogonal to $(\dot{p}(\alpha),\ddot{p}(\alpha))$.

Without loss of generality, suppose $\dot{p}(0) = 0$ and assume that

(H1) Every k-periodic solution of the homogeneous equation

$$\ddot{z} + g'(p)z = 0$$

is a constant multiple of $\dot{p}(t)$.

We now apply the method of Liapunov-Schmidt to equation (4). Let $P_k{}^r$ be the space of r-times continuously differentiable periodic functions with period k with $|f|_r = \sup\{|f^{(i)}(t)| : i = 0,1,...,r, t \ \varepsilon \ [o,k)\}$. For any $y \ \varepsilon \ P_k{}^2$ let

$$Ay = \ddot{y} + g'(p)y$$
$$Ny = -\lambda\dot{y} - \lambda\dot{p} + \mu f_\alpha(t) + G(t,y)$$

where f_α and $G(t,y)$ are the same as in (4). Then A is a continuous linear operator from $P_k{}^2 \to P_k{}^0$ and N is a continuous operator from

$P_k{}^2 \to P_k{}^0$. (H1) implies that the null space of A is one-dimensional. Define $P : P_k{}^0 \to P_k{}^0$ by

$$Py = \eta \dot{p} \int_0^k \dot{p}y\,dt$$

where

$$\eta = (\int_0^k \dot{p}^2 \, dt)^{-1}.$$

Then P is a continuous projection.

Let $b \in \mathbb{R}$ be a parameter and $p(t,b)$ be the periodic solution of (2) with least period $T(b)$. Hence,

$$p(t + T(b),b) = p(t,b)$$

and $p(t,b_0) = p(t)$. Let

$$(5) \qquad q(t) = \frac{\partial}{\partial b} p(t,b)|_{b=b_0} .$$

Hence, q satisfies (5) with initial condition $q(0) = 1$, $\dot{q}(0) = 0$. We have the following

Theorem 1. Assume (H1) holds. Let $X(t)$ be the fundamental matrix of $Ay = 0$. For any $\phi \in P_k{}^0$, define $G : (I - P)P_k{}^0 \to P_k{}^0$ by the first component of

$$(6) \qquad X(t)\begin{bmatrix} w_0 \\ 0 \end{bmatrix} + X(t) \int_0^t X^{-1}(s) \begin{bmatrix} 0 \\ \phi(s) \end{bmatrix} ds$$

where

$$w_0 = \frac{1}{-\dot{q}(k)} \int_0^k q(s)\phi(s)ds$$

and $q(t)$ is given by (5). Then G is a continuous linear operator, and $G\phi(t)$ is a solution of

$$\begin{cases} \ddot{z} + g'(p)z = \phi(t) \\ z \text{ is } k\text{-periodic} \\ \dot{z}(0) = 0. \end{cases}$$

Hence, G gives the generalized inverse of A and is dependent on k = 1,2,⋯ .

A fundamental question is to determine whether the norm of G is uniformly bounded in k. In the following, we will give two examples to show that uniformity depends on the equation. These results are new.

The readers may find the books by Guckenheimer and Holmes [2] and Chow and Hale [1] useful for more applications.

Example I. Consider the equation

$$(7) \qquad \ddot{x} = \begin{cases} x, & x \leqslant 1 \\ \\ x - 2, & x \geqslant 1 \end{cases}$$

Let G_k be the operator as in Theorem 1. We have the following.

Theorem 2. The Green's function G_k for equation (7) satisfies

$$\| G_k \| \geqslant \cosh(\frac{k}{4}) - 1$$

for all k = 1,2,⋯ .

Proof: Let $\phi(t)$ be any k-periodic function, it follows that $G_k\phi$ is the solution of

$$\begin{cases} \ddot{x} - x = \phi(t) \\ x \text{ is k-periodic} \\ \dot{x}(0) = 0 . \end{cases}$$

Since $\ddot{x} - x = 0$ has cosh t and sinh t as linearly independent solutions, we obtain

$$r(t) = \begin{cases} \sinh t & 0 \leqslant t \leqslant \frac{k}{4} \\ \\ -\sinh(t - \frac{k}{2}) & \frac{k}{4} \leqslant t \leqslant \frac{3k}{4} \\ \\ \sinh(t - k) & \frac{3k}{4} \leqslant t \leqslant k \end{cases}$$

and

$$q(t) = \begin{cases} \cosh t & 0 \leq t \leq \frac{k}{4} \\ -\cosh(t - \frac{k}{2}) - c_1 \sinh(t - \frac{k}{2}) & \frac{k}{4} \leq t \leq \frac{3k}{4} \\ \cosh(t - k) + c_2 \sinh(t - k) & \frac{3k}{4} \leq t \leq k \end{cases}$$

where

$$c_1 = \frac{2 \cosh \frac{k}{4}}{\sinh \frac{k}{4}} , \quad c_2 = \frac{4 \cosh \frac{k}{4}}{\sinh \frac{k}{4}} .$$

Let $\phi = 1$ and $0 \leq t \leq \frac{k}{4}$. Then

$$G_k(t) = \frac{\cosh t}{-c_2} \left(\int_0^{\frac{k}{4}} \cosh ds \right.$$

$$+ \int_{\frac{k}{4}}^{\frac{3k}{4}} (-\cosh(s - \frac{k}{2}) - c_1 \sinh(s - \frac{k}{4})) ds$$

$$+ \int_{\frac{3k}{4}}^{k} (\cosh(s - k) + c_2 \sinh(s - k)) ds)$$

$$- \cosh t \int_0^t (\sinh s \, ds + \sinh t \int_0^t \cosh s \, ds)$$

$$= \frac{\cosh t}{-c_2} (\sinh \frac{k}{4} - 0 - \sinh(\frac{k}{4}) + \sinh(-\frac{k}{4}) - c_1 \cosh(\frac{k}{4})$$

$$+ c_1 \cosh(-\frac{k}{4}) + 0 - \sinh(-\frac{k}{4}) + c_2 - c_2 \cosh \frac{k}{4})$$

$$- \cosh t(\cosh t - 1) + \sinh t \cdot \sinh t$$

$$= -\cosh t(1 - \cosh \frac{k}{4}) + \cosh t - 1$$

$$= \cosh t \cosh \frac{k}{4} - 1 .$$

Setting $t = 0$, $\| G_k \phi \| \geq \cosh \frac{k}{4} - 1$. Hence,

$$\| G_k \| \geq \cosh \frac{k}{4} - 1 .$$

This completes the proof.

<u>Example II.</u> Consider the following equation

$$(8) \qquad \ddot{x} = \begin{cases} -x & x \leqslant 1 \\ x - 2 & x \geqslant 1 \end{cases}.$$

For equation (8), the equilibrium point $(0,0)$ is a saddle and the other equilibrium point $(2,0)$ is a center. Also, the global stable and unstable manifolds of $(0,0)$ coincide. This implies that (8) has a homoclinic orbit which crosses the x-axis at $(0,0)$ and $(4,0)$.

Let $p_k(t)$ be the k-periodic solution of (8) with $p_k(0) = b_k$ and $p_k(k/2) = c_k$. Equation (8) admits the first integral

$$E = \tfrac{1}{2}\, \dot{x}^2 + \int^{x} g(s)ds$$

$$= \begin{cases} \tfrac{1}{2}\, \dot{x}^2 - \tfrac{1}{2}\, x^2 & \text{if } x \leqslant 1 \\ \tfrac{1}{2}\, \dot{x}^2 + \tfrac{1}{2}\, x^2 - 2x & \text{if } x \geqslant 1 \end{cases}.$$

Therefore the period of the periodic orbit $p_k(t)$ is given by the formula

$$T(b_k) = 2\Big(\int_{b_k}^{1} \frac{dx}{\sqrt{x^2 - b_k^2}} + \int_{1}^{c_k} \frac{dx}{\sqrt{-(x-2)^2 + 4 - b_k^2}} \Big)$$

where $c_k = 2 + \sqrt{4 - b_k^2}$. Hence,

$$(9) \qquad T(b_k) = 2 \ln \frac{1 + \sqrt{1 - b_k^2}}{b_k} + \pi - 2 \sin^{-1} \Big(\frac{-1}{\sqrt{4 - b_k^2}} \Big)$$

$$= 2\tau_k + 2\sigma_k$$

$$= k .$$

<u>Lemma 3.</u> $a = \dot{q}(k) \to 2$, as $k \to \infty$.

<u>Proof:</u> Note that $\dot{q}(k) = T'(b_k) \cdot g(b_k)$. Hence

$$\dot{q}(k) = \left(\frac{-2b_k}{\sqrt{1 - b_k^2}\,(1 + \sqrt{1 - b_k^2})} - \frac{2}{b_k} - \frac{4b_k}{\sqrt{4 - b_k^2}\,\sqrt{3 - b_k^2}} \right)(-b_k)$$

$$= b_k^2 \left(\frac{2}{\sqrt{1 - b_k^2}\,(1 + \sqrt{1 - b_k^2})} + \frac{4}{\sqrt{4 - b_k^2}\,\sqrt{3 - b_k^2}} \right) + 2 \; .$$

As $b_k \to 0$, $\dot{q}(k) \to 2$. This completes the proof.

Note that $p_k(t)$ is given precisely by the following formula

$$(10) \qquad p_k(t) = \begin{cases} b_k \cosh t & 0 \leqslant t \leqslant \tau \\ \dfrac{\sinh \tau}{\cosh \tau \sin \sigma} \cos(t - \frac{k}{2}) + 2 & \tau \leqslant t \leqslant k - \tau = \tau + 2\sigma \\ b_k \cosh(t - k) & k - \tau \leqslant t \leqslant k \end{cases}$$

where, $b_k \cosh \tau = 1$.

__Lemma 4.__ $G_k\phi$ is uniformly bounded in k, where $\phi \in P = \{\phi$ is a k-periodic characteristic function and is symmetric with respect to $k/2\}$.

__Proof:__ For simplicity, we will drop the subscript k. First choose k large enough, such that $3\pi/4 < \sigma < 5\pi/6$, where σ is given by (9). Let $r(t)$, $q(t)$ be the solutions of the linearlized equation

$$\begin{cases} \ddot{x} - x = 0 & 0 \leqslant t \leqslant \tau \\ \ddot{x} + x = 0 & \tau \leqslant t \leqslant k - \tau \\ \ddot{x} - x = 0 & k - \tau \leqslant t \leqslant k. \end{cases}$$

It follows from (10) and $\dot{p}(t) = br(t)$ that

$$(11) \qquad r(t) = \begin{cases} \sinh t & 0 \leqslant t \leqslant \tau \\ F(t) & \tau \leqslant t \leqslant k - \tau \\ \sinh(t - k) & k - \tau \leqslant t \leqslant k \end{cases}$$

where $F(t) = \dfrac{\sinh \tau}{-\sin \sigma} \sin(t - \frac{k}{2})$ and

$$(12) \qquad q(t) = \begin{cases} \cosh t & 0 \leqslant t \leqslant \tau \\ Q(t) & \tau \leqslant t \leqslant k - \tau \\ \cosh(t - k) + a \sin h(t - k) & k - \tau \leqslant t \leqslant k \end{cases}$$

where $a = \dot{q}(k)$ and

$$Q(t) = \frac{2 \cosh \tau - a \sinh \tau}{2 \cos \sigma} \cos(t - \frac{k}{2}) - \frac{a \sinh \tau}{\sin \sigma} \sin(t - \frac{k}{2}).$$

It follows from Theorem 1, that

$$(13) \qquad G\phi(t) = \frac{q(t)}{-a} \int_0^k q(s)\phi(s)ds - q(t) \int_0^t r(s)\phi(s)ds$$

$$+ r(t) \int_0^t q(s)\phi(s)ds .$$

For $\phi \varepsilon P$, there exist constants $\beta_1, \beta_2, \beta_3$ such that

$$\phi(t) = \begin{cases} 1 & \beta_1 \leqslant t \leqslant \beta_2 \qquad k - \beta_2 \leqslant t \leqslant k - \beta_1 \\ 1 & \beta_3 \leqslant t \leqslant k - \beta_3 \\ 0 & \text{otherwise} . \end{cases}$$

Therefore we only have to consider the following forms of ϕ

$$(14) \qquad \phi(t) = \begin{cases} 1 & \beta_1 \leqslant t \leqslant \beta_2 , \quad k - \beta_2 \leqslant t \leqslant k - \beta_1 \\ \\ 0 & \text{otherwise} \end{cases}$$

where $\beta_1 < \beta_2 \leqslant \tau$, and

$$(15) \qquad \phi(t) = \begin{cases} 1 & \beta_3 \leqslant t \leqslant k - \beta_3 \\ \\ 0 & \text{otherwise} \end{cases}$$

where $\tau \leqslant \beta_3 \leqslant \frac{k}{2}$.

Substituting (11), (12) and (14) into (13), we obtain the following, for $0 \leqslant t \leqslant \beta_1$

$$G\phi(t) = \frac{\cosh t}{-a} \left[\int_{\beta_1}^{\beta_2} \cosh s \, ds \right.$$

$$+ \int_{k-\beta_2}^{k-\beta_1} (\cosh(s - k) + a \sinh(s - k)) ds$$

$$= \frac{\cosh t}{-a} \left[(2 \sinh \beta_2) - a \cosh \beta_2 - 2 \sinh \beta_1 \right.$$

$$+ \cosh \beta_1]$$

$$= \frac{\cosh t}{-a} \left[(2 - a)e^{\beta_2} - (a + 2)e^{-\beta_2} + (a - 2)e^{\beta_1} \right.$$

$$+ (a + 2)e^{-\beta_1}].$$

It follows from Lemma 3 that $a - 2 = 0(e^{-2\tau})$ and $t \leqslant \beta_1 < \beta_2 \leqslant \tau$. Hence, there exists $M_1 > 0$ such that $|G\phi(t)| \leqslant M_1$ for $0 \leqslant t \leqslant \beta_1$. Next, for $\beta_1 \leqslant t \leqslant \beta_2$, we have

$$G\phi(t) = \frac{\cosh t}{-a} \left[2 \sinh \beta_2 - a \cosh \beta_2 - 2 \sinh \beta_1 \right.$$

$$+ a \cosh \beta_1] - \cosh t(\cosh t - \cosh \beta_1)$$

$$+ \sinh t(\sinh t - \sinh \beta_1)$$

$$= \frac{1}{-a} \left[(2 - a)e^{t+\beta_2} + (a - 2)e^{t+\beta_1} - (2 + a)e^{t-\beta_2} \right.$$

$$+ (2 - a)e^{\beta_2-t}$$

$$+ (a - 2)e^{\beta_1-t} - (a + 2)e^{-(t+\beta_2)}$$

$$+ (a + 2)e^{-(t+\beta_1)}]$$

$$+ (\frac{2}{-a} + 1)e^{t-\beta_1} + 2e^{\beta_1-t} .$$

Since $a - 2 = 0(e^{-2\tau})$ and $\beta_1 \leqslant t \leqslant \beta_2 \leqslant \tau$, there exists $M_2 > 0$, such that $|G\phi(t)| \leqslant M_2$ for $\beta_1 \leqslant t \leqslant \beta_2$.

Consider now the interval $\beta_2 \leqslant t \leqslant k - \beta_2$. We have

$$G\phi(t) = Q(t) \left[\frac{2}{-a}(\sinh \beta_2 - \sinh \beta_1) + (\cosh \beta_2 - \cosh \beta_1) \right]$$

$$- Q(t)(\cosh \beta_2 - \cosh \beta_1)$$

$$+ F(t)(\sinh \beta_2 - \sinh \beta_1)$$

$$= [(2 \cosh \tau - a \sinh \tau) \frac{\cos(t - \frac{k}{2})}{-a \cos \sigma}$$

$$- (2 \sinh \tau - a \sinh \tau) \frac{\sin(t - \frac{k}{2})}{\sin \sigma}]$$

$$\cdot (\sinh \beta_2 - \sinh \beta_1)$$

since $a - 2 = 0(e^{-2\tau})$, $3\pi/4 \leqslant \sigma \leqslant 5\pi/6$ and $-\sigma \leqslant t - k/2 \leqslant \sigma$, there exists $M_3 \geqslant 0$, such that $|G\phi(t)| \leqslant M_3$ for $\beta_2 \leqslant t \leqslant k - \beta_2$.

For $k - \beta_2 \leqslant t \leqslant k - \beta_1$ and $k - \beta_1 \leqslant t \leqslant k$, the computations are similar to the first two cases by replacing $\cosh t$ by $\cosh(t - k) + a \sinh(t - k)$ and $\sinh t$ by $\sinh(t - k)$. We obtain for some $M_4 \geqslant 0$

$$|G\phi(t)| \leqslant M_4, \quad k - \beta_2 \leqslant t \leqslant k.$$

One can see that the constants M_1, M, M_3 and M_4 can be chosen independent of k.

Now, repeat the above procedure for ϕ defined by (15), for $0 \leqslant t \leqslant \beta_3$ and $\beta_3 \geqslant \tau$, we have

$$G\phi(t) = \frac{q(t)}{-a} (\int_{\beta_3}^{k-\beta_3} Q(s)ds)$$

$$= \begin{cases} \cosh t(\dfrac{2 \cosh \tau - a \sinh \tau}{-2a \cos \sigma}(-2 \cos \sigma)) & t \leqslant \tau \\[2em] Q(t) (\dfrac{2 \cosh \tau - a \sinh \tau}{a}) & \tau \leqslant t \leqslant \beta_3 . \end{cases}$$

Again, since $a - 2 = 0(e^{-2\tau})$, there exists $N_1 > 0$, such that

$$|G\phi(t)| \leqslant N_1, \quad \text{for } 0 \leqslant t \leqslant \beta_3 .$$

For $\beta_3 \leqslant t \leqslant k - \beta_3$, we obtain

$$G\phi(t) = Q(t)(\frac{2 \cosh \tau - a \sinh \tau}{a}) - Q(t) \int_{\beta_3}^{t} F(s)ds$$

$$+ F(t) \int_{\beta_3}^{t} Q(s)ds$$

$$= Q(t)(\frac{2 \cosh \tau - a \sinh \tau}{a}) + (\frac{2 \cosh \tau - a \sinh \tau}{2 \cos \sigma})$$

$$(\frac{\sinh \tau}{\sin \sigma}) [-1 + \cos(t - \frac{k}{2})\cos(\beta_3 - \frac{k}{2})$$

$$+ \sin(t - \frac{k}{2})\sin(\beta_3 - \frac{k}{2})] \ .$$

Since $a - 2 = 0(e^{-2\tau})$, $-\sigma \leqslant t - k/2 \leqslant \sigma$ and $3\pi/4 \leqslant \sigma \leqslant 5\pi/6$, there exists $N_2 > 0$, such that $|G\phi(t)| \leqslant N_2$ for $\beta_3 \leqslant t \leqslant k - \beta_3$. For $k - \beta_3 \leqslant t \leqslant k$, by using similar arguments, we can choose $N_3 \geqslant 0$, independent of k, such that $G\phi(t)$ is bounded by N_3.

It follows that there exists $M \geqslant 0$ independent of k such that

$$\max_{0 \leqslant t \leqslant k} |G\phi(t)| \leqslant M \ .$$

Since

$$G\dot{\phi}(t) = \dot{q}(t)[\frac{1}{-a} \int_0^k q(s)\phi(s)ds] - \dot{q}(t) \int_0^t r(s)\phi(s)ds$$

$$+ \dot{r}(t) \int_0^t q(s)\phi(s)ds \ ,$$

one can see by similar arguments that there exists $N \geqslant 0$ which is independent of k such that

$$\max_{0 \leqslant t \leqslant k} |G\dot{\phi}(t)| \leqslant N \ .$$

Since

$$G\ddot{\phi} = -g'(p)G\phi + \phi$$

we have

$$\max_{0 \leqslant t \leqslant k} |G\ddot{\phi}| \leqslant \max_{0 \leqslant t \leqslant k} |g'(p)| \max_{0 \leqslant t \leqslant k} |G\phi| + 1$$

$$\leqslant JM + 1$$

where $J = \max\limits_{0 \leqslant t \leqslant k} |g'(p)|$. Since $p(t)$ is uniformly bounded, therefore J can be chosen independent of k.

We have shown that

$$\|G\phi\| \leq \max(N,M,jM + 1) = K_o .$$

This completes the proof.

The above two lemmas imply the following theorem.

Theorem 5. The operator G_k is uniformly bounded in P_s, where G_k is defined by Theorem 1 and $P_s = \{\phi$ is a continuous k-periodic function and is symmetric with respect to $k/2\}$.

REFERENCES

[1] S.N. Chow and J.K. Hale, Methods of Bifurcation Theory, Springer-Verlag N.Y. 1982.

[2] J. Guckenheimer and P.J. Holmes, Nonlinear Oscillations, Dynamical Systems and Bifurcation Vector Fields, Springer-Verlag, N.Y. 1983.

EXISTENCE OF RADIALLY SYMMETRIC SOLUTIONS
OF STRONGLY DAMPED WAVE EQUATIONS

Hans Engler
Department of Mathematics
Georgetown University
Washington, D.C. 20057

1.INTRODUCTION

The aim of this note is a study of the quasilinear third order partial differential equation

$$(1.1) \qquad u_{tt}(x,t) - \Delta_x u_t(x,t) - \operatorname{div}_x(g(\nabla_x u(x,t))) = f(x,t) \quad (x \in \Omega \subset \mathbb{R}^n, \, 0 < t < T)$$

in the special case where $\Omega = B = B_1(0) \subset \mathbb{R}^n$, $u(x,t)$ depends only on $|x| = r$ and t, and $g: \mathbb{R}^n \longrightarrow \mathbb{R}^n$ is isotropic, i.e. $g(\xi) = g(|\xi|^2) \cdot \xi$ for some scalar function g. In (1.1) Δ_x is the n-dimensional Laplacian, ∇_x is the gradient, div_x denotes the divergence operator, and subscripts denote differentiation. In addition to (1.1), initial data $u(\cdot,0) = u_0$, $u_t(\cdot,0) = u_1$ and zero boundary data $u(x,t) = 0$ for $|x| = 1$ are to be given. We want to give conditions under which (1.1) has unique global "regular" solutions for arbitrary regular data u_0, u_1, f.

Equation (1.1) is an example for a strongly damped nonlinear second order wave equation; such equations are discussed in more generality in [7]. It can also be viewed as a perturbed heat equation for $v = u_t$ (with an integral perturbation).

Global solutions for (1.1) have been found in various settings:

In the case where $\Omega \subset \mathbb{R}$ is an interval, classical solutions were constructed in [1], [5], [8], [9], and recent discussions focus on the asymptotic behavior of solutions, in particular for non-monotone g which can give rise to equilibrium states with co-existing phases ([2],[12]). Classical global solutions in $\Omega = \mathbb{R}^2$ were found for special (superlinear) g in [11] and in arbitrary $\Omega \subset \mathbb{R}^n$ for small initial data in [6]. On the other hand, global weak solutions of (1.1) (in which $\nabla_x u_t \in L^2(0,T;L^2(\Omega))$, $g(\nabla_x u) \in L^1(0,T;L^1(\Omega))$, and (1.1) holds in the sense of distributions) can be constructed in fairly general situations, see [4] for a concrete example and [13] for an abstract approach; however uniqueness and regularity is an open problem for these solutions.

Here, we use the notion of "mild" solutions as defined in [7]: Let $(T(t))_{t \geqslant 0}$ be the heat semigroup for zero boundary data in, say, $L^p(B)$, $1 < p < \infty$; then a mild solution u is required to be continuous with values in $D(\Delta_x) = W^{2,p} \cap W_o^{1,p}$ and to satisfy the integral equation

$$(1.2) \qquad u(t) = u_0 + (T(t) - I)(\Delta_x)^{-1}u_1 + \int_0^t (T(t-s) - I)(\Delta_x)^{-1}(f(s) + B(u(s)))ds$$

where $B(u) = div_x(g(\nabla_x u))$ (x-arguments are omitted). If $p > n$, then B will be Lipschitz-continuous from $D(\Delta_x)$ into $L^p(B)$, uniformly on bounded sets, due to Sobolev imbedding theorems, as soon as g is suitably smooth; so local mild solutions will always exist for $u_0 \in D(\Delta_x)$, $u_1 \in L^p(B)$, $f \in L^1(0,T;L^p(B))$ (see [7]), and the question arises if they can be continued for arbitrary initial data.

In Section 2, we give conditions under which this can be guaranteed for radially symmetric solutions (Theorems 2.1, 2.2), the key assumptions being a monotonicity assumption for g (up to affine functions) and a growth condition, if the space dimension n exceeds 2, namely $|g(\xi)| = o(|\xi|^{(n+2)/(n-2)})$. Since in this case the motion is essentially one-dimensional away from x = 0 and (1.1) is known to have classical solutions for smooth data in one space dimension, one would expect singularities to begin to form at the origin. Our result shows that strong dissipation will prevent this. As a direct consequence, solutions under these general conditions will automatically be smooth, if the data permit this. We also give a result for the existence of "almost everywhere" solutions (Theorem 2.3) under no growth restrictions for g. For such solutions, the representation formula (1.2) still holds (in some L^r, r close to 1), but they cannot be obtained using the results in [7], their uniqueness is not clear, nor is it obvious that they will be smooth if the data are.

Section 3 contains various a priori estimates for mild solutions of (1.1), and in section 4, proofs are completed.

We write $\nabla_x^2 u = \nabla^2 u$ for the matrix of second spatial derivatives. Lebesgue and Sobolev spaces are denoted by their usual symbols; L^p, $W^{k,p}$ etc. are spaces of radially symmetric functions on $B = B_1(0)$. Constants that are used during the proofs and that may change from line to line are denoted by the same letter C, in contrast to constants that appear in assumptions ($c, C_0, C_1, K, \ldots$). For a number or an expression z and k > 0, we write $z^k = (z)^k := z \cdot |z|^{k-1}$.

2. MAIN RESULTS

Throughout this paper, we assume that $g: \mathbb{R}^+ \longrightarrow \mathbb{R}$ is locally Lipschitz continuous together with its first derivative. We define $G(z) = \int_0^z g(\zeta)\, d\zeta$ for any $z \geqslant 0$. Some of the following assumptions will be needed:

(2.1) for some $C_0 \geqslant 0$, $G(z) + C_0 \cdot (1 + z) \geqslant 0$ for all $z \in \mathbb{R}^+$

(2.2) for some $L \in \mathbb{R}$, $g_0(z) = g(z) + L \geqslant 0$ for all $z \geqslant 0$

(2.3) $0 \leqslant g_0(z) + (2+\delta) \cdot z g_0'(z) \leqslant C_1 \cdot g_0(z)$ for some $\delta, C_1 > 0$ and all $z \in \mathbb{R}^+$

(2.4) $|g(z)| \leqslant c(1 + z^q)$ for some $q > 0, c > 0$ and all $z \in \mathbb{R}^+$.

Theorem 2.1: Assume (2.1) and that $n = 2$, $p > 2$. Then for any $u_0 \in \underline{W}^{2,p} \cap W_o^{1,2}$, $u_1 \in \underline{L}^p$ and $f \in L^1(0,T; \underline{L}^p)$ there is a unique radially symmetric mild solution $u \in C([0,T], \underline{W}^{2,p} \cap W_o^{1,2})$ of

$$(2.5) \quad u_{tt} - \Delta u_t - \text{div}_x(g(|\nabla_x u|^2)\nabla_x u) = f \quad \text{on } B \times [0,T];$$

$$(2.6) \quad u(\cdot,0) = u_0 , u_t(\cdot,0) = u_1 .$$

Theorem 2.2: Assume that (2.2), (2.3) with $\delta = 2$, and (2.4) hold, that $p > n \geqslant 3$ and $q < 2/(n-2)$. Then for any $u_0 \in \underline{W}^{2,p} \cap W_o^{1,2}$, $u_1 \in \underline{L}^p$ and $f \in L^2(0,T; \underline{L}^p)$, there is a unique radially symmetric mild solution $u \in C([0,T], \underline{W}^{2,p} \cap W_o^{1,2})$ of (2.5), (2.6).

Theorem 2.3: Assume that $n \geqslant 3$ and that (2.3) holds. Then for all $u_0 \in \underline{W}^{2,2} \cap W_o^{1,2}$ for which $\int_B G(|\nabla_x u_0|^2) < \infty$, for all $u_1 \in \underline{L}^2$ and $f \in L^{1+\epsilon}(0,T; \underline{L}^2)$ with $\epsilon > 0$ there is a radially symmetric solution u of (2.5), (2.6) that vanishes on $\partial B \times [0,T]$ and for which $u_{tt}, \nabla_x^2 u_t , \text{div}_x(g(|\nabla_x u|^2)\nabla_x u) \in L^r(0,T; \underline{L}^r)$ for some $r > 1$, and (2.5) holds almost everywhere in $B \times [0,T]$.

Comments:

(i) No uniqueness is claimed in Theorem 2.3.

(ii) For increasing δ, condition (2.3) becomes stronger, and (2.2) implies (2.1). The inequality $zg_0'(z) \leqslant C_1 \cdot g_0(z)$ implies the polynomial growth behavior (2.4) with $q = C_1$, and (2.3) with $\delta = 0$ is equivalent to assuming that $z \longrightarrow g_0(z^2) \cdot z$ is increasing in z or that the function $\xi \longrightarrow G_0(|\xi|^2)$ is convex on $\mathbb{R}^n$, where $G_0' = g_0$.

(iii) One can show that Theorem 2.3 holds in fact in arbitrary bounded smooth domains in any space dimension.

(iv) Since the mild solutions found in Theorems 2.1 and 2.2 are continuous curves in $\underline{W}^{2,p}$ for $p > n$, their spatial gradients are uniformly pointwise bounded on $B \times [0,T]$. Then one obtains by standard regularity arguments for parabolic equations that Hölder-continuous data (i.e. $\nabla_x^2 u_0, u_1, f \in C^\alpha$) imply that u_{tt} and $\Delta_x u_t$ will be in any $L^p(0,T; L^p(B))$. Thus, $B(u) = \text{div}_x(g(|\nabla_x u|^2)\nabla_x u)$ is, in fact, a second order operator with Hölder-continuous coefficients. Repeating the existence argument in a C^α-class (see [10]), and recalling the uniqueness of mild solutions implies that u_{tt} and $\nabla_x^2 u_t$ are also Hölder-continuous, and that (2.5) holds in the classical sense. On the other hand, one cannot expect the solution to be smoother than the initial data. In the case of $n = 1$ space dimensions, one can show that for weak solutions, jump discontinuities of the derivatives of the initial data u_0 will persist also for the solutions (and remain stationary); see [12]. This possibility is also suggested by the integral equation defining a mild solution (1.2).

3. A PRIORI ESTIMATES

If $f \in L^2(0,T;\underline{L}^p)$ with $r > 1$, then any local mild solution $u \in C([0,T_0], \underline{w}^{2,p} \cap w_o^{1,2})$ of (2.5) with $p > n$ actually satisfies (2.5) almost everywhere, and all members of the left hand side of (2.5) are in $L^2(\delta,T; \underline{L}^p)$ for any $\delta > 0$ ([10]); also, u_t and $\Delta_x u$ are continuous with values in $\underline{L}^p$ by construction. All estimates below are stated for such mild solutions on any existence interval $[0,T_0]$, assuming that $u_0 \in \underline{w}^{2,p} \cap w_o^{1,p}$ and $u_1 \in \underline{L}^p$. The slightly more general assumption $f \in L^1(o,T;\underline{L}^p)$ of Theorem 2.2 will be handled by an approximation argument.

We define, as in Section 2, $G: \mathbb{R}^+ \longrightarrow \mathbb{R}$ by $d/dz\, G(z) = g(z)$, $G(0) = 0$, and note that $d/dz\, G(|z|^2) = 2g(|z|^2)z$, $G_0(|z|^2) = G(|z|^2) + C_0\cdot(1 + |z|^2) \geqslant 0$.

Lemma 3.1: For any $0 \leqslant t \leqslant T_0$, if g satisfies (2.1), then

$$(3.1) \quad E_1(t) = 1/2\, \{(\|u_t(\cdot,t)\|_2)^2 + \int_B G_0(|\nabla_x u(\cdot,t)|^2)\} + \int_0^t (\|\nabla_x u_t(\cdot,s)\|_2)^2 ds$$

$$\leqslant C = C(\|u_1\|_2, \int_B G_0(|\nabla_x u_0(\cdot)|^2), \int_0^t \|f(\cdot,s)\|_2 ds, C_0, T).$$

Proof: Multiply (2.5) with u_t, integrate over $B \times [0,t]$, and integrate by parts. Then

$$(3.2) \quad 1/2\, (\|u_t(\cdot,t)\|_2)^2 + \int_0^t (\|\nabla_x u_t(\cdot,s)\|_2)^2\, ds + \int_B G(|\nabla_x u(\cdot,t)|^2)/2 = (\|u_1\|_2)^2/2 +$$

$$+ \int_B G(|\nabla_x u_0(\cdot)|^2)/2 + \int_0^t <u_t(\cdot,s),f(\cdot,s)>\, ds,$$

where $\|\cdot\|_2$ denotes the norm and $<\cdot,\cdot>$ the scalar product of L^2. Thus,

$$(3.3) \quad 1/2\, (\|u_t(\cdot,t)\|_2)^2 + \int_0^t (\|\nabla_x u_t(\cdot,s)\|_2)^2\, ds + \int_B G_0(|\nabla_x u(\cdot,t)|^2)/2 \leqslant (\|u_1\|_2)^2/2 +$$

$$+ \int_B G(|\nabla_x u_0(\cdot)|^2)/2 + \int_0^t |<u_t(\cdot,s),f(\cdot,s)>|ds + C_0\cdot(1 + (\|\nabla_x u(\cdot,t)\|_2)^2).$$

If $C_0 > 0$, then the last member on the right hand side can be eliminated using

$$(3.4) \quad \{(\|\nabla_x u(\cdot,t)\|_2)^2 - (\|\nabla_x u_0\|_2)^2\} \leqslant 2\int_0^t \|\nabla_x u_t(\cdot,s)\|_2 \|\nabla_x u(\cdot,s)\|_2\, ds \leqslant$$

$$\leqslant 2\{\int_0^t \int_0^s \|\nabla_x u_t(\cdot,\tau)\|_2 d\tau\, \|\nabla_x u_t(\cdot,s)\|_2\, ds + \int_0^t \|\nabla_x u_0\|_2 \|\nabla_x u(\cdot,s)\|_2 ds\}$$

$$\leqslant \varepsilon \int_0^t (\|\nabla_x u_t(\cdot,s)\|_2)^2 ds + C_\varepsilon\{(\|\nabla_x u_0\|_2)^2 + \int_0^t \int_0^s (\|\nabla_x u_t(\cdot,\tau)\|_2)^2 d\tau\}$$

for any $\varepsilon > 0$. This gives an integral inequality of the form

$$(3.5) \quad E_1(t) \leqslant C\cdot(E_1(0) + \int_0^t E_1(s)\, ds + \int_0^t \|u_t(\cdot,s)\|_2 \cdot \|f(\cdot,s)\|_2\, ds)$$

with C depending on C_0 and T. Gronwall-Bihari's inequality then implies (3.1).

□□□

Lemma 3.2: Let g_0 satisfy (2.3). Then for all $u \in \underline{W}^{2,2} \cap W^{1,2}$,

(3.7) $\quad <\Delta_x u, B(u)> \;\geqslant\; \varepsilon(\|\nabla_x(\sqrt{g_0}(|\nabla_x u|^2)\nabla_x u)\|_2)^2$,

where $B(u) = \text{div}_x(g_0(|\nabla_x u|^2)\nabla_x u)$ and $\varepsilon > 0$ depends on n and the constant δ in (2.3).

Proof: We prove the assertion in the case of C^2-smooth u; the general case follows by approximation. Recall that for radially symmetric u,

(3.8) $\quad \Delta_x u = u_{rr} + (n-1)r^{-1}u_r = r^{1-n}(r^{n-1}u_r)_r$

and, writing $\sigma(z) = g_0(|z|^2)z$,

(3.9) $\quad B(u) = \sigma'(u_r)u_{rr} + (n-1)r^{-1}\sigma(u_r) = r^{1-n}(r^{n-1}\sigma(u_r))_r$.

Then, denoting the area of the sphere S^{n-1} by c_n,

(3.10) $\quad <\Delta_x u, B(u)> = c_n \int_0^1 (u_{rr}+(n-1)r^{-1}u_r)(\sigma'(u_r)u_{rr}+(n-1)r^{-1}\sigma(u_r))r^{n-1}dr.$

Add and subtract the integral of $d/dr\,(c_n(n-1)r^{n-2}\sigma(u_r)u_r) =$

$c_n(n-1)(\sigma(u_r) + \sigma'(u_r)u_r)\,u_{rr}r^{n-2} + c_n(n-1)(n-2)\sigma(u_r)u_r r^{n-3}$. This implies

(3.11) $\quad <\Delta_x u, B(u)>/c_n = (n-1)\sigma(u_r(1))\cdot u_r(1) + \int_0^1 \{\sigma'(u_r)(u_{rr})^2 +(n-1)\sigma(u_r)u_r r^{-2}\}r^{n-1}dr.$

One calculates next that $|\nabla_x(\sqrt{g_0}(|\nabla_x u|^2)\nabla_x u)|^2 = (n-1)g_0(|u_r|^2)|u_r|^2 r^{-2} +$

$(g_0(|u_r|^2) + g_0'(|u_r|^2)|u_r|^2)(u_{rr})^2$. Then (2.3) implies that

(3.12) $\quad \sigma'(z) = g_0(|z|^2) + 2|z|^2 g_0'(|z|^2) \;\geqslant\; \varepsilon(g_0(|z|^2) + |z|^2 g_0'(|z|^2))$

with $\varepsilon = \delta/(1+\delta)$, which together with (3.11) proves (3.7).

$\qquad\qquad\qquad\qquad\qquad\qquad\qquad\qquad\qquad\qquad\qquad\qquad$ □□□

Lemma 3.3: Let u be a mild local solution of (2.5), (2.6). If (2.3) holds, then for any $0 \leqslant t \leqslant T_0$,

(3.13) $\quad E_2(t) = \|\Delta_x u(\cdot,t)\|_2 + \int_0^t (\|\nabla_x(\sqrt{g_0}(|\nabla_x u(\cdot,s)|^2)\nabla_x u(\cdot,s)\|_2)^2 \, ds \leqslant C$,

the constant C depending on $\|\Delta_x u_0\|_2$, $\sup_{s\leqslant t} E_1(s)$, $\int_0^t \|f(\cdot,s)\|_2 \, ds$, δ, and L.

Proof: Rewrite (2.5) as

$$u_{tt} - \Delta_x u_t - \text{div}(g_0(|\nabla_x u|^2)\nabla_x u) = f + L\cdot\Delta_x u,$$

multiply with $(-\Delta_x u)$ and integrate over $B \times [0,t]$. After integrating by parts, this gives the identity

$$(3.14) \quad (\|\Delta_x u(\cdot,t)\|_2)^2/2 + \int_0^t <\Delta_x u(\cdot,s), B(u)(\cdot,s)> ds = (\|\Delta_x u_0\|_2)^2/2 - <u_t(\cdot,t),\Delta_x u(\cdot,t)>$$

$$+ \int_0^t (\|\nabla_x u_t(\cdot,s)\|_2)^2 ds + L \int_0^t (\|\Delta_x u(\cdot,s)\|_2)^2 ds - \int_0^t <f(\cdot,s),\Delta_x u(\cdot,s)> ds.$$

We estimate the second member on the left hand side of (3.14) from below, using Lemma 3.2, note that $|<u_t(\cdot,t),\Delta_x u(\cdot,t)>| \leq 1/4 \,(\|\Delta_x u(\cdot,t)\|_2)^2 + E_1(t)$, and use Gronwall-Bihari's inequality to arrive at (3.13).

□□□

Lemma 3.4: Assume that g_0 satisfies (2.3) with $\delta = 2$ and that $2 \leq k < p \cdot n$. Then for all $u \in \underline{W}^{2,p} \cap W_0^{1,2}$, writing $B(u) = div_x(g_0(|\nabla_x u|^2)\nabla_x u)$,

$$(3.15) \quad <(\Delta_x u)^{k-1}, B(u)> \; \geq \; \epsilon \cdot (\|\nabla_x(g_0^{1/k}(|\nabla_x u|^2)\nabla_x u)\|_k)^k,$$

where $\|\cdot\|_k$ is the $\underline{L}^k$-norm and $\epsilon > 0$ depends on n,k, and the constant C_1 in (2.3).

Proof: We again use the abbreviation $\sigma(z) = g_0(|z|^2)z$. Then

$$(3.16) \quad <(\Delta_x u)^{k-1}, B(u)> \; = \; c_n \int_0^1 (u_{rr} + (n-1)r^{-1}u_r)^{k-1}(\sigma'(u_r)u_{rr} + (n-1)r^{-1}\sigma(u_r))r^{n-1}dr.$$

Add and subtract the integral of $d/dr \, (c_n \cdot (n-1)^{k-1}r^{n-k}\sigma(u_r)(u_r)^{k-1}) =$

$$= c_n \cdot (n-1)^{k-1}\{((k-1)\sigma(u_r) + \sigma'(u_r)u_r)|u_r|^{k-2}u_{rr}r^{n-k} + (n-k)\sigma(u_r)(u_r)^{k-1}r^{n-k-1}\}.$$ Since $|u_r| = O(r^{1-1/p})$ near $r = 0$ one has $r^{n-k}\sigma(u_r)(u_r)^{k-1} = o(1)$, which implies

$$(3.17) \quad <(\Delta_x u)^{k-1}, B(u)> = c_n \cdot (n-1)^{k-1}\sigma(u_r(1))(u_r(1))^{k-1} +$$

$$+ c_n \int_0^1 g_0(|u_r|^2) \cdot C(\rho,u_{rr},(n-1)r^{-1}u_r)r^{n-1}dr,$$

with $C(\rho,x,y) = (x + y)^{k-1}((1+\rho)x + y) - (n-k)(n-1)^{-1}|y|^k - y^{k-1}(1+\rho)x - (k-1)xy^{k-1}$

$= |x+y|^k + \rho x(x+y)^{k-1} - (\rho+k)xy^{k-1} - (n-k)(n-1)^{-1}|y|^k$, and $x = u_{rr}$, $y = (n-1)r^{-1}u_r$,

$\rho = 2 \cdot g_0'(|u_r|^2)|u_r|^2/g_0(|u_r|^2)$. Using now (2.3), Lemma A.1 implies that this integrand is pointwise bounded from below by $\epsilon \cdot ((n-1)^{-1}|y|^2 + (1+\rho/k)|x|^2)^{k/2}$, which is equal to $\epsilon \cdot |\nabla_x(g_0^{1/k}(|\nabla_x u|^2)\nabla_x u)|^k$.

□□□

Lemma 3.5: Assume that $n \geq 3$, let u be a local mild solution of (2.5), (2.6) for $0 \leq t \leq T_0$, and assume that (2.3) and (2.4) hold.

a) If for some $2 < k_0 \leqslant k < n$,

(3.18) $u_t \in L^\infty(0,T;\underline{L}^k(B))$ and $u_t \cdot \Delta_x u_t \in L^1(0,T;\underline{L}^{k/2}(B))$

with norms K_1, K_2 in these spaces, and if in (2.4) $q \leqslant 2 \cdot (n-2)^{-1}$, then also

(3.19) $\Delta_x u \in L^\infty(0,T;\underline{L}^k(B))$,

(3.20) $u_t \in L^\infty(0,T;\underline{L}^r(B))$ and $u_t \cdot \Delta_x u_t \in L^1(0,T;\underline{L}^{r/2}(B))$,

with norms depending on $K_1, K_2, r, k_0 - 2$, and the data, where $r^{-1} \geqslant k^{-1} - c$, and c is a fixed positive number.

b) If for some $n < k \leqslant p$,

(3.21) $u_t \in L^\infty(0,T;\underline{L}^k(B))$ and $u_t \cdot \Delta_x u_t \in L^1(0,T;\underline{L}^{k/2}(B))$

with norms K_1 resp. K_2, then also $\Delta_x u \in L^\infty(0,T;\underline{L}^k(B))$, the norm depending only on K_1, K_2, and the data.

<u>Proof:</u> Let $\delta \in (0,T)$ be small enough that $\|u(\cdot,t)\|_{W^{2,p}} \leqslant 2 \cdot \|u_0\|_{W^{2,p}}$, $\|u_t(\cdot,t)\|_{L^p} \leqslant 2 \cdot \|u_1\|_{L^p}$ for $0 \leqslant t \leqslant \delta$. Rewrite (2.5) as in the proof of Lemma 3.2, multiply with $(-\Delta_x u)^{k-1}$, integrate over $B \times [\delta,t]$, and integrate the term containing u_{tt} by parts. The result is

(3.22) $1/k \left(\|\Delta_x u(\cdot,t)\|_k \right)^k + \int_\delta^t <(\Delta_x u(\cdot,s))^{k-1}, B(u)(\cdot,s)> ds =$

$= 1/k \left(\|\Delta_x u(\cdot,\delta)\|_k \right)^k - <u_t(\cdot,t), (-\Delta_x u(\cdot,t))^{k-1}> + <u(\cdot,\delta),(-\Delta u(\cdot,\delta))^{k-1}> -$

$- (k-1)\int_\delta^t <u_t(\cdot,s) \cdot \Delta_x u_t(\cdot,s),|(-\Delta_x u(\cdot,s))|^{k-2}> ds +$

$\int_\delta^t \{L \cdot (\|\Delta_x u(\cdot,s)\|_k)^k + <f(\cdot,s),(-\Delta_x u(\cdot,s))^{k-1}>\} ds .$

Using (3.15), we get an estimate

(3.23) $1/2k \cdot (\|\Delta_x u(\cdot,t)\|_k)^k + \varepsilon \int_\delta^t \|\nabla_x(g_0^{1/k}(|\nabla_x u|^2)\nabla_x u)\|_k)^k(s) ds$

$\leqslant C(u_0, u_1, K_1) + C \cdot \int_\delta^t \{L \cdot \|\Delta_x u(\cdot,s)\|_k)^k + \|f(\cdot,s)\|_k \cdot (\|\Delta_x u(\cdot,s)\|_k)^{k-1} +$

$+ \|u_t(\cdot,s) \cdot \Delta_x u_t(\cdot,s)\|_{k/2} \cdot (\|\Delta_x u(\cdot,s)\|_k)^{k-2} \} ds .$

An estimate for the left hand side of (3.23) then follows in terms of K_1, K_2, and of the data, for $\delta \leqslant t \leqslant T$. By the choice of δ and by imbedding theorems, such an estimate is even true if the integration is extended over $[0,t]$. This proves (3.19) and part b).

To show (3.20), we define $u_t = v$. Then v satisfies the equation $v_t - \Delta_x v = h$, where $h = f - \text{div}_x(g(|\nabla_x u|^2) \cdot \nabla_x u)$ and $v(\cdot,0) = u_1 \in \underline{L}^p$. Now, by assumption $f \in L^2(0,T;\underline{L}^p)$, and

(3.23) gives an estimate for $g_0^{1/k}(|\nabla_x u|^2)\nabla_x^2 u$ in $L^k(0,T;\underline{L}^k)$ in terms of the data and of K_1, K_2. Also, the estimate for $\Delta_x u$ in $L^\infty(0,T;\underline{L}^k)$ and standard imbedding theorems imply that $\nabla_x u$ is bounded a priori in $L^\infty(0,T;\underline{L}^{k'})$ with $k' = nk/(n-k) < \infty$. Thus, $g_0^{(k-1)/k}(|\nabla_x u|^2)$ is estimated a priori in $L^\infty(0,T;\underline{L}^{k''})$ with $k'' = k \cdot k'/((k-1) \cdot 2q)$, where q is the growth exponent in (2.4). Therefore, $g(|\nabla_x u|^2) \cdot \nabla_x^2 u$ is estimated in $L^k(0,T;\underline{L}^s)$ with $s^{-1} = k^{-1} + (k'')^{-1}$, and h is estimated in the same space. Applying now Lemma A.2, we obtain that v and $v \cdot \Delta_x v$ can be estimated a priori in $L^\infty(\delta,T;\underline{L}^r)$, resp. in $L^1(\delta,T;\underline{L}^r)$, where the "regularity gain" is $r^{-1} - k^{-1} = (k'')^{-1} - n^{-1}$ resp. $(2r')^{-1} - k^{-1}$.

A simple calculation, using the restriction on q, shows that these two quantities are both bounded from above by some negative constant depending on q and n, as long as $2 \leqslant k \leqslant n$. This proves (3.20).

□□□

Lemma 3.6: Let u be a mild solution of (2.5) on the interval $[0,T_0)$. If $\nabla_x u \in L^\infty(B \times (0,T_0))$, then $u \in C([0,T_0], \underline{w}^{2,p})$, and therefore, u can be continued past T_0 as a solution of (2.5).

Proof: The function u is a mild solution of $u_{tt} - \Delta_x u_t = f + Du$, where $Du = \mathrm{div}_x(g(|\nabla_x u|^2)\nabla_x u) = g(|\nabla_x u|^2)\Delta_x u + 2g'(|\nabla_x u|^2)\nabla_x u \cdot \nabla_x^2 u \cdot \nabla u$ is a quasilinear second order operator with coefficients that are essentially bounded on $B \times [0,T_0)$ (due to the bound on $|\nabla_x u|$). In [7] it is shown that such equations always have unique global mild solutions (in the sense given in section 1); thus $u \in C([0,T_0], \underline{w}^{2,p})$. In particular, since u is uniformly bounded in $\underline{w}^{2,p}$, it can be continued past T_0, again by results on continuability of mild solutions from [7].

□□□

4. PROOFS OF THE MAIN RESULTS

The assumptions of Th. 2.1 and Th. 2.2 imply, together with the results in [7], that (2.5), together with the initial conditions (2.6), has a unique local mild solution $u(\cdot)$ in the sense defined in section 1. Also, by [7] it is sufficient to show that $\Delta_x u$ is uniformly bounded on its maximal existence interval $[0,T_0)$ with values in $\underline{L}^p$ in order to guarantee that the solution exists on all $[0,T]$.

Proof of Th. 2.1: By Lemma 3.1, for all $t \in [0,T_0)$, $\|u_t(\cdot,t)\|_2 \leqslant K$, if $f \in L^2(0,T;L^2)$, and K depends on the $L^1(0,T;L^2)$-norm of f. A standard approximation argument shows that the same estimate is also valid for mild solutions with $f \in L^1(0,T;L^2)$ (for which (2.5) need not hold almost everywhere). Rewriting the equation (2.5) in polar coordinates, and abbreviating $\sigma(\zeta) = g(|\zeta|^2)\zeta$, we obtain

$$(4.1) \qquad r \cdot u_{tt}(r,t) - (r \cdot (u_{rt}(r,t) + \sigma(u_r)))_r = r \cdot f(r,t).$$

We now use a device introduced in [1]: Since $u_r(0,t) = 0$, we can integrate (4.1) over $[0,r]$. Defining $p(r,t) = u_r(r,t) - r^{-1}\int_0^r s \cdot u_t(s,t)ds$ and $f_1(r,t) = r^{-1}\int_0^r s \cdot f(s,t)ds$, $q(r,t) = p(r,t) - u_r(r,t)$, we obtain the ordinary differential equation

$$(4.2) \qquad p_t(r,t) + \sigma(p(r,t) - q(r,t)) = f_1(r,t).$$

By Cauchy-Schwarz' inequality,

$$|q(r,t)| \leqslant r^{-1} \cdot \left(\int_0^r s \cdot |u_t(s,t)|^2 ds\right)^{1/2} \cdot \left(\int_0^r s ds\right)^{1/2} \leqslant K$$

due to the previous estimate for $\|u_t\|_2$. By the same argument, since $f \in L^1(0,T;L^2)$, $|f_1(r,t)| \leqslant m(t)$ with $m \in L^1(0,T;\mathbb{R})$. Also, $|p(r,0)|$ is bounded in terms of the initial data for all r, since $u_0 \in \underline{W}^{2,p}$ with $p > 2$. Now (2.1) implies that as $|\xi| \longrightarrow \infty$, $\lim \sup(\sigma(\xi) + (C+1)\xi) \cdot \text{sign}(\xi) = \infty$, and thus for the bound K established above, there is a constant C_2 such that for all $|\eta| \leqslant K$ and all ξ, $\sigma(\xi + \eta) \cdot \text{sign}(\xi) + C_2(1 + |\xi|) \geqslant 0$. Multiply (4.2) with $\text{sign}(p(r,t))$ and integrate from 0 to t. Then, due to the bound on q and the choice of C_2,

$$(4.3) \qquad |p(r,t)| \leqslant |p(0,t)| + \int_0^t \{C_2(1 + |p(r,\tau)|) + m(\tau)\} d\tau.$$

Gronwall's inequality implies a uniform bound for p, from which a uniform bound for u_r, i.e. for $\nabla_x u$ on $B \times [0,T_0)$ follows. Lemma 3.6 then implies that the solution is uniformly bounded on $[0,T_0)$ in $\underline{W}^{2,p}$, and hence can be continued on $[0,T]$.

□□□

Proof of Th. 2.2: Let u be a solution of (2.5) on $B \times [0,T_0)$. By Lemma 3.1 and 3.3, $\|\Delta_x u(\cdot,t)\|_2$ and $\int_0^t (\|\nabla(\sqrt{g_0}(|\nabla u|^2) \cdot \nabla u)\|_2)^2$ are bounded for any $t < T_0$ by constants depending only on the data. Since this implies that $\nabla^2 u(\cdot,t)$ is bounded in $\underline{L}^2$, uniformly in t, imbedding theorems and the growth restriction for g imply that $(|g(\nabla_x u(\cdot,t)|^2)|^{1/2}$ is bounded in $\underline{L}^{n+\varepsilon}$, uniformly in $0 \leqslant t < T_0$, for some $\varepsilon > 0$. The L^2-bound on $\nabla(\sqrt{g_0}(|\nabla u|^2) \cdot \nabla u)$, together with (2.3), imply then that $\text{div}_x(g(|\nabla_x u|^2)\nabla_x u)$ is bounded in $L^2(0,T_0;\underline{L}^s)$ with $s^{-1} < 2^{-1} + n^{-1}$.

Lemma A.2, applied to u_t, then shows that $u_t \in L^\infty(0,T_0;\underline{L}^r)$, $u_t \cdot \Delta_x u \in L^1(0,T_0;\underline{L}^{r/2})$ with $r > 2$. Employing now Lemma 3.5, a priori estimates for $\Delta_x u$ and thus for $\nabla^2 u$ in $L^\infty(0,T_0;\underline{L}^r)$, for u_t in $L^\infty(0,T_0;\underline{L}^{r'})$, and for $u_t \cdot \Delta_x u$ in $L^1(0,T_0;\underline{L}^{r'/2})$ follow, with $(r')^{-1} = r^{-1} - c$, where $c > 0$ is bounded away from 0. Iterating this argument finitely often, we obtain an a priori bound for $\nabla^2 u$ in $L^\infty(0,T_0;\underline{L}^{n+\epsilon})$, where $\epsilon > 0$. Therefore, $\nabla_x u \in L^\infty(0,T_0;\underline{L}^\infty)$. By Lemma 3.6, the solution can be continued past T_0 and must therefore exist for $t \in [0,T]$.

$\square\square\square$

Proof of Th. 2.3: For any $N > 0$, we truncate g to obtain smooth functions g^N that agree with g on $[0,N]$, and are constant on $[N+1, \infty)$. Replacing g with g^N in (2.5), and approximating the data u_0, u_1, f, by smooth data u_{No}, u_{N1}, and f_N, we then obtain approximating mild solutions u_N on $B \times [0,T]$, using e.g., Theorem 2.2. By Lemmas 3.1 and 3.3, $u_{N,t}$ and $\Delta_x u_N$ are bounded in $L^\infty(0,T_0;\underline{L}^2)$, and $\nabla_x u_{N,t}$ and $\nabla(\sqrt{g_{N0}}(|\nabla u_N|^2)\cdot\nabla u_N)$ are bounded in $L^2(0,T;\underline{L}^2)$, with bounds that depend on $\|u_1\|_2$, $\|u_0\|_{\underline{W}^{2,2}}$, on $\int_B G(|_x u_0|^2)$, and on $\int_0^T \|f(\cdot,s)\|_2 ds$. By Sobolev's imbedding theorem, the latter estimate implies a bound for $(\sqrt{g_{N0}}(|\nabla u_N|^2)\cdot\nabla u_N$ in $L^2(0,T;\underline{L}^{2n/(n-2)})$. Now, (2.3) implies that g grows at most polynomially and that $\sqrt{|g_N(z^2)|} \leqslant C \cdot (1 + \sqrt{g_{N0}}(z^2)\cdot z)^{1-\gamma}$, with some $C, \gamma > 0$, for all N. Therefore, $\sqrt{|g_N(|\nabla u_N|^2)|}$ is bounded in some $L^{2+\epsilon}(0,T;\underline{L}^{2+\epsilon})$, $\epsilon > 0$, and we obtain an N-independent bound for $\nabla_x(g_{N0}(|\nabla u_N|^2)\nabla_x u_N)$ in some $L^r(0,T;\underline{L}^r)$ with $r > 1$. We can assume that also f is in this space. By standard regularity results for parabolic equations (see [10]), this implies bounds for the $u_{N,tt}$ and for the $\Delta_x u_{N,t}$ in the same space, and $\nabla_x u_N$ and $u_{N,t}$ are in relatively compact subsets of $L^2(0,T;\underline{L}^2)$. Taking suitable subsequences, a weak solution of (2.5) is obtained which still satisfies the a priori estimates of all the u_N. Therefore u is a solution as stated in the Theorem.

$\square\square\square$

APPENDIX

Recall that we abbreviate $z^r = (z)^r = z \cdot |z|^{r-1}$ for numbers or terms z.

Lemma A.1: Let $k \geqslant 2$, $-1/2 \leqslant \rho \leqslant C$, $n \geqslant 2$. Then there exists $\epsilon = \epsilon(n,k,C) > 0$ such that for all $x, y \in \mathbb{R}$,

(A.1) $|x + y|^k + \rho x(x+y)^{k-1} - (\rho + k)xy^{k-1} - (n-k)(n-1)^{-1}|y|^k$

$\geqslant \epsilon\{(n-1)^{-1}|y|^2 + (1 + \rho/k)|x|^2)^{k/2}.$

50

<u>Proof:</u> Assume first that $y \neq 0$. Dividing by $|y|^k$ and defining $s = x/y$, we have to show

(A.2) $\{|1+s|^k + \rho s(1+s)^{k-1} - (\rho+k)s - (n-k)(n-1)^{-1}\} \geq \varepsilon \{(n-1)^{-1} + (1+\rho/k)|s|^2\}^{k/2}$

for all $s \in \mathbb{R}$. Clearly it suffices to show that

(A.3) $|1+s|^k + \rho s((1+s)^{k-1} - 1) - ks - 1 \geq 0$,

and since $s((1+s)^{k-1} - 1) \geq 0$, we only have to show (A.3) for $\rho = -1/2$. Then (A.3) is equivalent to

(A.4) $f(t) = |t|^k + t^{k-1} + (1-2k)t + (2k-3) \geq 0$ for all $t \in \mathbb{R}$,

where $t = 1+s$. One easily checks that f is convex on $(-\infty,-1] \cup [0,\infty)$, and

(A.5) $f'(-1) < 0 < f(-1)$, $f(1) = f'(1) = 0$.

Thus, (A.4) is true on $(-\infty,-1] \cup [0,\infty)$. For $-1 \leq t \leq 0$,

(A.6) $f(t) \geq -1 + 2k - 3 \geq 0$,

by the assumption on k. For $y = 0$, it is clear that an ε can be found such that (A.1) holds, which proves the Lemma.

$\square\square\square$

Lemma A.2: Let $n > 2$, $k > 2$, $p, s > 1$, and let $v : B \times [0,T] \longrightarrow \mathbb{R}$ be a weak solution of

(A.7) $v_t - \Delta_x v = h$,

that vanishes on $\partial B \times [0,T]$, and for which $v(\cdot,0) = v_0 \in L^p(B)$ and $h \in L^k(0,T;L^s(B))$ with norms bounded by K in these spaces. Let $\delta > 0$.

a) If $s \leq n/2$, then $v \in L^\infty(\delta,T;L^r(B))$, and $v \cdot \Delta_x v \in L^1(\delta,T;L^{r/2}(B))$ with $r^{-1} > s^{-1} - n^{-1}$.

b) If $n/2 < s$, then $v \in L^\infty(\delta,T;L^r(B))$, and $v \cdot \Delta_x v \in L^1(\delta,T;L^{r/2}(B))$ with $r > n$.

The norms of v and $v \cdot \Delta_x v$ depend on δ, T, n, k, s, r, p and K.

<u>Proof:</u> We write $v = v_1 + v_2$, where v_1 solves a homogeneous heat equation with initial data $v(\cdot,0)$, and v_2 solves (A.7) with zero initial data. Then $v_1 \in C^\infty([\delta,T] \times B)$ for any positive δ. By well-known maximal regularity results for parabolic equations (see [10]), $v_{2\,t} \in L^k(0,T;L^s(B))$, and $v_2 \in L^k(0,T;W^{2,s}(B))$ with norms depending on K. Thus, $v_2 \in W^{\theta,k}(0,T; [W^{2,s}(B), L^s(B)]_{\theta,k})$ for $0 < \theta < 1$, $\theta \neq 1/k$, using standard results from interpolation theory (see [3]). For $\theta > 1/k$, θ sufficiently close to $1/k$, we have continuous imbeddings from $W^{\theta,k}(0,T; X)$ into $L^\infty(0,T; X)$, if X is a Banach space, and from $[W^{2,s}(B), L^s(B)]_{\theta,k}$ into $L^r(B)$, if $r^{-1} > s^{-1} - n^{-1}$, since $k \geq 2$. Thus, $v_2 \in L^\infty(0,T; L^r(B))$, which proves the assertions about $v = v_1 + v_2$ in both cases.

In case a), v_2 is also bounded in $L^k(0,T; L^m(B))$, $m^{-1} > s^{-1} - 2n^{-1}$ Using the bound for $\Delta_x v_2$ in $L^k(0,T;L^s(B))$, and the previously established regularity properties of v_1, we obtain the assertion for $v \cdot \Delta_x v$ in this case. In case b), v_2 is bounded in $L^k(0,T;L^\infty(B))$, which gives the assertion about $v \cdot \Delta_x v$ due to $k \geqslant 2$.

□□□

References

1. G. Andrews, On the existence of solutions to the equation $u_{tt} = u_{xxt} + \sigma(u_x)_x$. J. Diff. Eqns. $\underline{35}$ (1980), 200 - 231.
2. G. Andrews, J. M. Ball, Asymptotic behaviour and changes of phase in one-dimensional nonlinear viscoelasticity. J. Diff. Eqns. $\underline{44}$ (1982), 306 - 341.
3. J. Bergh, J. Löfström, Interpolation spaces. Springer: Berlin - Heidelberg - New York 1976.
4. J. Clements, Existence theorems for a quasilinear evolution equation. SIAM J. Appl. Math. $\underline{26}$ (1974), 745 - 752.
5. C. M. Dafermos, The mixed initial-boundary value problem for the equations of nonlinear one-dimensional viscoelasticity. J. Diff. Eqns. $\underline{6}$ (1970), 71 - 86.
6. Y. Ebihara, On some nonlinear evolution equations with strong dissipation. J. Diff. Eqns. $\underline{34}$ (1979), 339 - 352.
7. H. Engler, F. Neubrander, J. Sandefur, Strongly damped second order equations. These proceedings.
8. J. M. Greenberg, R. C. MacCamy, V. J. Mizel, On the existence, uniqueness, and stability of the equation $\sigma'(u_x)u_{xx} + \lambda u_{xxt} = \rho_0 u_{tt}$. J. Math. Mech. $\underline{17}$ (1968), 707 - 728.
9. J. M. Greenberg, On the existence, uniqueness, and stability of the equation $\rho_0 X_{tt} = E(X_x)X_{xx} + \lambda X_{xx}$. J. Math. Anal. Appl. $\underline{25}$ (1969), 575 - 591.
10. O. A. Ladyzenskaya, V.A. Solonnikov, N.N. Uraltseva, Linear and quasilinear equations of parabolic type. Am. Math. Soc. Providence 1968.
11. H. Pecher, On global regular solutions of third order partial differential equations. J. Math. Anal. Appl. $\underline{73}$ (1980), 278 - 299.
12. R. L. Pego, Phase transitions: Stability and admissibility in one dimensional nonlinear viscoelasticity. IMA Preprint # 180, Sept. 1985.
13. N. Yamada, Note on certain nonlinear evolution equations of second order. Proc. Japan Acad. $\underline{55}$ Ser. A (1979), 167 - 171.

STRONGLY DAMPED SEMILINEAR SECOND ORDER EQUATIONS

H.Engler, F.Neubrander and J.Sandefur
Department of Mathematics
Georgetown University
Washington,   D.C. 20057

1. Introduction

In this note we try to give a unified treatment of the initial value problem

$$(1) \qquad u''(t) + (aA+bI)u'(t) + (cA+dI)u(t) = f(t,u(t),u'(t)),$$
$$u(0) = u_o \ , \ u'(0) = u_1,$$

in a Banach space E , which has been studied by each of the authors (see [1],[2],[6] and [8]) . Key assumptions that we make are that $a > 0$, -A generates a strongly continuous semigroup $(T(t))$, and $f(\cdot)$ is locally Lipschitz continuous from $\mathbb{R}_+ \times [D(A)] \times E$ into E , where $[D(A)]$ denotes the Banach space $D(A)$ endowed with the graph norm $|x|_A = |x| + |Ax|$. Examples of such problems are, among others, the strongly damped nonlinear Klein Gordon equation, and the vibrating beam equation. Replacing A by $\tilde{A} = aA+eI$ for an appropriate constant e and adjusting the constants a,b,c,d , one can assume without loss of generality that every $\mu \in \mathbb{C}$ with $\mathrm{Re}\mu > -w$, $w > 0$, is contained in the resolvent set $\rho(-A)$ of the generator -A . Moreover, it is easy to realize that one can absorb the terms $bu'(t)$, $cAu(t)$, and $du(t)$ into the nonlinearity without changing its character; i.e., the new nonlinearity will remain Lipschitz continuous from $\mathbb{R}_+ \times [D(A)] \times E$ into E . Therefore, as far as local existence and uniqueness of solutions of (1) is concerned, it is sufficient to study the initial value problem

$$(2) \qquad u''(t) + Au'(t) = f(t,u(t),u'(t)),$$
$$u(0) = u_o \ , \ u'(0) = u_1,$$

where -A generates a strongly continuous semigroup $(T(t))$ on a Banach space E, and $\{ \mu \in \mathbb{C} : \mathrm{Re}\mu > -w \} \subset \rho(-A)$ for some $w > 0$. Under these assumptions, we give conditions on $f(\cdot)$ that imply local existence and uniqueness of mild solutions to the equation (2) . We also indicate how global existence can be derived in certain situations.

2. Local Results

We use the following assumptions on $-A$:

(i) $-A$ generates a strongly continuous semigroup $(T(t))$ on a Banach space E and $\{ \mu \varepsilon \mathbb{C} : \operatorname{Re}\mu > -w \} \subset \rho(-A)$ for some $w > 0$.

(i') $-A$ generates an analytic semigroup $(T(t))$ on a Banach space E and $\{ \mu \varepsilon \mathbb{C} : \operatorname{Re}\mu > -w \} \subset \rho(-A)$ for some $w > 0$.

Under assumption (i') one can define the fractional powers $(-A)^{\alpha}$ in the usual way (see, for example, [5] or [7]), and we denote by $[D((-A)^{\alpha})]$ the Banach space obtained by endowing $D((-A)^{\alpha})$ with the "graph"-norm $|x|_{\alpha} = |(-A)^{\alpha}x|$ (which is equivalent to the usual graph norm $|x| + |(-A)^{\alpha}x|$). Recall that the analyticity of the semigroup $(T(t))$ implies that

(3)
$$|(-A)^{\alpha}T(t)| \leq C(\alpha,T)t^{-\alpha} \quad \text{for} \quad 0 < t \leq T .$$

We use the following assumptions on f :

(ii) Let $T > 0$, let $\alpha = 0$ if $-A$ fulfills assumption (i) , and let $0 \leq \alpha < 1$ if $-A$ fulfills assumption (i') .
   (a) $f: [0,T) \times D(-A) \times D((-A)^{\alpha}) \longrightarrow E$.
   (b) for every $(u_0,u_1) \varepsilon D(-A) \times D((-A)^{\alpha})$ we have that $f(\cdot,u_0,u_1)$ is strongly measurable and that there exists a constant C_0 such that $|f(t,u_0,u_1)| \leq C_0$ for all $t \varepsilon [0,T)$.
   (c) for every $(u_0,u_1) \varepsilon D(-A) \times D((-A)^{\alpha})$ there is an $R > 0$ and $c_R(T) > 0$ such that $|x_i - u_0|_1 + |y_i - u_1|_{\alpha} \leq R$ $(i = 1,2)$ implies that $|f(t,x_1,y_1) - f(t,x_2,y_2)| \leq$ $c_R(T)(|x_1 - x_2|_1 + |y_1 - y_2|_{\alpha})$ for all $t \varepsilon [0,T)$.

<u>Definition</u>   Consider the following system of integral equations:

$$u(t) = (T(t)-I)(-A)^{-1}u_1 + u_o + \int_o^t (T(t-s)-I)(-A)^{-1}f(s,u(s),v(s))ds$$

(2')

$$v(t) = T(t)u_1 + \int_o^t T(t-s)f(s,u(s),v(s))\,ds \quad .$$

The first component $u(\bullet)$ of a solution $(u(\bullet),v(\bullet))$ of the system (2') is called a <u>mild</u> solution of the initial value problem (2) . Note that if $(u(\bullet),v(\bullet))$ is a solution of (2') , then $u(t) = u_o + \int_o^t v(s)\,ds$.

<u>Theorem 2.1</u>   Assume that the assumptions (i') and (ii) hold. Then, for every $(u_o,u_1) \in D(-A) \times D((-A)^\alpha)$, there exists a $t_o > 0$ such that (2') has a unique local solution $(u(\bullet),v(\bullet)) \in C([0,t_o],[D(-A)] \times [D((-A)^\alpha)])$.

<u>Proof:</u> We define a mapping S on $C([0,t_o],E \times E)$ by $S(w_1(\bullet),w_2(\bullet)) = (\hat{w}_1(\bullet),\hat{w}_2(\bullet))$ where

$$\hat{w}_1(t) = (T(t)-I)u_1 - Au_o + \int_o^t (T(t-s)-I)f(s,(-A)^{-1}w_1(s),(-A)^{-\alpha}w_2(s))\,ds$$
and
$$\hat{w}_2(t) = T(t)(-A)^\alpha u_1 + \int_o^t (-A)^\alpha T(t-s)\,f(s,-A^{-1}w_1(s),(-A)^{-\alpha}w_2(s))\,ds,$$

and look for a fixed point $(\bar{w}_1(\bullet),\bar{w}_2(\bullet))$ of S for sufficiently small t_o . Obviously, $(u(\bullet),v(\bullet)) := ((-A)^{-1}\bar{w}_1(\bullet),(-A)^{-\alpha}\bar{w}_2(\bullet))$ will then be a solution of (2') . Also, any solution of (2') with the regularity properties in the statement of the theorem defines by $(\bar{w}_1(\bullet),\bar{w}_2(\bullet)) := ((-A)u(\bullet),(-A)^\alpha v(\bullet))$ a fixed point of S .

To find a fixed point of S, we apply Banach's fixed point theorem in the complete metric space

$$X(t_o) = \{(w_1(\bullet),w_2(\bullet)) \in C([0,t_o],E \times E) :$$
$$w_1(0) = -Au_o \,, \; w_2(0) = (-A)^{-\alpha}u_1,$$
$$|w_1(t)-(-A)u_o| + |w_2(t)-(-A)^\alpha u_1| \le R \quad \text{for all } 0 \le t \le t_0 \}$$

with R as in (ii) and a suitable small t_o . Since any fixed point of S will belong to some $X(t_o)$ (for t_o small), we only have to find a fixed point in a space of this form.

We equip $X(t_o)$ with the standard metric

$$d((w_1(\bullet),w_2(\bullet)),(z_1(\bullet),z_2(\bullet)))$$
$$= \sup \{ \ |w_1(t)-z_1(t)| \ + \ |w_2(t)-z_2(t)| \ : \ 0 \leq t \leq t_o \ \} \ .$$

Then, by the boundedness of $T(t)$ on compact intervals, by (3), and the assumption (ii) , we obtain

$$d(S(w_1(\bullet),w_2(\bullet)),S(z_1(\bullet),z_2(\bullet))) \ \leq$$

$$\sup \{ \ \int_o^t \ (|T(t-s)-I| \ + \ |(-A)^\alpha T(t-s)|)|f(s,-A^{-1}w_1(s),(-A)^{-\alpha}w_2(s)) \ -$$

$$f(s,-A^{-1}z_1(s),(-A)^{-\alpha}z_2(s))| \ : \ 0 \leq t \leq t_o \ \}$$

$$\leq \ \int_o^{t_o} C(1 + (t_o-s)^{-\alpha})(|w_1(s)-z_1(s)| \ + \ | \ w_2(s)-z_2(s)|) \ ds$$

for some constant C depending on $c_R(T)$ in (ii) and $C(\alpha,T)$ in (3) . From this we get

$$d(S(w_1(\bullet),w_2(\bullet)),S(z_1(\bullet),z_2(\bullet))) \ \leq$$

$$C(t_o+t_o^{1-\alpha}(1-\alpha)^{-1}) \ d((w_1(\bullet),w_2(\bullet)),(z_1(\bullet),z_2(\bullet))) \ \leq$$

$$\tfrac{1}{2} \ d((w_1(\bullet),w_2(\bullet)),(z_1(\bullet),z_2(\bullet))) \qquad \text{for an appropriately chosen } t_o \ .$$

Also, S maps $X(t_o)$ into itself since

$$d(S(w_1(\bullet),w_2(\bullet)),(-Au_o,(-A)^\alpha u_1)) \ \leq$$

$$d(S(w_1(\bullet),w_2(\bullet)),S(-Au_o,(-A)^\alpha u_1)) \ + \ d(S(-Au_o,(-A)^\alpha u_1),(-Au_o,(-A)^\alpha u_1))$$

$$\leq \ \tfrac{R}{2} + \sup\{|(T(t)-I)u_1| \ + \ |(T(t)-I)(-A)^\alpha u_1|$$
$$+ \int_o^t C(1+(t-s)^{-\alpha})|f(s,u_o,u_1)| \ ds \ : \ 0 \leq t \leq t_o\}$$

$$\leq \tfrac{R}{2} + \sup\{|(T(t)-I)u_1| \ + \ | \ (T(t)-I)(-A)^\alpha u_1| \ : \ 0 \leq t \leq t_o\} \ +$$
$$C'(t_o+t_o^{1-\alpha}(1-\alpha)^{-1}),$$

where C' depends on C_o . Thus, by picking t_o small enough, we obtain

$$d(S(w_1(\bullet),w_2(\bullet)),(-Au_o,(-A)^\alpha u_1)) \ < \ R \ .$$

It follows that S has a unique fixed point which gives the desired solution $(u(\bullet),v(\bullet))$. Since $u'(\bullet) = v(\bullet)$, we have that $(-A)^\alpha u'(\bullet)$ will still be continuous ∎

Comments: (i) This result improves a result of Fitzgibbon [3] , in which he obtained local existence and uniqueness of mild solutions of (2') for f(•) being locally Lipschitz continuous on $R_+ \times [D((-A)^\alpha)] \times [D((-A)^\alpha)] \longrightarrow E$ for $\alpha \in [0,1)$. As our examples will demonstrate, this is an important improvement.

(ii) If $c_R(•)$ of condition (ii) and $|f(•,u_o,u_1)|$ are in $L^p([0,T),R)$ with $p(1-\alpha) > 1$, then the statement of the theorem remains valid, and the proof requires only some modifications.

(iii) If the nonlinearity f(•) does not depend on u'(•) , then α can be any number between 0 and 1 . In this case our result contains a regularity statement: if $u_1 \in D((-A)^\alpha)$, then u'(•) is continuous with values in $D((-A)^\alpha)$.

(iv) The result is false if one allows $\alpha = 1$. A counterexample is simply f(t,u(•),v(•)) = 2Av(•), which gives rise to the 'backward' equation u''(•) = Au'(•). In general, this equation is not well-posed.

By taking $\alpha = 0$ in the above theorem, we can deduce a result on mild solutions of (2) if (-A) only generates a strongly continuous semigroup. Also, in this case it is possible to allow f(•) to depend in an L^1-manner on t . We state the corresponding result. Its proof follows the one of Theorem 2.1 .

<u>Corollary 2.2</u> Assume the assumptions (i) and (ii) hold with $\alpha = 0$, and with $c_R(T)$, C_o replaced by $c_R(t)$, $C_o(t)$ where $\int_o^T c_R(t) + C_o(t)\,dt < \infty$. Then there exists a unique local solution (u(•),v(•)) of (2') for which Au(•) and v(•) = u'(•) are continuous.

3. Examples

<u>Example 3.1</u>   General linear second order equations

Let (-A) be the generator of a strongly continuous semigroup (T(t)) on a Banach space E . Let $(B(t))_{o \leq t \leq T}$ be a family of closed linear operators with $D(-A) \subset D(B(t))$ for all $0 \leq t \leq T$ such that B(•)x is strongly measurable for all $x \in D(-A)$, and $|B(t)x| \leq c(t)|x|_1$ for all $0 \leq t \leq T$ and $x \in D(-A)$, where $\int_o^T c(t)\,dt < \infty$. Then the initial value problem

$$u''(t) + Au'(t) + B(t)u(t) = f_o(t), \quad (0 < t \le T),$$
$$u(0) = u_o, \quad u'(0) = u_1,$$

has a unique mild solution $u(\cdot)$ for every $u_o \in D(-A)$, $u_1 \in E$, and $f_o(\cdot) \in L^1([0,T],E)$. Also, $u'(\cdot) \in C([0,T],E)$.

This can be seen by letting $f(t,u(t),v(t)) = f_o(t) - B(t)u(t)$ in Corollary 2.2. A closer look at the proof shows that the existence interval $[0,t_o]$ can be found independently of u_o and u_1. Repeating the existence argument on $[t_o,2t_o]$ etc., we see that the solution can be continued on $[0,T]$. In fact, if $B(t)$ is defined for $t \ge 0$, and if $c(\cdot)$ is bounded on compact sets of $\mathbb{R}_+$, then we have unique global mild solutions. It is therefore seen that this technique uses semigroup methods to give existence and uniqueness results for linear time-dependent differential equations, where the operators A and $B(\cdot)$ do not necessarily commute, and also $D(B(\cdot))$ need not be constant.

A specific example would be the case in which $E = L^2(\Omega)$, where $\Omega \subset \mathbb{R}^n$ is open, bounded, with smooth boundary, and $-A = \Delta$ with domain $D(-A) = W^{2,2}(\Omega) \cap W_o^{1,2}(\Omega)$.

Let $a_{ij}(\cdot) : [0,\infty) \times \Omega \longrightarrow \mathbb{R}$ $(1 \le i,j \le n)$ be measurable such that

$$\int_o^T \text{ess sup}_\Omega |a_{ij}(t,\cdot)| \, dt < \infty$$

for all $T > 0$, and define $B(t)$ to be the closure of the operator

$$B(t)u(x) = \Sigma_{i,j=1}^n a_{ij}(t,x) \, \partial^2/\partial x_i \partial x_j \, u(x)$$

in $L^2(\Omega)$. Then $(B(t))_{t \ge o}$ satisfies all assumptions above. This example shows that the t-dependence of $B(\cdot)$ can be quite irregular.

<u>Example 3.2</u>   Strongly damped quasilinear wave equations

Let $E = L^p(\Omega)$, where $\Omega \subset \mathbb{R}^n$ is open, bounded, with smooth boundary, and $p > n$. Define $-A = \Delta$, $D(-A) = W^{2,p}(\Omega) \cap W_o^{1,2}(\Omega)$. Let $g: \mathbb{R}^n \longrightarrow \mathbb{R}^n$ be locally Lipschitz continuous together with its matrix of first derivatives, and define $f(u)(x) = \text{div}_x \, g(\nabla_x u)(x)$ for $u \in D(-A)$. Since for $p > n$ the imbedding of $D(-A)$ in $W^{1,\infty}(\Omega)$ holds, it is straightforward to show that $f(\cdot)$ satisfies all assumptions of Theorem 2.1. Consequently, for every $u_o \in W^{2,p}(\Omega) \cap W_o^{1,2}(\Omega)$ and every $u_1 \in L^p$, there exists a unique local mild solution $u(\cdot)$ of

$$u_{tt}(t,x) - \Delta u_t(t,x) = \text{div } g(\nabla u)(t,x) \quad ,$$
$$u(0,x) = u_o(x) \quad , \quad u_t(0,x) = u_1(x) \quad ,$$
$$u(t,x) = 0 \quad \text{for} \quad x \in bd(\Omega) \quad .$$

Moreover, in this case $u(t,\cdot)$ and $u_t(t,\cdot)$ are continuous as L^p-valued functions. If $u_1 \in W_o^{1,p}(\Omega)$, then, since $D((-A)^{1/2}) = W_o^{1,p}(\Omega)$ (see [5] and [9]), we obtain $u_t(\cdot) \in C([0,t_o],W_o^{1,p}(\Omega))$.

Example 3.3   The strongly damped Klein-Gordon equation

Consider the equation

$$u''(t) + aAu'(t) + bu'(t) + cAu(t) - du(t) = - \gamma |u(t)|^{p-1}u(t)$$

with $A = -\Delta$, $a > 0$, and $p \geq 1$ on a bounded domain $\Omega \subset \mathbb{R}^n$, $n = 1,2,3$, with smooth boundary $bd(\Omega)$, and $u(t,x) = 0$ for $x \in bd(\Omega)$. Letting $E = L^2(\Omega)$, the closure of $(-A)$ generates an analytic semigroup and $0 \in \rho((-A))$.
By the Sobolev imbedding theorems, $|u|_\infty \leq C|(-A)u|$ and $|u|_p = (\int_\Omega |u|^p dx)^{1/p} \leq C|(-A)u|$, we get conditions (i') and (ii), and therefore existence of $u(\cdot)$ and $u'(\cdot)$ on $[0,t_o)$ for some $t_o > 0$.

By having zero Neumann conditions on $bd(\Omega)$, or by letting $\Omega = \mathbb{R}^n$, and then letting $A = -a\Delta + eI$ for $e > 0$, we again get local existence and uniqueness, since the spectrum $\sigma(-A)$ is bounded away from zero; i.e., $\text{Re } \sigma(-A) \leq -e$.

4. Global Existence of Solutions

Theorem 4.1   Assume (i'). Also assume that condition (ii) holds for all $T > 0$ and all $R > 0$. Let $L > 0$ be such that there exists a solution $(u(\cdot),v(\cdot))$ to (2') on $[0,L]$ for which $Au(\cdot)$ and $(-A)^\alpha v(\cdot)$ are continuous. If $(u(\cdot),v(\cdot))$ cannot be continued beyond $[0,L]$, then either $L = +\infty$ or $\limsup_{t \to L-} (|u(t)|_1 + |v(t)|_\alpha) = \infty$.

Proof: Suppose $L < \infty$ and $|u(t)|_1 + |v(t)|_\alpha \leq M$ for all $t \in [0,L)$. We will show first that $\lim_{t \to L-} (-A)u(t)$ exists.

Define $w_1(t) = -Au(t)$, $w_2(t) = (-A)^\alpha v(t)$, and $f(s) = f(s,(-A)^{-1}w_1(s),(-A)^{-\alpha}w_2(s))$. Then $(w_1(\bullet),w_2(\bullet))$ satisfies

$$w_1(t) = (T(t)-I)u_1 - Au_0 + \int_0^t (T(t-s)-I)f(s)\, ds$$

$$= [(T(t)-I)u_1 - Au_0 - \int_0^t f(s)\, ds] + \int_0^t T(t-s)f(s)\, ds ,$$

$$w_2(t) = (-A)^\alpha T(t)u_1 + \int_0^t (-A)^\alpha T(t-s)f(s)\, ds .$$

We have that

$$|f(s)| \leq |f(s,(-A)^{-1}w_1(s),(-A)^{-\alpha}w_2(s)) - f(s,u_0,u_1)| + |f(s,u_0,u_1)|$$
$$\leq c_R(L)(1 + |u(s) - u_0|_1 + |v(s) - u_1|_\alpha) \leq C$$

for $s < L$ and some constant C . We want to show that $\lim_{t \to L-} (w_1(t),w_2(t))$ exists.

Clearly, due to the bound on $f(\bullet)$, the expressions in braces in the formula for $w_1(\bullet)$ and the first term in the formula for $w_2(\bullet)$ have limits as $t \to L-$.

To show this for the remaining terms , we take $0 < t < t_1 < t_2 < L$. Then, considering the second term in the expression for $w_2(\bullet)$,

$$|\int_0^{t_1}(-A)^\alpha T(t_1-s)f(s)ds - \int_0^{t_2}(-A)^\alpha T(t_2-s)f(s)ds| \leq$$

$$|\int_0^{t_1} (-A)^\alpha T(t_1-s)[I-T(t_2-t_1)]f(s)ds| + |\int_{t_1}^{t_2} (-A)^\alpha T(t_2-s)f(s)ds| = I_1 + I_2,$$

and $I_1 \leq |T(t_1-t)[I-T(t_2-t_1)]\int_0^t (-A)^\alpha T(t-s)f(s)\, ds|$

$$+ |[I-T(t_2-t_1)]\int_t^{t_1}(-A)^\alpha T(t_1-s)f(s)\, ds| = I_3 + I_4 .$$

By the analyticity of $(T(t))$, $I_4 \leq C(t_1-t)^{1-\alpha} < \varepsilon/4$ for t close to L . Then, letting $x = \int_0^t (-A)^\alpha T(t-s)f(s)\, ds$,

$I_3 \leq |T(t_1-t)| \, |[I-T(t_2-t_1)]x| < \varepsilon/4$ for t_2 and t_1 close enough to L . Likewise, by the analyticity of $(T(t))$, $I_2 \leq C(t_2-t_1)^{1-\alpha} < \varepsilon/4$ for t_1 and t_2 close enough to L . Thus, $\lim_{t \to L-} w_2(t)$ exists.

The argument for $w_1(\bullet)$ is the same (taking $\alpha = 0$).

Now define $u(L) = (-A)^{-1}w_1(L)$, $v(L) = (-A)^{-\alpha}w_2(L)$. Then $(u(L),v(L)) \in D(-A)xD((-A)^{\alpha})$ and

$$(u(\bullet),v(\bullet)) \in C([0,L],[D(-A)]x[D((-A)^{\alpha})]) .$$

Using $\tilde{u}_o = u(L)$ and $\tilde{u}_1 = v(L)$, we can extend the solution to an interval $[0,L+\delta)$, contradicting our original assumption ∎

Corollary 4.2   Under the assumptions of Theorem 4.1 except that $(T(t))$ is not analytic, Theorem 4.1 holds with $\alpha = 0$.

Corollary 4.3   Let $f : [0,\infty)xD(-A) \longrightarrow E$. Assume that for $u_o \in D(-A)$ and any given $R > 0$, there exists a function $c_R(\bullet)$ on $[0,\infty)$ with $\int_o^T c_R(t) dt < \infty$ for $T > 0$ such that $|f(t,u_o)| \leq c_1(t)$ for all $t \geq 0$, and that $|x_i - u_o| \leq R$ ($i = 1,2$) implies that $|f(t,x_1) - f(t,x_2)| \leq c_R(t)|x_1 - x_2|$ for $t \leq T$. Also assume (i) . Let $L > 0$ be such that there exists a solution to

$$u(t) = (T(t)-I)(-A)^{-1}u_1 + u_o + \int_o^t (T(t-s)-I)(-A)^{-1}f(s,u(s))ds$$

on $[0,L]$ for which $Au(\bullet)$ is continuous, but $u(\bullet)$ cannot be continued beyond $[0,L]$.
Then, either $L = +\infty$ or $\limsup_{t->L-} |u(t)|_1 = \infty$.

Proof: The proof that $\lim_{t->L-} (-A)u(t)$ exists is identical to the above. It is also straightforward to show that

$$v(t) = T(t)u_1 + \int_o^t T(t-s)f(s,u(s)) ds$$

is bounded and Cauchy (in the sense of the proof of Theorem 4.1) on $[0,L]$. Therefore, $\lim_{t->L-} v(t)$ also exists. ∎

We now give an example to show how these results can be used to obtain global existence.

Example 4.4   Consider the strongly damped Klein-Gordon equation of example 3.3 where $\Omega \subset \mathbb{R}^n$ is bounded with smooth boundary $bd(\Omega)$, and $u(t,x) = 0$ for $x \in bd(\Omega)$. Also assume that $a > 0$, $b \geq 0$, $c > 0$, $p > 1$, and $\gamma > 0$. We have that $(u(\bullet),v(\bullet))$ exists on $[0,L)$ for some $L > 0$. Define

$$E_1(t) = -c<\Delta u(t),u(t)> + |u'(t)|^2 - d|u(t)|^2 + \frac{2\gamma}{p+1}|u(t)^{(p+1)/2}|^2 .$$

Then $E_1'(t) = -2a|\nabla u'(t)|^2 -2b|u(t)|^2 \leq 0$, which implies that $E_1(\cdot)$ is nonincreasing. There is no problem in showing that $E_1(t)$ and consequently each of the terms in $E_1(t)$ is bounded if $d \leq 0$. In the case that $d > 0$, it is easy to show that if $|u(t)|^2 \longrightarrow +\infty$, then $- d|u(t)|^2 + \frac{2\gamma}{p+1}|u(t)^{(p+1)/2}|^2 \longrightarrow +\infty$ (see [1]).
Thus we get that $|u'(t)|^2$, $|\nabla u(t)|^2$, $|u(t)|^2$ and $|u(t)^{(p+1)/2}|^2$ are uniformly bounded in t .

Since we used $D(-\Delta)$ to derive local existence, we must now pay the price by showing that $|\Delta u(t)|^2$ does not go to infinity in finite time. To do this we consider

$$E_2(t) = \frac{1}{2}(E_1(t)/a + b|\nabla u(t)|^2 + a|\Delta u(t)|^2 - 2\langle u'(t), \Delta u(t)\rangle) .$$

We then have

$$E_2'(t) = -\frac{b}{a}|u'(t)|^2 - c|\Delta u(t)|^2 + d|\nabla u(t)|^2 - \gamma p \int_\Omega |u|^{p-1}|\nabla u|^2 dx .$$

Since $|\nabla u(t)|^2$ is bounded , $E_2'(t) \leq C$ for some constant $C \geq 0$. Thus , $E_2(t) \leq Ct + C'$ for some constant C' . Using the inequality $2|\langle\Delta u(t),u'(t)\rangle| \leq \delta|\Delta u(t)|^2 + |u'(t)|^2/\delta$ for δ arbitrarily small, and since $|u(t)|$ and $|u'(t)|$ are bounded, we have that $|\Delta u(t)|^2$ is bounded on finite intervals . By Theorem 4.1 with $\alpha = 0$, we have global existence of solutions for the strongly damped Klein-Gordon equation, see [1] for more details.

Acknowledgement: We would like to thank Professor Andrew Vogt for helpful suggestions.

References

[1]   P.Aviles, J.Sandefur: Nonlinear second order equations with applications to partial differential equations, Journal of Differential Equations, Vol.58, No.3 (1985), 404-427 .

[2]   H.Engler: Existence of regular solutions for semilinear parabolic integro-differential equations, Ann. Mat. Pura Appl. (to appear).

[3]   W.E.Fitzgibbon: Strongly damped quasilinear evolution equations, Journal of Mathematical Analysis and Applications 79 (1981), 536-550.

[4]   D.Fujiwara: On the asymptotic behavior of the Green operators
        for elliptic boundary problems and the pure imaginary powers of
        some second order operators, J.Math.Soc.Japan 21 (1969),
        481-522.

[5]   J.A.Goldstein: Semigroups of Linear Operators and Applications,
        Oxford University Press, 1985.

[6]   F.Neubrander: Well-posedness of higher order abstract Cauchy
        problems, Transactions AMS (to appear).

[7]   A.Pazy: Semigroups of Linear Operators and Applications to
        Partial Differential Equations, Springer , 1983.

[8]   J.Sandefur: Existence and Uniqueness of solutions of second
        order nonlinear differential equations, SIAM J.Math.Anal.14
        (1983), 477-487.

[9]   R.Seeley: Interpolation in L^p with boundary conditions. Studia
        Math.44 (1972), 47-60.

NONLINEAR SEMIGROUP THEORY AND VISCOSITY SOLUTIONS OF HAMILTON-JACOBI PDE

Lawrence C. Evans
Department of Mathematics
University of Maryland
College Park, MD 20742

1. Introduction

My purpose in this expository paper is to explain how the abstract theory of nonlinear semigroups has led, at least indirectly, to the development by M.G. Crandall, P.L. Lions and - to a lesser extent - myself and others of the concrete theory of viscosity solutions for Hamilton-Jacobi-type PDE. I hope thereby to make two points: first, that abstract considerations can indeed guide the developing investigation of specific nonlinear PDE, but second, that the abstractions alone rarely provide sufficiently detailed knowledge. In other words, my moral will be that in trying to verify for specific PDE the hypothesis of an abstract nonlinear theorem, it's often the case that we have to develop such expertise about the PDE itself that in the long run we don't really need the abstractions at all. The organization of the paper reflects this point: I begin with a description of nonlinear operators on Banach spaces and eventually conclude with a discussion of on-going research in PDE, the abstract considerations having evaporated along the way.

The following draws largely upon my two previous expository papers [16], [18], the first on nonlinear semigroups and the second on PDE from control theory. What's new is mainly my attempt to explain the connections between these subjects. I describe also various applications of the theory and list in the next-to-last section some important open and open-ended problems.

Any interested readers should look also at Crandall-Souganidis [15] for another expository introduction to these matters from a different viewpoint.

2. Nonlinear Semigroups; Accretive Operators

This section is a brief outline of some facts about nonlinear semigroups and accretive operators: see, for example, [16] or [6] for details and more references.

Nonlinear semigroup theory concerns the solving of initial value problems of this form:

$$(\text{IVP}) \quad \begin{cases} \dfrac{du}{dt}(t) + A(u(t)) = 0 \quad (t>0) \\[2ex] u(0) = x_0. \end{cases}$$

We are given here a Banach space X, a nonlinear operator A mapping some domain $D(A) \subset X$ into X, and an element $x_0 \in D(A)$; the unknown is $u: [0,\infty) \to X$.

We attempt to solve (IVP) by studying the discrete approximations

$$(\text{IVP})_\lambda \quad \begin{cases} \dfrac{x_k^\lambda - x_{k-1}^\lambda}{\lambda} + A(x_k^\lambda) = 0 \quad (k=1,2,..) \\[2ex] x_0^\lambda = x_0, \end{cases}$$

for $\lambda > 0$. Now the first line of $(\text{IVP})_\lambda$ reads

$$x_k^\lambda + \lambda A(x_k^\lambda) = x_{k-1}^\lambda;$$

and so

$$x_k^\lambda = J_\lambda(x_{k-1}^\lambda) \qquad\qquad (k=1,2,..),$$

provided the resolvent

$$J_\lambda \equiv (I+\lambda A)^{-1}$$

exists. If we hope to show solutions of $(\text{IVP})_\lambda$ converge to some kind of solution of (IVP), it is therefore natural to assume that the J_λ's exist and are somehow well-behaved. We will therefore suppose that

(1) $$\text{Range}(I+\lambda A) = X$$

and

(2) $$\| J_\lambda x - J_\lambda \hat{x} \| \leq \| x - \hat{x} \| \qquad (x, \hat{x} \in X, \lambda > 0).$$

Assumption (1) implies the J_λ's to be everywhere defined on X and assumption (2) the J_λ's to be single-valued and contractions. An operator satisfying (2) is called <u>accretive</u>; it is <u>m-accretive</u> if (1) holds as well.

Generation Theorem (Crandall-Liggett [9]) Assume A is m-accretive on X, $x_0 \in \overline{D(A)}$.   Then the functions
$u^\lambda(t)$ ($\equiv x_k^\lambda$ for $k\lambda \leqslant t < (k+1)\lambda$)   converge as $\lambda \to 0$ uniformly on compact subsets of $[0,\infty)$ to a limit function $u(t)$.

We sometimes write $u(t) = S(t)x_0$ to display explicitly the dependence on the initial condition; the family $\{S(t)\}_{t \geqslant 0}$ is the nonlinear contraction semigroup generated by A. Clearly $u(t) = S(t)x_0$ is a major contender for a solution of (IVP), but a precise interpretation of this is difficult (since, for example, $u(t)$ may be nowhere differentiable).

I omit any discussion of this point (cf. [6]) and instead turn our attention to hypotheses (1),(2), in the hopes of finding some specific examples. First notice that we can rewrite (2) to read

(2)* $$\|y - \hat{y}\| \leqslant \|y - \hat{y} + \lambda(A(y) - A(\hat{y}))\| \qquad (y, \hat{y} \in D(A), \lambda > 0)$$

and thus

(2)** $$[y - \hat{y}, A(y) - A(\hat{y})]_+ \geqslant 0 \qquad (y, \hat{y} \in D(A)),$$

where we set

$$[u,v]_+ \equiv \lim_{\lambda \to 0_+} \frac{\|u + \lambda v\| - \|u\|}{\lambda}$$
$$= \inf_{\lambda > 0} \frac{\|u + \lambda v\| - \|u\|}{\lambda} \qquad (u, v \in X).$$

The pairing $[,]_+$, defined for any Banach space, is a kind of "approximate inner product;" we use this observation later in §4. A related definition is this:

$$[u,v]_- \equiv \lim_{\lambda \to 0_-} \frac{\|u + \lambda v\| - \|u\|}{\lambda}$$
$$= \sup_{\lambda < 0} \frac{\|u + \lambda v\| - \|u\|}{\lambda}.$$

An operator A is strongly accretive provided

(3) $$[y - \hat{y}, A(y) - A(\hat{y})]_- \geqslant 0 \qquad (y, \hat{y} \in D(A)).$$

For later applications let us consider the special case that $X \equiv C(\overline{\Omega})$, for $\Omega \subset \mathbf{R}^n$ open, bounded, smooth. It then turns out that the brackets $[,]_\pm$ have these specific forms:

$$\begin{cases} [u,v]_+ = \max\ v(x_0)\ \text{sgn} u(x_0) \quad (u \not\equiv 0), \\ \qquad |u(x_0)| = \|u\| \\[2em] [u,v]_- = \min\ v(x_0)\ \text{sgn} u(x_0) \quad (u \not\equiv 0), \\ \qquad |u(x_0)| = \|u\| \\[3em] [u,v]_\pm = \|v\| \quad (u \equiv 0), \end{cases}$$

(4)

for
$$\|u\| \equiv \max_{x_0\ \overline{\Omega}} |u(x_0)|.$$

Using these explicit characterizations of $[,]_\pm$ we easily check that

(5)
$$\begin{cases} \text{if } A^\gamma(\gamma \in \Gamma) \quad \text{are strongly accretive on } ,X = C(\overline{\Omega}), \\ \qquad \text{then so is } A \equiv \max_{\gamma \in \Gamma} A^\gamma. \end{cases}$$

Here Γ is a, say, finite index set and $D(A) = \bigcap_{\gamma \in \Gamma} D(A^\gamma)$.

3. Control Theory and Dynamic Programming

I now present a wide class of interesting but badly nonlinear PDE arising from control theory, which formally at least, falls within the foregoing abstract theory.

Let Γ denote some index set, assumed for simplicity to be finite. Suppose that for each $\gamma \in \Gamma$, we are given a homogeneous Markov process X_t^γ taking values in $\overline{\Omega}$. The associated _Feller semigroup_ is defined by

$$[S^\gamma(t)u](x) = \int_{\overline{\Omega}} u(y) P^\gamma(t,x,dy),$$

$P^\gamma(t,x,A)$ ($= \text{Prob}(X_t^\gamma \in A | X_0^\gamma = x)$) denoting the transition function. Then, given some appropriate hypotheses on $P^\gamma(.,.,.)$, it turns out that $\{S^\gamma(t)\}_{t>0}$ is a _linear contraction semigroup_ on $X = C(\overline{\Omega})$, with _generator_

$$A^\gamma(u) \equiv \lim_{t \to 0+} \frac{u - S^\gamma(t)u}{t},$$

$D(A^\gamma)$ comprising those $u \in C(\overline{\Omega})$ for which this limit exists in $C(\overline{\Omega})$. (Details for all this are to be found in Lamperti [30]). Furthermore, abstract theory implies

(6)   each A^γ is strongly accretive on $X = C(\overline{\Omega})$.

We construct an associated control theory problem by allowing now for the possibility of switching according to a "control" $\gamma(\cdot)$ among the parameters γ as time evolves. Given an appropriate such $\gamma(\cdot)$, we define the corresponding expected cost function

$$C_x[\gamma(\cdot)] = E_x(\int_0^\infty f(X_s^{\gamma(\cdot)})ds),$$

$X_t^{\gamma(\cdot)}$ denoting the process resulting from our switching according to $\gamma(\cdot)$ among the processes $X_t^\gamma(\gamma \in \Gamma)$ as time evolves. Here $f(\cdot)$ is a state-dependent "running cost". We are interested in minimizing this cost among all controls. The method of dynamic programming addresses this problem by considering the value function

$$u(x) \equiv \inf_{\gamma(\cdot)} C_x[\gamma(\cdot)],$$

and showing - formally - that u solves the general Hamilton-Jacobi-Bellman equation:

(HJB) $$\max_{\gamma \in \Gamma} [A^\gamma u] = f \quad \text{in} \quad \Omega.$$

The idea then is to solve (HJB) (plus possibly some boundary conditions) and then to use our solution, if it is nice enough, to design an optimal feedback control. Consult, for example, Krylov [29] or Fleming-Rishel [22] for a further explanation.

Now all of this is quite delicate in practice, entailing various complications I will not discuss: my point is simply that in light of assertions (5) and (6) above, the nonlinear operator on the left hand side of (HJB) is accretive on $X = C(\overline{\Omega})$. This is the case should the processes X_t^γ be diffusions, pure jump processes, deterministic flows, or any combination thereof. In other words I claim, very loosely speaking, that underline{"most" dynamic programming PDE give rise to accretive operators on $X = C(\overline{\Omega})$} (or a similar space), a statement which is strictly speaking absolutely false, but which nevertheless indicates that $C(\overline{\Omega})$ and related function spaces are the natural settings in which to study (HJB)-type PDE.

4. Minty's Device; Viscosity Solutions

After the wild claims concluding the previous section, I now address the practical problem of how to use the foregoing abstract considerations really and rigorously to get information about specific PDE.

I begin, however, by remaining a bit longer in the abstract mode to investigate the problem of solving the stationary problem

$$(S) \qquad\qquad A(u) = f,$$

where A is some kind of nonlinear operator on a Banach space X, $f \in X$ is given, and $u \in D(A)$ is the unknown. A reasonable plan for solving (S) consists in finding "nicer" approximate operators A^ε solving the approximate problems

$$(S)_\varepsilon \qquad\qquad A^\varepsilon(u^\varepsilon) = f,$$

and then showing the u^ε converge somehow to a solution of (S). In practice, however, this is rather tricky since usually, good estimates are not available for the u^ε.

Minty's method often resolves such difficulties if $X = H$, a Hilbert space, and if A is monotone; that is, if

$$(7) \qquad\qquad (y-\hat{y}, A(y)-A(\hat{y})) \geq 0 \qquad (y, \hat{y} \in D(A)).$$

(This turns out to be equivalent to condition $(2)^{**}$). Let us thus suppose u^ε satisfy $(S)_\varepsilon$ but we know only $u^\varepsilon \rightharpoonup u$ weakly in H. Suppose the A^ε satisfy (7). Does u solve (S)? Following Minty (cf. J.-L. Lions [31]) we fix $\phi \in D(A)$ and use (7), $(S)_\varepsilon$, to write

$$(u^\varepsilon-\phi, f-A^\varepsilon(\phi)) = (u^\varepsilon-\phi, A^\varepsilon(u^\varepsilon)-A^\varepsilon(\phi)) \geq 0.$$

Let $\varepsilon \to 0$ and suppose $A^\varepsilon(\phi) \to A(\phi)$ strongly in H. Then

$$(u-\phi, f-A(\phi)) \geq 0.$$

Write $\phi = u - t\psi$ for $t > 0$ to find

$$(\psi, f-A(u-t\psi)) \geq 0.$$

Send $t \to 0$ and suppose $A(u-t\psi) \to A(u)$ in H:

$$(\psi, f-A(u)) \geq 0.$$

This inequality holding for a sufficiently wide class of ψ forces $A(u) = f$, and so u solves (S).

Now, Minty's technique works quite well for Hilbert space problems, but - as explained in $\S 3$ - $X = C(\overline{\Omega})$ is the natural space for the dynamic programming PDE of concern here. Furthermore, the operators we're interested in are accretive in $C(\overline{\Omega})$, and so satisfy

$$(8) \qquad [u-\hat{u}, A(u)-A(\hat{u})]_+ \geqslant 0 \qquad (u, \hat{u} \in D(A)).$$

The basic observation now is that this is an analogue of the monotonicity condition (7), and thus there's a possibility Minty's method will work in this setting as well.

Let us try this, following [17], by supposing A^ϵ, A are nonlinear operators on $C(\overline{\Omega})$ satisfying (8) and that $u^\epsilon \in D(A^\epsilon)$ solve $(S)_\epsilon$. Fix a smooth function ϕ and use (8), $(S)_\epsilon$, to write

$$[u^\epsilon - \phi, f - A^\epsilon(\phi)]_+ = [u^\epsilon - \phi, A^\epsilon(u^\epsilon) - A^\epsilon(\phi)]_+ \geqslant 0.$$

Let $\epsilon \to 0$ and suppose $u^\epsilon \to u$, $A^\epsilon(\phi) \to A(\phi)$, in $C(\overline{\Omega})$. As the mapping $[,]_+$ is upper semicontinuous, we find

$$(9) \qquad [u-\phi, f-A(\phi)]_+ \geqslant 0 \text{ for all smooth functions } \phi.$$

At this point it's tempting to write $\phi = u - t\psi$ and mimic the rest of Minty's argument, but for various reasons (for example, we don't know u is smooth) this does not work.

<u>We instead decide to define u to be a generalized solution of (S) provided (9) holds.</u>

Then, trivially, we obtain a solution of (S) as a limit of solutions of $(S)_\epsilon$; but does this approach have any real content? This answer is "yes" at least for operators A corresponding to nonlinear first-order PDE of <u>Hamilton-Jacobi</u> type. This is, in effect, the basic observation of Crandall-Lions [10], as reformulated in Crandall-Evans-Lions [8]. The idea is to assume $A(u) = H(Du)$, where $H: \mathbf{R}^n \to R$ is continuous, and to interpret (9) in light of the definition of $[,]_+$ on $C(\overline{\Omega})$ given by (4). Since boundary value problems for first order PDE are usually ill-posed on bounded domains Ω, we furthermore switch attention to $\mathbf{R}^n$.

All these considerations motivate the following definition, where $BUC(\mathbf{R}^n)$ denotes the space of bounded, uniformly continuous functions:

<u>Definition</u> ([10],[8]). Assume $f \in BUC(\mathbf{R}^n)$, $H: \mathbf{R}^n \to R$ is continuous. Then $u \in BUC$ is called a <u>viscosity solution</u> of

$$(HJ) \qquad\qquad H(Du) = f$$

provided for all $\phi \in C_0^\infty(\mathbf{R}^n)$,

(i) if $u - \phi$ has a local maximum at $x_0 \in \mathbf{R}^n$,

$$H(D\phi(x_0)) \leqslant f(x_0);$$

and

(ii) if $u-\phi$ has a local minimum at $x_0 \in \mathbf{R}^n$,

$$H(D\phi(x_0)) \geqslant f(x_0).$$

Granted this definition is inspired by abstract considerations; but is it useful? The answer again is "yes", for it turns out that we have here a very useful notion of solution for PDE of Hamilton-Jacobi type.

We record as follows some of the principle properties of viscosity solutions:

Theorem ([10],[8]).

(i) Assume u is a viscosity solution of (HJ). If $Du(x_0)$ exists at some point x_0, then $H(Du(x_0)) = f(x_0)$.

(ii) Suppose u,v are viscosity solutions of

$$(10) \qquad\qquad u + \lambda H(Du) = f,$$

$$v + \lambda H(Dv) = g.$$

Then

$$\| (u-v)^\pm \| \leqslant \| (f-g)^\pm \|.$$

(iii) For each $f \in BUC(\mathbf{R}^n)$ there exists a (unique) viscosity solution of

$$u + \lambda H(Du) = f.$$

The most important of these assertions is (ii), which implies the uniqueness of a viscosity solution for (10). Thus is the major discovery of Crandall-Lions [10].

5. Applications

a. Dynamic programming PDE

I return now to the considerations from $\oint 3$ and explain P. L. Lions' nice observation ([32]) that the value function satisfies the (HJB) equation in the viscosity sense. This is true in the full stochastic setting (see P. L. Lions [34]), but I restrict myself to deterministic control problems, for which the problem reads

$$C_x[\gamma(\cdot)] = \int_0^\infty f(x^{\gamma(\cdot)}(s))\,ds,$$

the function $x^{\gamma(\cdot)}(\cdot) = x(\cdot)$ solving

$$\begin{cases} \dot{x}(s) = a(x(s), \gamma(s)) & (s>0) \\ x(0) = x \end{cases}$$

for a measurable control $\gamma: [0,\infty) \to \Gamma$. Then it turns out that under some technical assumptions the value function rigorously solves in the viscosity sense the appropriate (HJB) equation, which for the case at hand reads

$$H(Du) \equiv \max_{\gamma \in \Gamma}(a(x,\gamma)\cdot Du) = f \quad \text{in } \mathbf{R}^n.$$

To prove this, choose any $\phi \in C_0^\infty(\mathbf{R}^n)$ and suppose that $u - \phi$ has a local max(min) at x_0; we must then prove

(11)
$$H(D\phi)) \leqslant f \quad (\geqslant f).$$

Inequalities (11) turn out to follow fairly easily from the optimality condition

(12)
$$u(x) = \inf_{\gamma(\cdot)} \left\{ \int_0^t f(x^{\gamma(\cdot)}(s))\,ds + u(x^{\gamma(\cdot)}(t)) \right\},$$

which is valid for all $x \in \mathbf{R}^n$ and $t > 0$. We leave as an exercise the derivation of (12) and of (11) from (12).

The foregoing observations have many, many variants, applying for example to finite horizon control problems, stopping time problems, costly switching problems, etc.; see, for example, Lions[32], Capuzzo Dolcetta-Evans[7], etc.

More interesting in my opinion are extensions to two-person, zero-sum differential games. Here the problem involves the competition between two players, each of whom can affect an ODE describing the dynamics of the controlled system. In this situation there are (at least) two value functions, called the upper and lower value functions, which turn out to solve in the viscosity sense certain nonlinear, first-order PDE called Isaacs' equations. I find such PDE particularly interesting as they entail max-min (or min-max) type nonlinearites. Conversely, it is not particularly hard to show that any, say, Lipschitz function $H: \mathbb{R}^n \to \mathbb{R}$ can be written as a max-min of affine mappings, whence Hamilton-Jacobi equations involving H can be regarded as an Isaacs' equation for some differential game. This is useful since the appropriate value function will be the (unique) viscosity solution of this H-J equation. The upshot is a kind of representation formula for the solution of the original fully nonlinear and nonconvex H-J equation; see [20], [36], [40], etc.

b. Large deviations

The ideas just now discussed apply also to the study of certain large deviation problems concerning stochastic differential equations subject to small random fluctuations of "size" ε. The idea is to consider an appropriate functional u_ε on the sample paths which should be very small as $\varepsilon \to 0$, and to prove

$$u_\varepsilon = \exp \left[\frac{-I + o(1)}{\varepsilon^2} \right]$$

as $\varepsilon \to 0$ for some appropriate quantity I. Following Fleming [21] we observe that $v_\varepsilon = -\varepsilon^2 \log u_\varepsilon$ solves some kind of second-order quasilinear elliptic or parabolic PDE. We next prove that $v_\varepsilon \to I$, the latter function being the viscosity solution of an appropriate H-J-type first order PDE. But then we can represent I as the value function of an associated control theory (in the convex case) or game theory (in the nonconvex case) problem, obtaining thereby what amounts to a constructive formula for I.

Such results are well-known in probability from the Ventcel-Freidlin or Donsker-Varadhan theories; my point is that the above approach - when it applies - is simplier and uses purely PDE techniques. See [19], Fleming - Souganidis [23], Bardi [3], Kamin [27], etc.

c. Hamilton-Jacobi PDE

In addition, there have recently been many papers written on Hamilton-Jacobi type equations in their own right. These include studies concerning the existence and uniqueness of solutions for

problems of more general type (see, for example, Souganidis [39],
Crandall-Lions [11], Barles [5], Ishii [24], etc.), numerical methods
(Crandall-Lions [12], Souganidis [40], etc.,); H-J equations in
infinite dimensions (Crandall-Lions [13], Barbu-DaPrato [1]);
representation formulas for solutions (P.L. Lions [32], Osher [37],
Bardi-Evans [4], Lions-Rochet [35], etc.), etc. One nice observation
here is that, by definition, viscosity solutions satisfy a kind of
maximum principle, whereas, on the other hand, maximum principles are
a main tool for studying the structure of nonlinear, second-order
elliptic and parabolic PDE. Based upon this analogy Bardi [2] has
utilized the "moving plane" method of elliptic equations to get
geometric information on solutions of certain H-J equations.

6. Open Questions

a. Control and game theory

As seen in $\oint$ 5, the notion of viscosity solutions provides a
rigorous meaning to the dynamic programming principle that the value
function satisfies some appropriate Hamilton-Jacobi-Bellman PDE. Do
viscosity solutions provide any further insight into the problem of
sythesizing optimal controls? Perhaps F. Clarke's generalized
gradient would be useful in this context.

b. Uniqueness for second order PDE

Although I have not done so here, it is easy to define viscosity
solutions of various fully-nonlinear elliptic PDE and furthermore, to
construct such viscosity solutions, as for example the value functions
of a related stochastic control or game theory problems. Are such
viscosity solution unique? (P.L. Lions in [34] has important results
in this direction for convex nonlinearites.)

c. Physics applications

As noted above, viscosity solution ideas are extremely useful for
applications in control and game theory. It seems to me striking
however, that there have been only a few papers written concerning
applications to physics (where, of course, we find Hamiltonians of all
sorts). Does the notion of viscosity solution provide any physical
insight? Less ambitiously, can viscosity solution ideas make rigorous
any of the formal calculations involving Hamiltonians in physics?

d. Singularities of viscosity solutions

A major advantage of the theory set forth above is that the viscosity solution of a time-dependent Hamilton-Jacobi PDE exists for all times, even after it looses smoothness when the classical characteristics cross. It is important for many purposes to understand the singularities in the gradient which thereafter develop, but except for some work of Jensen and Souganidis [26] not much is known for nonconvex H.

7. Conclusions

As announced in ϕ 1, I have discussed how abstract semigroup theory identifies the notion of a nonlinear accretive operators, how such arise from dynamic programming, and finally how these abstract notions lead naturally to the definition of viscosity solutions for first-order nonlinear PDE of Hamilton-Jacobi type. The reader should carefully note that, although I used the Crandall-Liggett Generation Theorem to motivate everything, I never at the end went back to apply this theorem to H-J type PDE. This could be done, but it would not be particularly useful: the full theory of viscosity solutions already provides more information. Recall from ϕ1 that this is a moral of the paper.

Finally, a historical note. I have presented the theory as it naturally develops, but not exactly as it was discovered. Indeed, Crandall and Lions' original paper [10] was primarily motivated by Kruzkov's work [28] on scalar hyperbolic conservation laws; the connections with [17]; and in particular, the advantages of starting with the definition of viscosity solution given here (as opposed to the original definition in [10]) were set forth only later, in [8]. (See however, Remark 1.18 on pp. 60-61 in [32]).

REFERENCES

1. V. Barbu and G. DaPrato, to appear.

2. M. Bardi, Geometric properties of solutions of Hamilton-Jacobi equations, to appear.

3. M. Bardi, An asymptotic formula for the Green's function of an elliptic operator, to appear

4. M. Bardi and L. C. Evans, On Hopf's formulas for solutions of Hamilton-Jacobi equations, to appear in J. Nonlinear Analysis.

5. G. Barles, Existence results for first order Hamilton-Jacobi equations, Ann. L'Institut H. Poincaré 1(1984), 325-340.

6. P. Benilan, M. G. Crandall and A. Pazy, forthcoming book.

7. I. Capuzzo Dolcetta and L. C. Evans, Optimal Switching for ordinary differential equations, SIAM J. Control Op. 22 (1984), 143-161.

8. M. G. Crandall, L. C. Evans and P. L. Lions, Some properties of viscosity solutions of Hamilton-Jacobi equations, Trans. Am. Math. Soc. 282 (1984), 487-502.

9. M. G. Crandall and T. Liggett, Generation of Semigroups of nonlinear transformations on general Banach spaces, Amer. J. Math 93(1971), 265-298.

10. M. G. Crandall and P. L. Lions, Viscosity solutions of Hamilton-Jacobi equations, Trans Am. Math. Soc. 277(1983), 1-42.

11. M. G. Crandall and P. L. Lions, On existence and uniqueness of solutions of Hamilton-Jacobi equations, to appear.

12. M. G. Crandall and P. L. Lions, Two approximations of solutions of Hamilton-Jacobi equations, to appear in Math Comp.

13. M. G. Crandall and P. L. Lions, Hamilton-Jacobi equations in infinite dimensions, Parts I and II, to appear.

14. M. G. Crandall and R. Newcomb, Viscosity solutions of Hamilton-Jacobi equations at the boundary, to appear.

15. M. G. Crandall and P. E. Souganidis, Developments in the theory of nonlinear first-order partial differential equations, Proc. of International Symposium on Differential Equations, Birmingham, Alabama, 1983, I. Knowles and R. Lewis, eds., North Holland.

16. L. C. Evans, Applications of nonlinear semigroup theory to certain partial differential equations, in Nonlinear Evolution Equations, ed. by M. G. Crandall, Academic Press, 1979.

17. L. C. Evans, On solving certain nonlinear partial differential equations by accretive operator methods, Israel J. Math 36(1980), 225-247.

18. L. C. Evans, Systems of nonlinear PDE in control theory and related topics, in Systems of Nonlinear PDE, ed. by J. Ball, Reidel, 1983.

19. L. C. Evans and H. Ishii, A PDE approach to some asymptotic problems concerning random differential equations with small noise intensities, Ann. L'Institut H. Poincaré 2(1985), 1-20.

20. L. C. Evans and P. E. Souganidis, Differential games and representation formulas for solutions of Hamilton-Jacobi-Isaacs equations, Indiana Univ. Math. J. 33(1984), 773-797.

21. W. H. Fleming, Exit probabilities and optimal stochastic control, Appl. Math. Optim. 4(1978), 329-346.

22. W. H. Fleming and R. W. Rishel, Deterministic and Stochastic Optimal Control, Springer-Verlag, New York, 1975.

23. W. H. Fleming and P. E. Souganidis, PDE viscosity solution approach to some problems of large deviations, to appear.

24. H. Ishii, Uniqueness of unbounded solutions of Hamilton-Jacobi equations, to appear.

25. H. Ishii, Viscosity solutions of Hamilton-Jacobi equations with discontinuous Hamiltonian and differential games, to appear.

26. R. Jensen and P. E. Souganidis, A regularity result for one dimensional stationary Hamilton-Jacobi equation, to appear.

27. S. Kamin, Singular perturbation problems and the Hamilton-Jacobi equation, to appear.

28. S. N. Kruzkov, Generalized solutions of first order nonlinear equations in several independent variables, I. Mat. Sb. 70 (1966), 394-415; II. Mat. Sb. 72(1967), 93-116.

29. N. V. Krylov, Controlled Diffusion Processes, Springer, 1980.

30. J. Lamperti, Stochastic Processes, Springer, 1977.

31. J. L. Lions, Quelques méthodes de resolution des problèmes aux limites non linéaires, Dunod, 1969.

32. P.-L. Lions, Generalized Solutions of Hamilton-Jacobi Equations, Pitman, Boston, 1982.

33. P.-L. Lions, Existence results for first order Hamilton-Jacobi equations, Ricerche Mat. Napoli, 1982-1983.

34. P.-L. Lions, Optimal control of diffusion processes and Hamilton-Jacobi-Bellman equations, Parts 1-3, Comm. P.D.E., 1983.

35. P. L. Lions and J. C. Rochet, to appear.

36. P.-L. Lions and P. E. Souganidis, Differential games, optimal control and directional derivatives of viscosity solutions of Bellman's and Isaacs' equations, I and II, to appear in SIAM J. Control and Op.

37. S. Osher, The Riemann problem for nonconvex scalar conservation laws and Hamilton-Jacobi equations, to appear.

38. H. Soner, Optimal control with state-space constraint, to appear.

39. P. E. Souganidis, Existence of viscosity solutions of Hamilton–Jacobi equations, J. Diff. Eqns. 56(1985), 345–390.

40. P. E. Souganidis, Approximation schemes for viscosity solutions of Hamilton–Jacobi equations with applications to differential games, J. Nonlinear Analysis.

Support by NSF Grant DMS-83-01265. The author is currently a member of the Institute for Physical Science and Technology, University of Maryland.

EVOLUTION EQUATIONS WITH NONLINEAR BOUNDARY CONDITIONS

Jerome A. Goldstein
Department of Mathematics
Tulane University
New Orleans, LA 70118

1. INTRODUCTION

Of concern are mixed problems of the form

$$
\text{(MP)} \quad
\begin{cases}
du(t)/dt = Au(t) + F & (t > 0) \\
u(0) = f \\
Bu(t) = g(u(t)) & (t > 0) .
\end{cases}
$$

Here A is a linear operator from $D = \text{Dom}(A) \subset X$ to X, (X being a Banach space), B is a linear operator from $D \subset X$ to a Banach space Y, f is a given vector in D, g is a given (nonlinear) operator from D to Y, and for the moment we take F to be zero (although later we will allow $F = F(t,u(t))$).

Many problems in the applications can be written in the form (MP). Examples include parabolic problems with nonlinear boundary conditions, feedback control problems, reaction-diffusion systems, problems in age-dependent population dynamics or other problems involving "memories" (in which $g(u(t))$ is replaced by $g(u(s) ; 0 < s < t)$), and so on.

A sampling of the related literature includes articles by Amann [1], Brezis [2], Desch, Lasiecka, and Schappacher [4], Dubois and Lumer [5], Greiner [8], and Henry [9].

This is a report on work begun in the summer of 1985 at the University of Graz in Austria. It is actually a preliminary report on research still in progress and being done jointly with Wolfgang Desch and Wilhelm Schappacher (cf. [3]).

The solution (local or global) of a problem such as (MP) will be achieved by (i) a linear analysis, and (ii) use of the linear analysis and a fixed point theorem to solve the nonlinear problem.

The methodology of (i) works in a general context. The methodology of (ii) is variable and depends on the applications one has in mind. After explaining (i) in some detail we shall briefly discuss (ii) in one special case, namely in the context of a parabolic equation or system.

2. THE LINEAR AUTONOMOUS PROBLEM

We consider the problem

$$
\text{(LA)} \quad
\begin{cases}
u'(t) = Au(t) & (t \in \mathbb{R}^+ = [0,\infty)) \\
u(0) = f & \\
Bu(t) = g & (t \in \mathbb{R}^+)
\end{cases}
$$

where $A : D \subset X \to X$ is a closed linear operator, $B : D \subset X \to Y$ is linear, $A = A|_{N(B)}$ (the restriction of A to the null space of B) is the infinitesimal generator of a (C_0) semigroup $T = \{T(t) : t \in \mathbb{R}^+\}$ on X (cf. Goldstein [7]), and B is "A-closed" in the sense that if $x_n \in D$, $x_n \to x$, $Ax_n \to y$, and $Bx_n \to z$, then $x \in D$ and $Bx = z$. (This does *not* imply that B is a closed operator. In the typical application, B will be A-closed but will not be closed.) Example: Think of $X = L^P(\Omega)$, $A = \Delta$, $D = W^{2,P}(\Omega)$, $Bu = u|_{\partial\Omega}$ with $\Omega \subset\subset \mathbb{R}^n$. We take $p > n$ so that $Bu = u|_{\partial\Omega}$ is a well-defined continuous function for $u \in D$. Then B is A-closed but is not closable.

We want (LA) to have a strict solution, i.e. a solution in $C^1(\mathbb{R}^+,X) \cap C(\mathbb{R}^+,D)$. This necessitates the *compatibility condition*

$$
\text{(CC)} \qquad\qquad Bf = g \ .
$$

HYPOTHESIS I. (LA) *is well-posed, i.e. there exists a unique strict solution of exponential growth if* (CC) *holds.*

Denote this solution by $u(t) = U(t;f,g)$. By exponential growth in the above hypothesis we mean that there exist constants $M > 1$, $\omega \in \mathbb{R}$ (independent of f,g) such that

$$
\|U(t;f,g)\| \le Me^{\omega t}(\|f\| + 1)
$$

holds for all $t > 0$, provided $Bf = g$.

3. THE LINEAR PROBLEM WITH A TIME-DEPENDENT BOUNDARY CONDITION

$$(\text{LTDBC}) \qquad \begin{cases} u'(t) = Au(t) + F(t) \qquad (t \in \mathbb{R}^+) \\ u(0) = f \\ Bu(t) = w(t) \end{cases}$$

We assume $f \in D$, $w \in C^1(\mathbb{R}^+, Y)$, $w(\mathbb{R}^+) \subset \text{Ran}(B)$, and $F \in C^1(\mathbb{R}^+, X) + C(\mathbb{R}^+, [D])$ are all given, where $[D]$ means D equipped with the graph norm of A. The appropriate compatibility condition corresponding to (CC) is $Bf = w(0)$.

Our goal is to find an explicit formula for (LTDBC) in terms of the solutions of (LA), i.e. using the semigroup T and the vectors $U(t; f, g)$. We begin by writing down the *provisional basic formula*

$$(\text{PBF}) \qquad \begin{cases} u(t) = \{T(t)f + \int_0^t T(t-s)F(s)ds\} \\ \qquad + [U(t; 0, w(0)) + \int_0^t U(t-s; 0, w'(s))ds] \\ \equiv \{u_1(t)\} + [u_2(t)] . \end{cases}$$

It is straightforward to check that, formally,

$$u_1'(t) = Au_1(t) + F(t) , \quad u_1(0) = f , \quad Bu_1(t) = 0 ,$$
$$u_2'(t) = Au_2(t) , \quad u_2(0) = 0 , \quad Bu_2(t) = w(t) .$$

The reason why $u = u_1 + u_2$ need not be a strict solution to (LTDBC) is because of the compatibility conditions. Indeed, applying the compatibility condition (CC) to the two terms in $u_2(t)$ gives $B0 = 0$, $B0 = w'(t-s)$, which together imply $w \equiv 0$.

We get around this problem by introducing a linear operator $G : Y \to X$ satisfying

$$AGg = 0 , \quad BGg = g$$

for all $g \in Y$.

If one thinks of $Bu = u|_{\partial\Omega}$, then $v = Gg$ is the solution of the "Dirichlet problem" $Av = 0$ in Ω, $Bv = g$ on $\partial\Omega$.

We now show how to construct G assuming HYPOTHESIS I. First of all we may, without loss of generality, replace A by $A - \lambda I$ for some fixed scalar λ. (This can be seen by noting the relation between the problems solved by $u(t)$ and $v(t) = e^{-\lambda t}u(t)$.) Take $\lambda > \omega$ where ω is as in HYPOTHESIS I and define

$$v = \lambda \int_0^\infty e^{-\lambda t} U(t;f,g)dt - \lambda(\lambda I - A)^{-1}f.$$

Then we get, using $A \supset \bar{A}$,

$$(\lambda I - \bar{A})v = \lambda \int_0^\infty e^{-\lambda t}(\lambda I - \bar{A})U(t;f,g)dt - \lambda f$$

$$= \lambda \int_0^\infty \frac{d}{dt} [e^{-\lambda t}U(t;f,g)]dt - \lambda f$$

$$= \lambda f - \lambda f = 0$$

since $\lim_{t \to \infty} e^{-\lambda t} U(t;f,g) = 0$ because $\lambda > w$. Since B is A-closed we obtain

$$Bv = \lambda \int_0^\infty e^{-\lambda t} BU(t;f,g)dt - B\lambda(\lambda I - A)^{-1}f$$

$$= \lambda \int_0^\infty e^{-\lambda t} g \, dt - 0 = g \ .$$

(Note that even though B may not be closable, we can treat B like a closed operator because of the A-closedness assumption.) It follows that G defined by $Gg = v$ has the desired properties (cf. Desch, Lasiecka, and Schappacher [4]).

It is now clear how to modify (PBF) to get our *basic formula*, viz.

$$(BF) \quad \begin{cases} u(t) = T(t)(f - Gw(0)) + U(t;Gw(0),w(0)) \\ \\ \qquad + \int_0^t T(t-s)(F(s) - Gw'(s))ds + \int_0^t U(t-s;Gw'(s),w'(s))ds. \end{cases}$$

THEOREM. *Assuming HYPOTHESIS I,* (LTDBC) *has a unique strict solution provided* $w \in C^1(\mathbb{R}^+,Y)$, $f - Gw(0) \in Dom(A)$, *and* $F - Gw' \in C^1(\mathbb{R}^+,X) + C(\mathbb{R}^+,[D])$.

The proof is by straightforward computations based on the preceding considerations. (See also Dubois and Lumer [5].)

Without assuming the three compatibility and regularity conditions stated at the end of the theorem, we still get a unique solution (in a weaker sense), but we prefer strict solutions in anticipation of application to nonlinear problems.

4. NONLINEAR PROBLEMS

Of concern is the nonlinear problem

$$
\text{(NP)} \quad
\begin{cases}
u'(t) = Au(t) + F(t,u(t)) & (0 < t < \tau) \\
u(0) = f \\
Bu(t) = N(u(t)) & (0 < t < \tau)
\end{cases}
$$

where $0 < \tau \leq \infty$. Here A, B are as before, $N : X \to Y$ and $F : \mathbb{R}^{+} \times X \to X$ are given nonlinear functions; and the compatibility condition on f is $Bf = N(f)$.

If u is a known solution of (NP), then the terms on the right hand sides of (NP) are known and may be regarded as inhomogeneous terms. Thus by (BF), u satisfies the integral equation

$$
\begin{aligned}
u(t) = (Su)(t) &\equiv T(t)(f - GN(f)) \\
&+ U(t;GN(f),N(f)) + \int_0^t T(t-s)[F(s,u(s)) - G\tfrac{d}{ds} N(u(s))]ds \\
&+ \int_0^t U(t-s;\ G\tfrac{d}{ds} N(u(s)),\ \tfrac{d}{ds} N(u(s)))ds
\end{aligned}
$$

for $0 < t < \tau$. We solve (NP) by finding a fixed point of S in a suitable closed subset Z of $C^1([0,\sigma],X)$ where $0 < \sigma < \tau$. We must work in C^1 rather than C because of the $(d/ds)N(u(s))$ terms, which involve $u'(s)$. (This is why we wanted strict solutions of (LTDBC).)

For concreteness we define $X_1 = [D]$ and take $Z \subset C([0,\sigma],X_1) \cap C^1([0,\sigma],X)$. We require F, N and N' to satisfy suitable local Lipschitz conditions on various spaces and assume that G is bounded from Y to X_2, where $X_2 \hookrightarrow X$.

One wants to apply Banach's fixed point theorem (alias the contraction mapping principle). In order to do this one must take σ to be small and estimate various terms, including for example,

$$
\| \int_0^t T(t-s)G \tfrac{d}{ds} (N(u_1(s)) - N(u_2(s)))ds \|_{X_1};
$$

this necessitates estimating

$$
\| A \int_0^t T(t-s)G \tfrac{d}{ds} (N(u_1(s)) - N(u_2(s)))ds \|_X .
$$

If A generates an analytic semigroup T , then $T(t)(X) \subset \text{Dom}(A)$
for all $t > 0$ and the above term is estimated by

$$\int_0^t \|AT(t-s)\|_{\mathcal{B}(X_2,X)} \|G\|_{\mathcal{B}(Y,X_2)} \|N'(u_1(s))u_1'(s) - N'(u_2(s))u_2'(s)\|_Y \, ds.$$

Let us look at the term involving the norm of $AT(t-s)$; this should

be integrable near $s = t$. The estimate $\|AT(t-s)\|_{\mathcal{B}(X,X)} < \dfrac{Me^{\omega(t-s)}}{(t-s)}$

is not satisfactory. But the estimate $\|AT(t-s)\|_{\mathcal{B}(X_2,X)} < \dfrac{Me^{\omega(t-s)}}{(t-s)^\alpha}$

for $\alpha < 1$ is useful; this can be achieved by taking X_2 to be the
domain of the fractional power $(-A + \mu I)^\beta$ for suitable positive β
and μ. This kind of estimation goes back to the work of Kato and
Fujita [10], [6] on the Navier-Stokes equations in the early sixties.

In the context of reaction-diffusion systems on $\Omega \subset\subset \mathbb{R}^n$ having
N components, our analysis can be carried out by choosing (with
$p > n$)

$$X = L^p(\Omega;\mathbb{R}^N) \ , \quad X_1 = W^{2,p}(\Omega;\mathbb{R}^N) \ ,$$

$$Y = W^{1-\frac{1}{p},p}(\partial\Omega;\mathbb{R}^N), \quad X_2 = W^{2+\varepsilon,p}(\Omega;\mathbb{R}^N)$$

for suitable $\varepsilon > 0$. This is similar to the earlier work of Amann [1].

The end result is a local (in time) existence and uniqueness
theorem. Global existence would follow if the assumed local Lipschitz
conditions were global or, more interestingly, if one could prove an a
priori estimate on the local solution which would preclude blow-up in
a finite time. This is rather standard so we can avoid giving details.

Other applications would necessitate other sorts of estimates.
That is, the hypotheses and proof for a particular case of (NP) should
be tailored to the specific problem along the broad lines indicated
above.

This work was partially supported by an NSF grant, which is
gratefully acknowledged.

REFERENCES

[1] Amann, H., Parabolic evolution equations with nonlinear boundary conditions, to appear.

[2] Brezis, H., Monotonicity methods in Hilbert space and some applications to nonlinear partial differential equations, in Contribution to Nonlinear Functional Analysis (ed. by E. Zarantonello), Academic Press, New York (1971), 101-156.

[3] Desch, W., J. A. Goldstein, and W. Schappacher, in preparation.

[4] Desch, W., I. Lasiecka, and W. Schappacher, Feedback boundary control problems for linear semigroups, Israel J. Math. 51 (1985), 177-207.

[5] Dubois, R. M. and G. Lumer, Formule de Duhamel abstraites, Arch. Math. 43 (1984), 49-56.

[6] Fujita, H. and T. Kato, On the Navier-Stokes initial value problem. I. Arch. Rat. Mech Anal. 16 (1984), 269-315.

[7] Goldstein, J. A., Semigroups of Linear Operators and Applications, Oxford U. Press, New York and Oxford, 1985.

[8] Greiner, G., Perturbing the boundary conditions of a generator, to appear.

[9] Henry, D., Some infinite-dimensional Morse-Smale systems defined by parabolic partial differential equations, J. Diff. Eqns., in press.

[10] Kato, T. and H. Fujita, On the nonstationary Navier-Stokes system, Rend. Sem. Mat. Univ. Padova 32 (1962), 243-260.

ASYMPTOTICALLY SMOOTH SEMIGROUPS AND APPLICATIONS

Jack K. Hale

Lefschetz Center for Dynamical Systems
Division of Applied Mathematics
Brown University
Providence, Rhode Island 02912 U.S.A.

1. INTRODUCTION.

Our purpose in these notes is to give some illustrations of infinite dimensional problems for which the techniques and ideas of dynamical systems have led to surprisingly new results in classical problems, to improvement of results that have been obtained by more classical approaches and to the formulation of more meaningful and significant problems.

All of the systems contain some type of dissipation in the sense that solutions to the corresponding evolutionary equations uniformly leave infinity. This eliminates the discussion of conservative systems. However, it does not exclude the consideration of hyperbolic problems. The abstract class of semigroups considered are called asymptotically smooth maps and were introduced in the early 1970's by Hale, LaSalle and Slemrod [1972], Hale and Lopes [1973]. The motivation at that time was to discuss retarded and neutral functional differential equations (FDE's) as well as partial differential equations. In the last few years, several rather important observations have been made concerning the scope of the class and its applicability. The purpose of these notes is to present a few types of problems that fall under the category of asymptotically smooth and to indicate directions of research that are being directly motivated by this approach. Some of the basic ideas of dynamics in infinite dimensions have already been expounded by Hale, Magalhães and Oliva [1984] for FDE's and the lecture notes of Hale [1985] for other problems. Because of the availability of these references, the ideas rather than proofs are emphasized here.

2. ASYMPTOTICALLY SMOOTH SEMIGROUPS AND COMPACT ATTRACTORS.

Let X be a Banach space, $R^+ = [0,\infty)$. A family of mappings $T(t): X \to X$, $t \geq 0$, is said to be a C^r-semigroup, $r \geq 1$, provided that

    (i) $T(0) = I$

   (ii) $T(t + s) = T(t) T(s)$, $t \geq 0$, $s \geq 0$,

 (iii) $T(t)x$ is continuous in t, x together with Fréchet derivatives up through order r for $(t,x) \in R^+ \times X$.

For any $x \in X$, the _positive orbit_ $\gamma^+(x)$ through x is defined as $\gamma^+(x) = \{T(t)x, t \geq 0\}$. A _negative orbit_ through x is a function $\phi:(-\infty,0] \to X$ such that $\phi(0) = x$ and, for any $s < 0$, $T(t)\phi(s) = \phi(t+s)$ for $0 \leq t \leq -s$. A _complete orbit_ through x is a function $\phi:R \to X$ such that $\phi(0) = x$ and, for any $s \in R$, $T(t)\phi(s) = \phi(t+s)$ for $t \geq 0$.

Since the range of $T(t)$ may not be all of X, to say there is a negative or complete orbit through x may impose restrictions on x. Also, since $T(t)$ need not be one-to-one it is not necessary for a negative orbit to be unique if it exists. Let _the negative orbit_ through x be defined as the union of all negative orbits through x. Then

$\gamma^-(x) = U_{t \geq 0} H(t,x)$,

$H(t,x) = \{y \in X:$ there is a negative orbit $\gamma^-(x)$ through x
     defined by $\phi:(-\infty,0] \to X$ with $\phi(0) = x$ and $\phi(-t) = y\}$.
The complete orbit $\gamma(x)$ through x is defined as $\gamma(x) = \gamma^-(x) U \gamma^+(x)$. When a negative (or complete) orbit through x is defined, we sometimes write $T(t)x$ for an element on the orbit for $t < 0$ ($t \in R$). For any subset $B \subset X$, let $\gamma^+(B) = U_{x \in B}\gamma^+(x)$, $\gamma^-(B) = U_{x \in B}\gamma^-(x)$, $\gamma(B) = U_{x \in B}\gamma(x)$ be respectively, the positive orbit, negative orbit, complete orbit through B if the latter exist.

For any set $B \subset X$, define the _ω-limit set_ $\omega(B)$ _of B_ and the _α-limit set_ $\alpha(B)$ _of B_ as

$\omega(B) = \cap_{s \geq 0} C\ell U_{t \geq s} T(t)x$

$\alpha(B) = \cap_{s \geq 0} C\ell U_{t \geq s} H(t,x)$.

Notice that $\omega(B)$ may not be equal to $U_{x \in B}\omega(x)$.

A set $S \subset X$ is said to be _invariant_ if, for any $x \in S$, there is a complete orbit $\gamma(x)$ through x such that $\gamma(x) \subset S$. It is not difficult to show that S invariant is equivalent to $T(t)S = S$ for $t \geq 0$. Saying that a set S is invariant may impose restrictions upon S. It is well known that an important class of invariant sets are the α-limit sets and ω-limit sets of orbits.

A compact invariant set A is said to be a _maximal compact invariant set_ if every compact invariant set belongs to A. An invariant set A is said to be a _compact attractor_ if A is a maximal compact invariant set and $\omega(B)$, for each bounded set $B \subset X$, is compact and $\omega(B) \subset A$.

The semigroup $T(t)$ is _point dissipative_ if there is a bounded set B such that, for any $x \in X$, dist $(T(t)x,B) \to 0$ as $t \to \infty$.

A set $B \subset X$ is said to <u>attract a set $C \subset X$ under $T(t)$</u> if dist $(T(t)C,B) \to 0$ as $t \to \infty$. The semigroup $T(t)$ is <u>asymptotically smooth</u> if, for any bounded set $B \subset X$, there is a compact set $J = J(B) \subset X$ such that, J attracts the set $L(B) = \{x \in B: T(t)x \in B$ for $t \geq 0\}$. As in the discrete case (Lemma 2.2.1), $T(t)$ is asymptotically smooth if and only if, for any closed bounded set $B \subset X$ such that $T(t)B \subset B$ for $t \geq 0$, there is a compact set $J \subset B$ such that J attracts B. This alternative definition implies

Lemma 2.1. <u>If $T(t)$ is asymptotically smooth and B is a nonempty set in X such that $\gamma^{+}(B)$ is bounded, then $\omega(B)$ is nonempty, compact, and invariant, and dist $(T(t), \omega(B)) \to 0$ as $t \to \infty$. If, in addition B is connected, then $\omega(B)$ is connected. In particular, if $\gamma^{+}(x)$ is bounded, then $C\ell\gamma^{+}(x)$ is compact and $\omega(x)$ is nonempty, compact, connected, and invariant</u>.

A basic result for asymptotically smooth semigroups is

Theorem 2.2. <u>If $T(t)$, $t \geq 0$, is asymptotically smooth, point dissipative, and orbits of bounded sets are bounded, then there exists a compact attractor</u> A.

Asymptotically smooth maps were introduced by Hale, LaSalle and Slemrod [1972] (see, also Hale and Lopes [1973]). It was shown that asymptotically smooth maps for which there is a compact set that attracts compact sets must have a maximal compact invariant set that attracts a neighborhood of any compact set. Cooperman [1978] and Massatt [1983a] clarified the attractivity properties of maximal compact invariant sets. The above Theorem 2.2 is essentially due to Massatt [1983a].

Theorem 2.3. <u>The following are particular cases of asymptotically smooth semigroups:</u>

(i) <u>There is a $t_1 \geq 0$ such that $T(t)$ is compact for $t > t_1$.</u>

(ii) <u>$T(t) = S(t) + U(t)$ where $U(t)$ is compact and $S(t)$ is linear with $|S(t)| \leq k \exp(-\alpha t)$ for all $t \geq 0$ and some constants $k \geq 1$, $\alpha > 0$.</u>

(iii) <u>$T(t) = S(t) + U(t)$ where $U(t)$ is compact and there is a continuous function $k: R^{+} \times R^{+} \to R^{+}$ such that $k(t,r) \to 0$ as $t \to \infty$ and $|S(t)x| \leq k(t,r)$ if $|x| \leq r$.</u>

It was observed by Hale and Lopes [1973] that (i), (ii) are asymptotically smooth. In fact, they are special cases of α-contractions (k-set contractions). Case (iii) contains α-contractions, but can be satisfied when $T(t)$ is not an α-contraction and so is new. Although the proof is not trivial, it is not too difficult and is omitted.

In the next sections, we give some examples of asymptotically smooth semigroups which arise in applications.

3. EXAMPLES OF ASYMPTOTICALLY SMOOTH SEMIGROUPS.

Several interesting applications lead to asymptotically smooth semigroups. In this section, we list some of these giving the appropriate references where proofs are given.

3.1. Systems of reaction diffusion equations.

$$u_t = D\Delta u + f(u) \quad \text{in} \quad \Omega \subset R^n,$$

$$u = 0 \quad \text{in} \quad \partial\Omega \tag{3.1}$$

or some other type of boundary condition (Neumann or mixed), where $u \in R^N$, Ω is a smooth bounded domain. If one can prove that solutions are globally defined and depend continuously on inital data in an appropriate Sobolev space, then the corresponding semigroup $T(t)$ will be compact for $t > 0$. The same remark applies to many sectorial evolutionary equations (see, for example, Henry [1981]).

3.2. Porous media equations.

Consider the equation

$$u_t = (u^m)_{xx} + f(u) \quad \text{in} \quad (-L,L) \times R^+,$$

$$u(\pm L,t) = 0 , \quad t \in R^+, \tag{3.2}$$

where $m > 1$ is constant and $f(u)$ is locally Lipschitz. If $f(0) = f(1) = 0$, then Aronson, Crandall and Peletier [1982] show that the initial value problem for (3.2) defines a C_0-semigroup on $L_\infty((-L,L),[0,1])$ which is compact for $t > 0$.

3.3. Retarded functional differential equations (RFDE's).

Let $C = C([-1,0],R^n)$, $f: C \to R^n$ and consider the equation

$$\dot{x}(t) = f(x_t) \tag{3.3}$$

where $x_t(\theta) = x(t+\theta)$, $-1 \leq \theta \leq 0$. If f is a C^1-function and the solutions $x(t,\phi)$ with initial value ϕ are globally defined, then $[T(t)\phi](\theta) \overset{\text{def}}{=} x(t + \theta,\phi), -1 \leq \theta \leq 0$, is a C^1-semigroup on C. Furthermore, $T(t)$ is compact for $t \geq 1$. One can even show that $T(t) = S(t) + U(t)$ where $U(t)$ is compact for $t \geq 0$. In fact, $S(t)$ is the semigroup corresponding to $\dot{x}(t) = 0$ on the space $C_0 = \{\phi \in C: \phi(0) = 0\}$. See Hale [1977].

For RFDE's with infinite delay on a Banach space X, the solution operator $T(t)$ will have the same decomposition as above with $S(t)$ being the semigroup defined by $\dot{x}(t) = 0$ on $X_0 = \{\phi \in X: \phi(0) = 0\}$. If the radius of the essential spectrum of $S(t)$ is $\leq \exp(-\delta t)$ for some constant $\delta > 0$, then $T(t)$ asymptotically smooths. This puts restrictions on the space X; for example, for fading memory spaces, the kernel must appraoch zero exponentially (see Hale and Kato [1978]).

3.4. Neutral FDE's.

With the notation as in Section 3.3, let $D: C \to R^n$ be the linear operator $D\phi = \phi(0) - g(\phi)$ with g nonatomic at zero for which the semigroup $T_D(t)$ generated by $Dx_t = 0$ on the set $C_D = \{\phi \in C: D\phi = 0\}$ satisfies $|T_D(t)| \le k \exp(-\delta t)$ for some $\delta > 0$. Then the semigroup $T(t)$ on C generated by

$$\frac{d}{dt}Dx_t = f(x_t) \tag{3.4}$$

is asymptotically smooth. In fact, $T(t) = S(t) + U(t)$ with $S(t)$, $U(t)$ as before (see, for example, Hale [1977]). A special case for D is $D\phi = \phi(0) - B\phi(-1)$, where the eigenvalues of B have moduli < 1.

3.5. Damped wave equation.

Consider the equation

$$u_{tt} + \partial \alpha u_t - \Delta u = -f(u) \quad \text{in} \quad \Omega,$$
$$u = 0 \quad \text{in} \quad \partial\Omega, \tag{3.5}$$

where $\alpha > 0$ is a constant, Ω is a smooth bounded domain, and let $X = H_0^1(\Omega) \times L^2(\Omega)$. Under appropriate conditions on f (growth conditions and sign conditions at ∞) and using energy estimates, Babin and Vishik [1983] proved that (3.5) generates a C^1-semigroup $T(t)$ on X. In Hale [1985], it is shown that $T(t) = S(t) + U(t)$ as above and therefore, asymptotically smooths.

3.6. Strongly damped nonlinear wave equation.

Consider the equation

$$u_{tt} - \alpha\Delta u_t - \Delta u = f(u) \quad \text{in} \quad \Omega$$
$$u = 0 \quad \text{in} \quad \partial\Omega$$

where $\alpha > 0$ is a constant and Ω is a smooth bounded domain in R^n. Under some growth and sign conditions in f, this equation defines a C^1-semigroup $T(t)$ on an appropriate Banach space X. (see Webb [1980], Massatt [1983b]). The decomposition $T(t) = S(t) + U(t)$ as above is also valid and so is asymptotically smooth. The function f can also depend on $u_t, \nabla u, \nabla u_t$ (see Massatt [1983b]).

3.7. Age dependent populations.

Even to write the equations for this type of application takes considerable space and discussion. Therefore, we only remark that many of the models considered in Webb [1985] generate asymptotically smooth semigroups $T(t)$ on a Banach space X. If the birth rates are linear, $T(t)$ has the decomposition $S(t) + U(t)$ with $S(t)$ linear and satisfying the above estimates. If the birth rates are nonlinear, then $S(t)$ satisfies the conditions of Theorem 2.3 (iii) (see Proposition 3.16, p. 106 of Webb [1985]).

4. FLOW ON THE ATTRACTOR.

In this section, we assume that our semigroup is always asymptotically smooth, point dissipative and orbits of bounded sets are bounded. Therefore, a compact attractor A exists. The problem is to study the flow on the attractor and see how this flow changes as parameters are varied.

For gradient systems; that is, ones for which there is a Liapunov function which implies that every solution approaches an equilibrium point, (for a more precise discussion, see Hale [1985]), point dissipativeness is a consequence of the set E of equilibrium points being bounded. Often the Liapunov function (energy estimates) also gives orbits of bounded sets bounded. Thus, any asympototically smooth gradient system has a compact attractor A. If each equilibrium point ϕ is hyperbolic and $W^u(\phi)$ is the unstable manifold of ϕ, then

$$A = U_{\phi \in E} \ W^u(\phi) \tag{4.1}$$

If A_λ depends on parameters λ, then only two types of bifurcations occur for gradient systems as λ varies:

(i) Bifurcation of equilibria (a local problem),

(ii) "Saddle connections" are created and broken;

that is, $W^u_\lambda(\phi)$ becomes nontransversal to $W^s_\lambda(\psi)$ at some point λ_0 and was transversal before and after λ. This type of bifurcation affects the global dynamics by creating drastic changes in the basins of attraction of equilibria.

For ordinary differential equations in the plane, both types of bifurcation (i) and (ii) can occur. A remarkable result of Henry [1985] says: If ϕ, ψ are hyperbolic equilibria for a scalar parabolic equation in one space dimension on a bounded interval, $W^s(\phi)$ is always transversal to $W^u(\psi)$; that is, no "saddle connections" exist. The infinite dimensional problem dynamically cannot be as complicated as a planar ode!

Problem: To what extent is the above transversality always true for scalar PDE's of other types which are asymptotically smooth?

We have no examples showing that non-transversal intersection of stable and unstable manifolds occur.

The above problem cannot have a positive solution for all scalar RFDE's. An example was given by Hale and Rybakowski [1982] where there was nontransversal intersection of stable and unstable manifolds.

Let us indicate how other natural questions arise when we restrict the discussion to the attractor.

Suppose $\varepsilon \geq 0$ and consider the equation

$$\varepsilon u_{tt} + u_t - \Delta u = f(u) \quad \text{in} \quad \Omega,$$
$$u = 0 \quad \text{in} \quad \partial\Omega, \tag{4.2}$$

and suppose f is such that there is a compact attractor A_ε in a space $X = X_1 \times X_2$ for $\varepsilon > 0$. This requires growth and sign conditions on f (see Babin and Vishik [1983] or Hale [1985]). One can also choose f so that the parabolic equation

$$u_t - \Delta u = f(u) \tag{4.3}$$

has a compact attractor A_0 in X_1. We can consider A_0 as imbedded in X as $\tilde{A}_0 = (A_0, 0)$.

Problem. Does $A_\varepsilon \to \tilde{A}_0$ in X as $\varepsilon \to 0$ and is the flow on A_ε equivalent to the flow on $\tilde{A}_0$?

For the case of one space variable and $X = H_0^1 \times L^2$, preliminary work by Hale, Lin and Rocha indicate that $A_\varepsilon \to \tilde{A}_0$ and the graphs of the flows are preserved. Our proof only shows this convergence in L^2 rather than H_0^1. It is clear that such questions should be discussed in great detail.

Another class of problems that are of interest concerns variations in the boundary conditions. Consider the system of reaction-diffusion equations

$$u_t = D\Delta u + f(u) \quad \text{in} \quad \Omega \subset R^n,$$
$$D\partial u/\partial n + \theta E(x)u = 0 \quad \text{in} \quad \partial\Omega, \tag{4.4}$$

where Ω is a smooth bounded domain, $u \in R^N$, $D = \text{diag}(d_1,\ldots,d_N)$, $d_j > 0$, $\theta \in [0,\infty)$, $E = \text{diag}(e_1,\ldots,e_N)$ is continuous, and f is a smooth function.

For $\theta = 0$; that is, Neumann boundary conditions, Conway, Hoff and Smoller [1978], assuming the existence of an invariant rectangle for the PDE, have shown that the solutions of the PDE (4.4) are in some sense close to the solutions of the ODE if $d\lambda$ is large where $d = \text{num } d_j$ and $-\lambda$ is the first eigenvalue of the Laplacian with Neumann conditions.

Hale [1985a] removed the restriction to invariant regions and showed that, if the ODE has a compact attractor, then the PDE has a compact attractor for $d\lambda$ large.

Hale and Rocha [1985] have studied the effects of variations in the boundary conditions when d is large. Assuming that the equation

$$\frac{dv}{dt} = \theta\zeta v + f(v), \quad \zeta = \int_{\partial\Omega} E \qquad (4.5)$$

has a compact attractor A_0, they showed that (4.4) has a compact attractor $A_{D,\theta}$ for d large, $A_{D,\theta} \to A_0$ as $d \to \infty$, and the structural properties of the flow of (4.4) on $A_{D,\theta}$ is determined by the structural properties of the flow of (4.5) on A_θ. The results are applicable for θ in any compact set and show that "seemingly small" perturbations in the boundary conditions lead to large changes in the flow. In more recent unpublished work, they show that reasonable hypotheses on f ensure that one can obtain a compact attractor $A_{D,\theta}$ for (4.4) for all $d \geq d_0 > 0$, sufficiently large and all $\theta \geq 0$. The flow on the attractor is governed by an ODE and, as $\theta \to \infty$, $A_{D,\theta} \to$ a singleton which is a solution of (4.4) with Dirichlet boundary conditions. One can follow the attractor with Neumann data homotopically to one with Dirichlet data. Bifurcations, of course, occur along the way.

A more interesting problem is to go from Neumann to Dirichlet along an attractor for which the flow does not change its topological structure. A special case of a scalar one-dimensional problem has been given by Gardner [1983] if we make use of the previously mentioned result of Henry on transversality.

Other natural perturbations of the attractor can be considered. Suppose one has an evolutionary equation which defines an asymptotically smooth semigroup for which there is a compact attractor. Suppose the evolutionary equation is approximated in some way by some type of projection or Galerkin procedure. Will the approximate equations have an attractor and will it approach the original one? How will the corresponding flows be related? Some results in this direction have been obtained by Hale, Lin and Raugel [1985].

References

Aronson, D., Crandall, M.G. and L.A. Peletier [1982], Stabilization of solutions of a degenerate nonlinear problem, Nonlinear Anal. 6(1982), 1001-1022.

Babin, A.V. and M.I. Vishik [1983], Regular attractors of semigroups and evolution equations, J. Math. Pures et Appl. 62(1983), 441-491.

Cooperman, G. [1978], α-condensing maps and dissipative systems, Ph.D. Thesis, Brown Univ., Providence, RI, 1978.

Gardner, R. [1983], Global continuation of branches of nondegenerate solutions. Preprint.

Hale, J.K. [1977], Functional Differential Equations. Appl. Math. Sci, Vol. 3, Springer-Verlag, 1977.

Hale, J.K. [1985], Asymptotic behavior and dynamics in infinite dimensions. p. 1-40 in Nonlinear Differential Equations, Research Notes in Math, Vol. 132, Pitman Publ. 1985.

Hale, J.K. [1985a], Large diffusivity and asymptotic behavior in parabolic systems, J. Math. Anal. Appl., to appear.

Hale, J.K., LaSalle, J.P. and M. Slemrod [1972], Theory of a general class of dissipative processes, J. Math. Anal. Appl. 39(1972), 177-191.

Hale, J.K., Lin, X.-B., and G. Raugel [1985], Approximation of the attractor in infinite dimensional systems. Preprint.

Hale, J.K. and G. Lopes [1973], Fixed point theorems and dissipative processes, J. Diff. Eqns. 13(1973), 391-402.

Hale, J.K., Magalhães, L. and W. Oliva [1984], An Introduction to Infinite Dimensional Dynamical Systems - Geometric Theory. Appl. Math. Sci. Vol. 47, Springer, 1984.

Hale, J.K. and J. Kato [1978], Phase space for retarded equations with infinite delay, Funk. Ekvac. 21(1978), 11-41.

Hale, J.K. and C. Rocha [1985], Varying boundary conditions with large diffusivity, J. Math. Pures et Appl., to appear.

Hale, J.K. and K. Rybakowski [1982], On a gradient-like integro-differential equation, Proc. Roy. Soc. Edinburgh 92A(1982), 77-85.

Henry, D. [1981], Geometric Theory of Semilinear Parabolic Equations. Lect. Notes in Math., Vol. 840, Springer, 1981.

Henry, D. [1985], Some infinite dimensional Morse-Smale systems defined by parabolic partial differential equations. J. Diff. Eqns. 59 (1985), 165-205.

Massatt, P. [1983a], Attractivity properties of α-contractions. J. Diff. Eqns. 48(1983), 326-333.

Massatt, P. [1983b], Limiting behavior for strongly damped nonlinear wave equations, J. Diff. Eqns. 48(1983), 334-349.

Webb, G. [1980], Existence and asymptotic behavior for a strongly damped nonlinear wave equation. Canad. J. Math. 32(1980), 631-643.

Webb, G. [1985], Theory of Nonlinear Age-Dependent Population Dynamics, Marcel-Dekker, New York and Basel, 1985.

THE PRINCIPLE OF SPATIAL AVERAGING
AND INERTIAL MANIFOLDS FOR REACTION-DIFFUSION EQUATIONS

John Mallet-Paret
Division of Applied Mathematics
Brown University
Providence, Rhode Island 02912

George R. Sell
School of Mathematics
University of Minnesota
Minneapolis, Minnesota 55455

1. INTRODUCTION

In this paper we want to examine the long-term behavior of nonlinear evolution equations of the form

$$u' + Au = N(u)$$

on a Hilbert space H. An example of such an equation arises with the reaction-diffusion equation

$$u_t = \nu \Delta u = f(x,u)$$

with suitable boundary conditions on a sufficiently regular bounded domain $\Omega \subseteq R^n$. In this case $A = -\nu \Delta$ (with the given boundary conditions) is a self-adjoint operator with compact resolvent.

One of the interesting recent developments in the study of nonlinear evolutionary equations was the observation that for dissipative equations there is a universal attractor Γ and that Γ is compact, invariant, and has finite Hausdorff dimension, cf. Mallet-Paret (1976) and other references cited in Hale-Magalhães-Oliva (1984) and Foias-Sell-Temam (1986). Such results, in the context of reaction-diffusion equations, are independent of the space-dimension n.

Just recently the new concept of an _inertial manifold_ was introduced into the study of the long-time behavior of solutions of nonlinear dissipative equations, cf. Foias-Sell-Teman (1985,1986). (Also see Conway-Hoff-Smoller (1978),

---

This research was done in part at the Institute for Mathematics and its Applications. Support came from several grants the National Science Foundation to Brown University and the University of Minnesota including DMS-8507056, DMS-8120789, and DMS-8501933.

Foias-Nicolaenko-Sell-Temam (1985,1986), Constantin-Foias-Nicolaenko-Temam (1986) and Mallet-Paret and Sell (1986a,b).) As they are now understood, the existence theorems for inertial manifolds depend heavily on the spectral properties of the linear operator A. As a result of this, one expects to find that the theories of inertial manifolds for reaction-diffusion equations will change as one changes the space dimension.

In our attempt to study reaction-diffusion equations in higher space dimensions we have discovered an important new feature of harmonic analysis, viz., the Principle of Spatial Averaging, or PSA, for short. The PSA is a very powerful tool and, as we will see in this paper, it can be used to demonstrate the existence of inertial manifolds for reaction-diffusion equations which were intractable with earlier techniques. The PSA also shows an interesting difference in the spectral properties of the Schrödinger operators in space dimensions 3 and 4.

II. THE PRINCIPLE OF SPATIAL AVERAGING

In order to formulate the PSA we let $N: H \to H$ be a (possibly nonlinear) vector field on a Hilbert space H. Let P be any bounded, linear projection on H. The vector field <u>induced</u> by N on the range $\mathcal{P}$ of P is defined to be $PN(Pu)$.

Consider now the Hilbert space $H = L^2(\Omega)$ where $\Omega \subseteq R^n$ is a bounded domain. For any $g \in L^\infty$ we let B_g denote the multiplication operator on L^2 defined by

$$(B_g u)(x) = g(x)u(x), \quad u \in L^2,$$

and let $\overline{g}$ denote the mean value

$$\overline{g} = (\text{vol } \Omega)^{-1} \int_\Omega g \, dx.$$

Let $-A = \Delta$ be the Laplacian on Ω with a given choice of boundary conditions, let $\{\lambda_m\}_{m=1}^\infty$ denote the eigenvalues of A ordered (with multiplicities) so that

$$0 < \lambda_1 < \lambda_2 < \lambda_3 < \ldots < \lambda_m \to \infty,$$

and let $\{e_m\}_{m=1}^{\infty} \subseteq L^2$ be a corresponding complete orthonormal set of eigenfunctions. For any $\lambda > 0$ set

$$\mathcal{P}_\lambda = \text{span}\{e_m \mid \lambda_m < \lambda\},$$

$$\mathcal{Q}_\lambda = \mathcal{P}_\lambda^\perp = \text{closure}(\text{span}\{e_m \mid \lambda_m > \lambda\}),$$

and let P_λ and $Q_\lambda = I - P_\lambda$ denote the orthogonal projections onto these subspaces. Of course, the space $\mathcal{P}_\lambda$ has finite dimension m_0 where $\lambda_{m_0} < \lambda < \lambda_{m_0+1}$. Let $\lambda > \kappa > 0$. The PSA, which we define next, gives a comparison between the vector fields induced by B_g and $\overline{g}I$ on the finite dimensional range of $(P_{\lambda+\kappa} - P_{\lambda-\kappa})$.

<u>Definition</u>. For a bounded domain $\Omega \subseteq R^n$, $n < 3$, and a given choice of boundary conditions for the Laplacian, we say the <u>Principle of Spatial Averaging</u>, or <u>PSA</u>, holds if there exists a quantity $\xi > 0$, such that for every $\delta > 0$ and $\kappa > 0$ there exists $\lambda > \kappa$, such that

$$\|(P_{\lambda+\kappa} - P_{\lambda-\kappa})(B_g - \overline{g}I)(P_{\lambda+\kappa} - P_{\lambda-\kappa})\|_{op} < \delta \|g\|_{H^2}$$

holds whenever $g \in H^2$; and such that

$$\lambda_{m+1} - \lambda_m > \xi,$$

where m is such that $\lambda_m < \lambda < \lambda_{m+1}$. Here $\|\cdot\|_{op}$ denotes the norm of an operator on L^2.

The main result which we use is that the PSA holds for certain domains. The following theorem will play a crucial role in the theory in the next section:

Theorem 1. The PSA holds for any rectangular domain

$$\Omega_2 = (0,a_1) \times (0,a_2),$$

for any $a_1 > 0$, $a_2 > 0$. Further, PSA holds for any parallelepiped

$$\Omega_3 = (0,a_1) \times (0,a_2) \times (0,a_3)$$

for which all quantities $(\dfrac{a_i}{a_j})^2$ are rational. In any case, either Dirichlet, Neumann, or periodic boundary conditions can be taken.

The fact that PSA is valid on the unit cube Ω_3 in R^3 is especially interesting. Since among any three consecutive integers there is one which is the sum of three squares, one has

$$\text{dim range } (P_{\lambda+\kappa} - P_{\lambda-\kappa}) > \frac{2\kappa}{3} - 1,$$

i.e., the dimension of range of $(P_{\lambda+\kappa} - P_{\lambda-\kappa})$ can be chosen to be arbitrarily large.

The reason for restricting the definition of PSA to regions of dimension $n \leqslant 3$ is so that $H^2 \subseteq L^\infty$; i.e. the operator B_g is bounded for $g \in H^2$. For dimension $n \geqslant 4$ one might consider restricting to $g \in H^s$, where $s > \frac{n}{2}$, so that $H^s \subseteq L^\infty$. However, this strategy is not successful. Consider the four-dimensional hypercube $\Omega_4 = (0,2\pi)^4$, with either Neumann or periodic boundary conditions, and fix $g(x) = \cos x_1$ (here $x = (x_1,x_2,x_3,x_4)$). It is not difficult to show that for any $\kappa \geqslant 2$ one has

$$\|(P_{\lambda+\kappa} - P_{\lambda-\kappa})(B_g - \overline{g}I)(P_{\lambda+\kappa} - P_{\lambda-\kappa})\| > \frac{1}{2}$$

for every $\lambda \geqslant \kappa$.

III. INERTIAL MANIFOLDS

Let us now consider the scalar reaction-diffusion equation

(1) $$u_t = \nu\Delta u + f(x,u), \quad x \in \Omega \subseteq R^n$$

in the bounded domain Ω, cf. Henry (1981). As boundary conditions we take either

$$u = 0 \quad \text{on} \quad \partial\Omega \quad \text{(Dirichlet), or}$$

$$\frac{\partial u}{\partial n} = 0 \quad \text{on} \quad \partial\Omega \quad \text{(Neumann).}$$

Periodic boundary conditions can also be taken when Ω is a cartesian product of intervals, say a rectangle Ω_2 in R^2 or a parallelepiped Ω_3 in R^3. In case the boundary $\partial\Omega$ is not smooth, the boundary conditions are interpreted in a weak sense: the domain $\mathcal{D} \subseteq L^2$ of the Laplace operator Δ is the set $\mathcal{D}^D = H^2 \cap H_0^1$ when Dirichlet conditions are taken, and the set

$$\mathcal{D}^N = \{u \in H^2 \mid \int_\Omega (\phi\Delta u + \nabla\phi\cdot\nabla u)dx = 0 \quad \text{for all} \quad \phi \in H^1\}$$

with Neumann conditions. Some regularity of $\partial\Omega$ will be required, specifically, that Ω should be bounded and possess the Cone Property. This means there exists a finite cone $C \subseteq R^n$, with nonempty interior, such that for each $x \in \overline{\Omega}$ one has $\tilde{C} \subseteq \overline{\Omega}$, where $\tilde{C}$ is a cone congruent to C with vertex at x. For simplicity we shall call such a domain Ω a _regular domain_.

The diffusion coefficient (i.e., diffusivity or viscosity) $\nu > 0$ of equation (1) is fixed, and the nonlinearity $f: \Omega \times R \to R$ is assumed sufficiently smooth. To be precise, we assume that

(2)
$$f \text{ is continuous in } \overline{\Omega} \times R, \text{ and}$$
$$D_u f \text{ exists and is } C^2 \text{ in } \overline{\Omega} \times R,$$

where $D_u f = \frac{\partial f}{\partial u}$ denotes the partial derivative. We also impose a sign condition on f for large $|u|$: for some $K_1 > 0$

(3)
$$uf(x,u) < 0 \quad \text{whenever} \quad |u| > K_1.$$

The purpose of this sign condition is to ensure that the dynamical system generated by (1) is dissipative, in which case (1) possesses a maximal compact attractor Γ in the phase space. Finally, we assume the growth conditions

(4)
$$|f(x,u)| \leq K_2|u| + K_3 \quad \text{and} \quad |D_u f(x,u)| \leq K_2 \quad \text{in} \quad \Omega \times R.$$

In practice there is no loss of generality in assuming (4): the sign condition (3) alone implies, by the maximum principle, that any solution $u(t,x)$ with initial condition

$$(5) \qquad u(0,x) = \phi(x),$$

where $\phi \in L^{\infty}$, satisfies

$$\|u(t,\cdot)\|_{L^{\infty}} < \max\{\|\phi\|_{L^{\infty}}, K_1\}$$

for all $t > 0$. Thus, modifying f, if necessary, for large $|u|$ in order to achieve (4) does not affect solutions starting in a given bounded set in L^{∞}.

With the above assumptions, the equation (1) generates a dynamical system in L^2. (Actually, less regularity is required of f; however, the assumptions (2) will be needed later.) A solution of the initial value problem (1), (5) in an interval $[0,\omega)$ with $\phi \in L^2$ is defined to be a continuous function $u: [0,\omega) \to L^2$ satisfying the integral equation

$$u(t) = T(t)\phi + \int_0^t T(t - s)\tilde{f}(u(s))ds.$$

Here $T(t) = e^{-At}$ is the semigroup in L^2 generated by the self-adjoint operator $-A = \nu\Delta$, and $\tilde{f}: L^2 \to L^2$ is the evaluation map given by

$$[\tilde{f}(u)](x) = f(x,u(x))$$

for $u \in L^2$. Any such solution exists for all $t > 0$; it is jointly continuous in t and ϕ (with values in L^2); and it is C^1 in t for $t > 0$ and satisfies the abstract equation

$$(6) \qquad u' = -Au + \tilde{f}(u).$$

Let $u(t) = S(t)\phi$ denote the solution of (1),(5). Then the regularity property $u(t) \in \mathcal{D} \cap L^{\infty}$ holds for $t > 0$; in fact, with a bit more regularity on f (i.e., a Hölder condition in x), the solution $u = u(t,x)$ is a classical solution of equation (1).

Equation (6) possesses a universal attractor $\Gamma \subseteq L^2$. This may be given as the nested intersection

$$\Gamma = \bigcap_{m=0}^{\infty} S(m\tau)B_R$$

where B_R denotes the ball in L^2 of sufficiently large radius R, and $\tau > 0$ is such that $S(\tau)B_R \subseteq B_R$ (for each large R such τ exists). The attractor Γ is independent of R and satisfies $S(t)\Gamma = \Gamma$ for all $t \geqslant 0$. In addition, Γ is nonempty, compact, connected, has finite Hausdorff dimension, and enjoys the L^{∞} bound

$$\|u\|_{L^{\infty}} \leqslant K_1 \quad \text{for all} \quad u \in \Gamma.$$

The dynamics on Γ describe the long-time asymptotics of equation (6), that is $u(t) \to \Gamma$ as $t \to \infty$, for each solution. Consequently, Γ is a principal object of study.

In view of the finite dimensionality of Γ, one approach to understanding the dynamics of (6) would be to embed Γ into a smooth, finite dimensional manifold $M \subseteq L^2$. If, in addition, M is invariant, then by restricting the vector field (6) to M one obtains an ordinary differential equation. In order to better understand this approach we make a formal definition.

<u>Definition.</u> A finite dimensional C^1 manifold $M \subseteq L^2$ is called an <u>inertial manifold</u> for the dynamical system generated by (6) if

    (a) M is locally positively invariant: i.e., $\phi \in M$ implies there is a

        $\delta = \delta(\phi)$ such that $S(t)\phi \in M$ for all $t \in [0,\delta)$;

    (b) the universal attractor $\Gamma \subseteq M$, and

    (c) M is normally hyperbolic over Γ.

In the neighborhood of any point u of an inertial manifold M, there exists a splitting $L^2 = \mathcal{P}_u \oplus \mathcal{Q}_u$ into subspaces, with $\mathcal{P}_u$ finite dimensional (with dimension independent of u), such that M is given locally as the graph $q = \Phi(p)$ of a C^1 function Φ. Writing $u \in L^2$ as $u = (p,q)$, one is led to the ordinary differential equation (note A is bounded on $\mathcal{P}_u$)

$$p' = -PAPp + P\tilde{f}(p,\Phi(p)),$$

which describes the dynamical system restricted to M. Here $P = P_u$ is the projection onto $\mathcal{P}_u$ along $\mathcal{Q}_u$. In particular, all solutions on the attractor Γ are obtained in this way. One thinks of the inertial manifold M as a global center manifold for the set Γ.

For the invariant manifolds obtained below, the subspace $\mathcal{P}$ will be the span of the first m eigenfunctions of the operator A for some m, and $\mathcal{Q}$ will be the complementary subspace. This decomposition is independent of u, and one can also write $u = (p,q) = p + q$. In addition, the projection P will commute with A and the dynamics restricted to M will have the form

$$p' = -Ap + P\tilde{f}(p + \Phi(p)).$$

In this sense the existence of M can be thought of as a dynamic version of Galerkin's Method.

Normal hyperbolicity of M is a condition involving the linear variational equation

$$(7) \qquad\qquad \mu' = -A\mu + D\tilde{f}(u(t))\mu$$

for $\mu \in L^2$, with $u(t)$ a solution of (6) and $D\tilde{f}(u)$ the multiplication operator

$$[D\tilde{f}(u)\mu](x) = D_u f(x,u(x))\mu(x)$$

on L^2. To say that M is normally hyperbolic over Γ means there exists a splitting $L^2 = T_u M \oplus N_u M$ into subspaces, depending continuously on $u \in \Gamma$, where $T_u M$ is the tangent space of M and $N_u M$ is some complementary space (not necessarily orthogonal). Furthermore, the tangent and normal bundles

$$T\Gamma = \{(u,\mu) \in L^2 \times L^2 | u \in \Gamma \text{ and } \mu \in T_u M\}$$

$$N\Gamma = \{(u,\mu) \in L^2 \times L^2 | u \in \Gamma \text{ and } \mu \in N_u M\}$$

are required to be positively invariant, with the flow satisfying uniform exponential estimates

$$\|\mu(t)\| < K_4 e^{-\beta t} \|\mu(0)\| \quad \text{in } N\Gamma, \quad \text{for } t > 0,$$

$$\|\mu(t)\| < K_4 e^{-\alpha t} \|\mu(0)\| \quad \text{in } T\Gamma, \quad \text{for } t < 0,$$

for some $K_4 > 0$ and $0 < \alpha < \beta$, for solutions $u(t) \in \Gamma$, cf. Hirsch-Pugh-Shub (1977) and Sacker-Sell (1980). (Invariance of $T\Gamma$ follows from the invariance of Γ.) Thus, the flow of (7) admits an exponential dichotomy over Γ. The idea is that M experiences a contraction in the normal direction which is stronger than that in the tangential direction. As a consequence of the normal hyperbolicity one expects that under small perturbations of the vector field in L^2, the manifold M should undergo a small perturbation. That is, M is robust under perturbations of the differential equation.

The theorems below give sufficient conditions on the domain Ω for equation (6) to possess an inertial manifold. These conditions, which involve spectral properties of the Laplacian, are fulfilled for certain low-dimensional domains. (See Foias-Sell-Temam (1985,1986).) We also state a result (Theorem 4) which concerns the nonexistence of an inertial manifold for a four dimensional domain and a certain nonlinearity.

In what follows, we let $\|\cdot\|_X$ denote the norm in a Banach space X, with $\|\cdot\|$ denoting the norm in the Hilbert space $X = L^2$. We use the notation of Section II and write $u \in L^2$ as $u = (p + q) \in \mathcal{P}_\lambda \oplus \mathcal{Q}_\lambda$, with $p = P_\lambda u \in \mathcal{P}_\lambda$ and $q = Q_\lambda u \in \mathcal{Q}_\lambda$. Equation (6) then becomes

$$p' = -Ap + P_\lambda \tilde{f}(p + q)$$

(8)

$$q' = -Aq + Q_\lambda \tilde{f}(p + q)$$

and the associated variational equation (7) can be written as

$$\rho' = -A\rho + P_\lambda D\tilde{f}(p + q)(\rho + \sigma)$$

(9)

$$\sigma' = -A\sigma + Q_\lambda D\tilde{f}(p + q)(\rho + \sigma),$$

where $\mu = (\rho + \sigma) \in \mathcal{P}_\lambda \oplus \mathcal{Q}_\lambda$.

Given $\lambda > 0$ and $u = (p + q) \in \mathcal{P}_\lambda \oplus \mathcal{Q}_\lambda$, we introduce the following Cone Condition for the variational system (9) at the point u:

> Define $V = \frac{1}{2} \|\sigma\|^2 - \frac{1}{2} \|\rho\|^2$ and $V' = \langle\sigma,\sigma'\rangle - \langle\rho,\rho'\rangle$, where σ' and ρ' are given by (9). Then one has $V' < 0$ whenever $\|\rho\| = \|\sigma\| \neq 0$ and $\sigma \in \mathcal{O} \cap \mathcal{Q}_\lambda$.

Geometrically, this means horizontal cone $V < 0$ in the tangent space $T_u L^2$ above $u \in L^2$ is positively invariant for (9). We shall not require this Cone Condition for all u, rather only for those u in a set containing the attractor Γ. Indeed, if for some $\lambda > 0$ the Cone Condition is fulfilled for each u near Γ, then one expects that Γ should lie in a C^1 inertial manifold $M \subseteq L^2$ which is in fact the graph $q = \Phi(p)$ of some C^1 function Φ, for p near the set $P_\lambda(\Gamma) \subseteq \mathcal{P}_\lambda$. We note here the formula

$$V' = \nu \int_\Omega (\sigma\Delta\sigma - \rho\Delta\rho)dx + \int_\Omega (\sigma^2 - \rho^2) \, g \, dx,$$

where

$$(10) \qquad g(x) = D_u f(x,u(x)).$$

For technical (but probably not essential) reasons, we need a slightly stronger Modified Cone Condition in order to obtain such an inertial manifold.

Theorem 2. Define $A_0 = I - \Delta$, and let $n \leq 3$. Assume that for each $R > 0$ there exist arbitrarily large $\lambda > 0$ such that

$$(11) \qquad -\nu\langle\sigma,A_0\sigma\rangle + \nu\|\rho\|\|A_0\rho\| + \int(\sigma^2 - \rho^2)g \, dx < 0$$

whenever $\|\rho\| = \|\sigma\| \neq 0$ and $u \in \mathcal{O}$ satisfies $\|A_0 u\| \leq R$, where g is as in (10). Then the dynamical system generated by equation (6) possesses an inertial manifold M which has the form $M = \text{Graph}(\Phi)$, where $\Phi: \mathcal{U} \subseteq \mathcal{P}_\lambda \to \mathcal{Q}_\lambda$ is a C^1-function and $\mathcal{U}$ is a neighborhood of $P_\lambda(\Gamma)$.

Actually, in Theorem 2 it is only necessary for the hypotheses to hold for some sufficiently large R and λ, which may depend on f. Also, note the inequality (11) implies that $V' < 0$.

Since the expression (11) in the statement of Theorem 2 is homogeneous in ρ and σ, it is enough to take $\|\rho\| = \|\sigma\| = 1$. With this, one has

$$-\nu\langle\sigma, A_0\sigma\rangle + \nu\|\rho\|\|A_0\rho\| \leq -\nu(\lambda_{m+1} - \lambda_m)$$

as A_0 is self-adjoint, and

$$\int_\Omega (\sigma^2 - \rho^2)g \, dx \leq 2(\text{vol } \Omega)K_2$$

from the bound (4). Thus, the inequality (11) is achieved if the spectrum of A_0 satisfies the gap condition

$$(12) \qquad\qquad \nu(\lambda_{m+1} - \lambda_m) > 2(\text{vol } \Omega)K_2.$$

For example, if

$$(13) \qquad\qquad \limsup_{m\to\infty} (\lambda_{m+1} - \lambda_m) = \infty$$

then for any $\nu > 0$ and any f, the gap condition (12) is satisfied for infinitely many m. As noted in Foias-Sell-Temam (1986) the gap condition (13) holds, in particular, for any rectangular domain $\Omega_2 = (0, a_1) \times (0, a_2)$ where $(a_1/a_2)^2$ is rational. In this case the eigenvalues of the Laplacian have the form

$$\pi^2(k_1^2 a_1^{-2} + k_2^2 a_2^{-2}),$$

where k_1, k_2 are integers. Among the numbers of the above form, there exist gaps of arbitrarily high length when $(a_1/a_2)^2$ is rational, cf. Richards (1982).

If (13) fails, as it does for the cube $\Omega_3 = (0, a)^3$ with Neumann or periodic boundary conditions (among any three consecutive positive integers, at least one is a sum of three perfect squares, cf. Hardy-Wright (1962)), then a more subtle approach is needed. In such a case, one tries to achieve the inequality (11) by showing the term $\int(\sigma^2 - \rho^2) \, g \, dx$ is close to zero, so is dominated in absolute value by the quantity $\nu(\lambda_{m+1} - \lambda_m)$. Indeed, the PSA achieves precisely that: One has the approximations

$$\int_\Omega \rho^2 g \, dx = \langle \rho, B_g \rho \rangle \approx \langle \rho, \overline{g}\rho \rangle = \left(\int_\Omega \rho^2 dx\right)\left(\int_\Omega g \, dx\right) = \int_\Omega g \, dx,$$

$$\int_\Omega \sigma^2 g \, dx = \langle \sigma, B_g \sigma \rangle \approx \langle \sigma, \overline{g}\sigma \rangle = \left(\int_\Omega \sigma^2 \, dx\right)\left(\int_\Omega g \, dx\right) = \int_\Omega g \, dx,$$

The idea is that ρ and σ are composed of rather high-frequency Fourier modes, so suffer a great deal of oscillation; by contrast, the function g is relatively smooth. Thus, the PSA implies the hypotheses of Theorem 2.

Theorem 3. Assume $\Omega \subseteq R^n$, with $n \leq 3$, is a regular domain satisfying PSA for some choice of boundary conditions (Dirichlet, Neumann, or periodic). Then the hypotheses of Theorem 2 hold, so (6) possesses an inertial manifold.

If the gap condition (13) holds, then PSA holds trivially: one need only select λ so that $\lambda - \kappa$ and $\lambda + \kappa$ both lie in an interval $[\lambda_m, \lambda_{m+1})$, for then $P_{\lambda+\kappa} - P_{\lambda-\kappa} = 0$. However, PSA can hold even if (13) fails. As noted in Theorem 1, PSA is valid on any rectangular domain $\Omega_2 = (0,a_1) \times (0,a_2)$ with no further assumptions on a_1 and a_2, and PSA is valid on any 3-dimensional parallelepiped provided all quantities $(\frac{a_i}{a_j})^2$ are rational. It is a conseqeunce of the failure of PSA in $\Omega_4 = (0,2\pi)^4$ that there exists a counterexample of the existence of an inertial manifold in this domain:

Theorem 4. With $\Omega_n = (0,2\pi)^n$, and $n \geq 4$ there exists an analytic function $f: \Omega_n \times R \rightarrow R$, satisfying the assumptions (2), (3), and (4), such that the equation (6), with $\nu = 1$ and Neumann boundary conditions, does not possess an inertial manifold in the phase space L^2.

We remark that the failure of an inertial manifold to exist for the equation of Theorem 4 is not due to a lack of smoothness (f is analytic); nor is it due to the attractor Γ being too large (Γ is compact, bounded in L^∞, and has finite Hausdorff dimension), nor is it due merely to the failure of a critical Sobolev inequality. What does happen is that the requirement of normal hyperbolicity, along with the occurence of eigenvalues of high multiplicity for the linearized problem, places severe restrictions on the possible dimension of the manifold M. The counterexample is constructed by arranging matters so that the normal hyperbolicity implies that dim M must vary from point to point on Γ; however, this is impossible since Γ is connected so it must lie in a single connected component of M. Thus, an inertial manifold M cannot exist.

The existence of this counterexample shows that the occurence of a normally hyperbolic inertial manifold in a delicate phenomenon and only occurs in low space dimension. The requirement that an inertial manifold be normally hyperbolic is, of course, not sacred. However, without this property one cannot expect the inertial manifolds to be robust or stable under small perturbations or the coefficients.

REFERENCES

1. E. Conway, D. Hoff, J. Smoller, (1978) Large time behavior of solutions of nonlinear reaction-diffusion equations. SIAM J. Appl. Math., 35, #11, p. 1-16.

2. P. Constantin, C. Foias, B. Nicolaenko, R. Temam, (1986) Integral manifolds and intertial manifolds for dissipative partial differential equations. To appear.

3. C. Foias, B. Nicolaenko, G.R. Sell, R. Temam, (1985) Variétés inertielles pour l'équation de Kuramoto-Sivashkinsky, C.R. Acad. Sc. Paris, Serie 1, 301, p. 285-288.

4. C. Foias, B. Nicolaenko, G.R. Sell, R. Temam, (1986) Inertial manifold for the Kuramoto Sivashinsky equation. To appear.

5. C. Foias, G.R. Sell, R. Temam, (1985) Variétés Inertielles des équations differentielles dissipatives, C.R. Acad. Sci. Paris, Serie 1, 301, p. 139-141.

6. C. Foias, G.R. Sell, R. Temam (1986) Inertial manifolds for nonlinear evolutionary equations, IMA Preprint No. 234.

7. J.K. Hale, L.T. Magalhaes, W.M. Oliva, (1984) An Introduction to Infinite Dimensional Dynamical Systems - Geometric Theory. Appl. Math. Sciences No. 47, Springer-Verlag, Berlin-Heidelberg - New York.

8. G.H. Hardy, E.M. Wright, (1962) An Introduction to the Theory of Numbers. Oxford Press.

9. D. Henry, (1981) Geometric Theory of Semilinear Parabolic Equations. Lecture Notes in Mathematics, No. 840, Springer-Verlag, Berlin-Heidelberg-New York.

10. M. Hirsch, C. Pugh, M. Shub, (1977) Invariant Manifolds. Lecture Notes in Mathematics, No. 583, Springer-Verlag, Berlin-Heidelberg - New York.

11. J. Mallet-Paret, (1976) Negatively invariant sets of compact maps and an extension of a Theorem of Cartwright, J. Diff. Eqns., 22, p. 331-348.

12. J. Mallet-Paret, G.R. Sell (1986a) Inertial manifolds for reaction-diffusion equations in higher space dimension. To appear.

13. J. Mallet-Paret, G.R. Sell (1986b) A counterexample to the existence of inertial manifolds. To appear.

14. J. Richards, (1982) On the gaps between numbers which are the sum of two squares, Adv. Math., vol. 46, pp. 1-2.

15. R.J. Sacker, G.R. Sell, (1980) The spectrum of an invariant submanifold, J. Diff. Eqns., vol. 38, p. 135-160.

APPLICATIONS OF SEMIGROUP THEORY
TO REACTION-DIFFUSION SYSTEMS

by

Robert H. Martin, Jr.
Department of Mathematics
North Carolina State University
Raleigh, NC 27650-8205

The purpose of this paper is to indicate how the theory of semilinear differential equations in Banach spaces can be applied to the study of the behavior of solutions to reaction-diffusion systems. The emphasis here is on _systems_ (i.e., more than one unknown function) and the approach is to use techniques from ordinary differential equations in Banach spaces for the analysis. In general our equations will have a linear term (corresponding to diffusion) and a nonlinear term (corresponding to reaction); however, as opposed to the usual situation, the linear term is frequently viewed as a perturbation of the nonlinear term. In particular we are often interested in the effect of adding diffusion to a nonlinear ordinary differential equation.

Throughout this paper it is assumed that Ω is a bounded region in $\mathbb{R}^N$ with $\partial\Omega$ (the boundary of Ω) smooth, that Δ denotes the Laplacian operator on Ω and that $\frac{\partial}{\partial\nu}$ denotes the outward pointing normal derivative on $\partial\Omega$. Also let m be a positive integer, $m \geq 2$, and consider the family of diffusion equations

(1)
(a) $\partial_t v_i(x,t) = d_i \Delta v_i(x,t)$, $\quad t > 0$, $x\in\Omega$, $i = 1,\ldots,m$

(b) $\alpha_i v_i(x,t) + (1-\alpha_i)\frac{\partial}{\partial\nu}v_i(x,t) = \beta_i$,

$\quad t > 0$, $x\in\partial\Omega$, $i = 1,\ldots,m$

(c) $v_i(x,0) = \gamma_i(x) \geq 0$, $\quad x\in\Omega$, $i = 1,\ldots,m$

where the following basic hypotheses are assumed to hold:

(H1) $d_i > 0$, $0 \leq \alpha_i \leq 1$ and $\beta_i \geq 0$ are constants for $i = 1,\ldots,m$. Also, $\beta_i = 0$ if $\alpha_i = 0$.

(H2) γ_i is measurable on Ω and there is an $M_o > 0$ such that $0 = \gamma_i(x) \leq M_o$ for $x\in\Omega$, $i = 1,\ldots,m$.

In addition, consider the nonlinear ordinary differential system

(2)
$z_i'(t) = f_i(z_1(t),\ldots,z_m(t))$, $t > 0$ $\quad i = 1,\ldots,m$

$z_i(0) = \eta_i \geq 0$, $i = 1,\ldots,m$

where the functions $f_i: \mathbb{R}^m \longrightarrow \mathbb{R}$ are assumed to satisfy:

(H3) f_i is locally Lipschitz continuous on $[0,\infty)^m$ and

(H4) if $k \in \{1,\ldots,m\}$ and $\xi_i \geq 0$ for $i = 1,\ldots,m$,

then $\xi_k = 0$ implies $f_k(\xi_1,\ldots,\xi_m) \geq 0$.

A function $f = (f_i)_1^m$ that satisfies (H4) is called quasi-positive, and it is easily seen that this implies that solutions to (2) remain nonnegative so long as they exist: if $z = (z_i)_1^m$ is the solution to (2) on $[0,b)$, then $z_i(t) \geq 0$ for all $t \in [0,b)$ and $i = 1,\ldots,m$.

Combining equations (1) and (2) leads to the following reaction-diffusion system:

(a) $\partial_t u_i = d_i \Delta u_i + f_i(u_1,\ldots,u_m)$, $t > 0$, $x \in \Omega$, $i = 1,\ldots,m$

(3) (b) $\alpha_i u_i + (1-\alpha_i)\frac{\partial}{\partial\nu}u_i = \beta_i$, $t > 0$, $x \in \partial\Omega$, $i = 1,\ldots,m$

(c) $u_i = \gamma_i \geq 0$, $t = 0$, $x \in \Omega$, $i = 1,\ldots,m$.

Of particular interest is the comparison of solutions to (3) with those of the ordinary differential equation (2) or with those of some matrix perturbation of (2).

§1. VARIATION OF CONSTANTS: ABSTRACT FORMULATION:

Suppose that X is a real Banach space with norm denoted by $|\,.\,|$, and that $T = \{T(t):t \geq 0\}$ is a C_o semigroup of bounded linear operators on X:

(a) $T(0)x = x$ and $T(t+s)x = T(t)T(s)x$ for all $x \in X$, $t,s \geq 0$.

(1.1) (b) $t \longrightarrow T(t)x$ is continuous on $[0,\infty)$ for each $x \in X$.

(c) $|T(t)x| \leq M e^{\omega t}|x|$ for all $t \geq 0$, $x \in X$, where M and ω are real constants, $M \geq 1$.

It is well-known that T has an infinitesimal generator A defined by

(1.2) $$Ax = \lim_{h \to 0+} \frac{T(h)x - x}{h} \text{ for all } x \in D(A),$$

where D(A) is precisely the set of all x in X where the limit in (1.2) exists. Moreover, D(A) is dense in X, the resolvent $(I-hA)^{-1}$ exists for all $h > 0$ with $h\omega < 1$, and A and T are related by the formulas

(a) $T(t)x = \lim_{n \to \infty}(I - \frac{t}{n}A)^{-n}x$ for all $t \geq 0$, $x \in X$

(1.3) (b) $(I-hA)^{-1}x = \frac{1}{h}\int_0^\infty e^{-t/h}T(t)xdt$ for all $x \in X$

and $h > 0$ such that $h\omega < 1$

(c) $T(t)x \in D(A)$ and $\frac{d}{dt}T(t)x = AT(t)x$ for all

$t > 0$ and $x \in D(A)$.

Suppose further that following are satisfied:

(a) D is a closed subset of X and $F:D \longrightarrow X$

(1.4) (b) for each $R > 0$ there is an $L(R) > 0$ such that

$|Fx - Fy| \leq L(R)|x-y|$ for all $x,y \in D$, $|x|,|y| \leq R$.

We consider the initial value problem

(1.5) $\qquad u'(t) = Au(t) + Fu(t), \; u(0) = z \in D, \; t > 0.$

A function $u:[0,b) \longrightarrow X$ is said to be a (strong) solution to (1.5) on [0,b) if u is continuous on [0,b), continuously differentiable on(0,b), satisfies $u(0) = z$ and $u(t) \in D(A) \cap D$ for all $t \in (0,b)$, and (1.5) holds for $t \in (0,b)$. In many cases, requiring the existence of strong solutions to (1.5) is too stringent, and so we apply the variation of constants formula to (1.5) to obtain the equation.

(1.6) $\qquad u(t) = T(t)z + \int_0^t T(t-s)Fu(s)ds, \; t \geq 0.$

A continuous function $u:[0,b) \longrightarrow D$ that satisfies (1.6) for $t \in [0,b)$ is called a mild solution to (1.5) on [0,b). Any strong solution to (1.5) is also a mild solution; however, the converse is not true in general. In order to take into account the nonhomogeneous boundary conditions in equation (3), we suppose that $z_0 \in X$ and consider the equation

(1.7) $\qquad u(t) = T(t)(z-z_0) + z_0 + \int_0^t T(t-s)Fu(s)ds, \; t \geq 0.$

A solution u to (1.7) would be a mild solution of the differential equation

(1.8) $\qquad u'(t) = A(u(t) - z_0) + Fu(t), \; u(0) = z, \; t > 0.$

If u is C^1 on [0,b) with $u(t) - z_0 \in D(A)$ and $u(t) \in D$ for all $t \in (0,b)$, then u is a solution to (1.7) only in case u is a solution to (1.8). However, a solution to (1.7) is not necessarily a strong solution to (1.8). We have the following basic existence result for (1.7):

<u>Theorem 1</u>.   In addition to (1.1) and (1.4), suppose that

$$(1.9) \qquad \lim_{h \to 0+} \frac{d(T(h)(x-z_o)+z_o+hFx;D)}{h} = 0 \text{ for all } x \in D$$

where $d(w;D) \equiv \inf\{|w-y|:y\in D\}$ for each $w \in D$. Then (1.7) has a unique
(noncontinuable) solution $u = u_z$ on $[0,b_z)$ for each $z \in D$. Furthermore, if $b_z < \infty$
then $|u_z(t)| \longrightarrow \infty$ as $t \longrightarrow b_z-$.

The proof of this theorem may be found in [3, Theorem 3.2 and Proposition 3.1].
It is interesting that condition (1.9) is actually necessary for (1.7) to have a
local solution for each $z \in D$. This is observed in Pavel [6].

The main objective here is to show that several standard techniques in the
theory of ordinary differential equations can be applied to study the behavior of
solutions to the integral equation (1.7). In many situations the proofs remain the
same, but sometimes   modifications are needed because of the lack of
differentiability of mild solutions. Our main abstract results use basic
perturbation techniques, and in order to include a reasonably wide variety of
equations, they are stated in terms of subadditive, positively homogeneous
functionals. In particular it is assumed that $V:X \longrightarrow [0,\infty]$ (the extended positive
numbers) satisfies :

(V1) $V[x+y] \leq V[x] + V[y]$ for all $x,y \in X$
(V2) $V[\alpha x] = \alpha V[x]$ for al $x \in X$, $\alpha > 0$
(1.10)   (V3) V is lower semicontinuous: $|x_n - x| \longrightarrow 0$

as $n \longrightarrow \infty$ implies $V[x] \leq \lim_{n \to \infty} \inf V[x_n]$.

(V4) $V[x] \geq \nu|x|$ for all $x \in X$ and some constant $\nu > 0$.

An important example of such a V is $V[x] \equiv |x|$. Another important example is when X
$= L^p(\Omega)$ $(1\leq p<\infty)$ and $V[x] \equiv |x|_\infty$. In this case $V[x] < \infty$ only in case $x \in L^\infty(\Omega)$ and V
is lower semicontinuous but not continuous.

Now suppose that $z^* \in D$ and the solution u_{z^*} to (1.7) satisfies $u_{z^*}(t) \equiv z^*$
[i.e., z^* is an equilibrium for (1.7)]. Substituting z^* for $u(t)$ in (1.7) implies
that

$$\theta = \frac{T(t)(z^*-z_o) - (z^*-z_o)}{t} + \frac{1}{t} \int_o^t T(t-s)Fz^* ds$$

for all $t > 0$, and letting $t \longrightarrow 0+$ shows that $z^* - z_o \in D(A)$ and $A(z^* - z_o) + Fz^* =$

θ. Thus we have the following:

If $z^* \in D$ is an equilibrium for (1.7) then

(1.11)
$$z^* - z_o \in D(A) \text{ and } A(z^* - z_o) + Fz^* = \theta.$$

Finally, we are able to state and prove our abstract perturbation results.

<u>Theorem 2</u>. Suppose that $V:X \longrightarrow [0,\infty]$ satisfies (1.10) and $z^* \in D$ is an equilibrium for (1.7). Suppose further that positive constants M, δ, α and R exist such that

(a) $V[T(t)x] \leq Me^{-\delta t}V[x]$ for all $t \geq 0$, $x \in X$.

(b) $V[Fx-Fz^*] \leq \alpha V[x-z^*]$ for $x \in D$ with $V[x - z^*] \leq 2R$.

(c) If $z \in D$ and $t \geq 0$ is such that $V[u_z(t) - z^*] \leq R$, then there is an $\eta(t,z) > 0$ such that $V[u_z(s) - z^*] \leq 2R$ for all $s \in [t,t + \eta(t,z)]$.

(d) $M\alpha \leq \delta$.

Then for each $z \in D$ with $V[z-z^*] \leq R/M$ the solution u_z to (1.7) exists on $[0,\infty)$ and satisfies

(1.12)
$$V[u_z(t) - z^*] \leq V[z-z^*]Me^{(M\alpha-\delta)t} \text{ for all } t \geq 0.$$

Thus, if $M\alpha < \delta$, $V[u_z(t) - z^*] \longrightarrow 0$ and $|u_z(t) - z^*| \longrightarrow 0$ as $t \longrightarrow \infty$.

<u>Remark</u>. Assumption (c) is always satisfied when V is continuous, and when V is not continuous (c) usually follows readily from (a) and (b).

<u>Proof of Theorem 2</u>. If $z \in D$ with $V[z - z^*] \leq R/M$, then

$$u_z(t)-z^* = T(t)(z-z_o) + z_o + \int_o^t T(t-s)Fu_z(s)ds$$
$$- [T(t)(z^*-z_o)+z_o + \int_o^t T(t-s)Fz^* ds$$

and we have

(1.13)
$$u_z(t)-z^* = T(t)(z-z^*) + \int_o^t T(t-s)(Fu_z(s) - Fz^*)ds.$$

By (a) and the subadditive property of V it follows that

$$V[u_z(t)-z^*] \leq Me^{-\theta t}V[z-z^*] + V[\int_o^t T(t-s)(Fu_z(s)-Fz^*)ds].$$

Using Jensen's inequality and the positive homogenuity of V,

$$V[\int_0^t T(t-s)(Fu_z(s) - Fz^*)ds] \le \int_0^t V[T(t-s)(Fu_z(s)-Fz^*)]ds.$$

Now set $J = \{t{\ge}0: V[u_z(s)-z^*] \le 2R$ for $0{\le}s{\le}t\}$ and observe that if $t \in J$

$$V[u_z(t) - z^*] \le Me^{-\delta t}V[z - z^*] + M\alpha\int_0^t e^{-\delta(t-s)}V[u_z(s) - z^*]ds$$

by (a), (b) and the two preceding inequalities. Setting $p(t) = e^{\delta t}V[u_z(t) - z^*]$ for all $t \in J$ we have

$$p(t) \le MV[z-z^*] + M\alpha\int_0^t p(s)ds$$

for all $t \in J$. Since p is lower semicontinuous and bounded on J, Gronwall's inequality shows that

$$V[u_z(t) - z^*] \le Me^{(M\alpha-\delta)t}V[z-z^*]\text{ for all }t \in J.$$

It is clear that $0 \in J$, that J is an interval, and that J is closed since $t \longrightarrow V[u_z(t)-z^*]$ is lower semicontinuous. Also, if $t \in J$ then since $M\alpha-\delta<0$ by (d),

$$V[u_z(t) - z^*] \le MV[z-z^*] \le R$$

and it follows from (c) that $t + \eta(t,z) \in J$. Thus $J = [0,\infty)$ and we see that (1.12) is true and the proof is complete.

Theorem 2 can be applied to linearization techniques for studying the behavior of solutions to (1.5). Suppose that the function F can be written in the form

(1.14) $Fx = Fz^* + B(x-z^*) + N(x-z^*)$ for all $x{\in}D$, where B is a bounded linear operator on X (and hence $x \longrightarrow N(x-z^*)$ is locally Lipschitz on D).

Since $B:X \longrightarrow X$ is a bounded linear operator, it follows that A + B is the generator of a C_0 linear semigroup $S = \{S(t): t \ge 0\}$, and that S is the solution to the integral equation

(1.15) $S(t)x = T(t)x + \int_0^t T(t-s)BS(s)xds$ for all $t \ge 0$, $x{\in}X$.

We have the following perturbation result:

<u>Theorem 3</u>. Suppose that (1.14) holds, that S is defined by (1.15), and that there are positive numbers M and δ such that

(a) $V[S(t)x] \leq Me^{-\delta t}V[x]$ for all $t \geq 0$, $x \in X$ and

(b) $V[N(x-z^*)] \leq \epsilon(\gamma)V[x-z^*]$ for all $x \in D$ with

$V[x-z^*] \leq \gamma$, where $\epsilon(\gamma) \longrightarrow 0+$ as $\gamma \longrightarrow 0+$.

(c) Assumption (c) in Theorem 2 holds.

If $R > 0$ is such that $M\epsilon(2R) < \delta$, then for each $z \in D$ with $V[z-z^*] \leq R/M$, the solution u_z to (1.7) exists on $[0,\infty)$ and

(1.16)
$$V[u_z(t) - z^*] \leq V[z-z^*]Me^{(M\epsilon(R)-\delta)t}$$

for all $t \geq 0$.

<u>Proof</u>. This theorem is an immediate consequence of Theorem 2 [with T replaced by S and $Fx - Fz^*$ by $N(x-z^*)$] once it is shown that the solution u_z to (1.7) satisfies

(1.17)
$$u_z(t) = S(t)(z-z^*) + z^* + \int_0^t S(t-s)N(u_z(s)-z^*)ds$$

for all t. To see that this is valid define

(1.18)
$$w(t) = S(t)(z-z^*) + z^* + \int_0^t S(t-s)N(u_z(s)-z^*)ds$$

for all t. Observe that if $h > 0$,

$$w(t+h) = S(h)[S(t)(z-z^*)] + z^* + S(h)\int_0^t S(t-s)N(u_z(s)-z^*)ds$$

$$+ \int_t^{t+h} S(t+h - s)N(u_z(s) - z^*)ds$$

$$= S(h)[w(t) - z^*] + z^* + \int_t^{t+h} S(t+h-s)N(u_z(s)-z^*)ds$$

$$= T(h)[w(t)-z^*] + hB(w(t) - z^*) + z^* + hN(u_z(t)-z^*) + o(h)$$

where $h^{-1}|o(h)| \longrightarrow 0$ as $h \longrightarrow 0+$. Similarly, using (1.13) and (1.14),

$$u_z(t+h) = T(h)(u_z(t) - z^*) + z^* + \int_t^{t+h} T(t+h)-s)(Fu_z(s) - Fz^*)ds$$

$$= T(h)(u_z(t) - z^*) + z^* + h[B(u_z(t)-z^*) + N(u_z(t)-z^*)] + o(h).$$

If $\omega > 0$ is such that $\| T(t) \| \le e^{\omega t}$ for all $t \ge 0$ we see that

$$\begin{aligned}|u_z(t+h) - w(t+h)| &= |T(h)[u_z(t)-w(t)] + B[u_z(t)-w(t)]| + o(h)\\ &\le e^{h\omega}|u_z(t)-w(t)| + h\| B \|\cdot|u_z(t)-w(t)| + o(h)\end{aligned}$$

and hence

$$\frac{|u_z(t+h)-w(t+h)| - |u_z(t)-w(t)|}{h}$$

$$\le (\frac{e^{h\omega}-1}{n} + \| B \|)|u_z(t)-w(t)| + \frac{o(h)}{h} .$$

Therefore, if $p(t) \equiv |u_z(t)-w(t)|$ for all t, the upper right Dini derivative $D^+p(t)$ satisfies

$$D^+p(t) \le (\omega + \| B \|)p(t) \quad \text{for all } t$$

since p is continuous and $p(0) = 0$ this implies $p(t) \equiv 0$. Thus $u_z(t) \equiv w(t)$ and so (1.17) is valid. This establishes the theorem.

Lyapunov – like methods may also be applied to analyze the bahavior of solutions to (1.7). In order to keep the techniques reasonably straight-forward, it is assumed that our functionals are locally Lipschitz continuous (as opposed to lower semicontinuous). This allows for obtaining estimates directly from equation (1.7) instead of indirectly by the construction of approximate solutions to (1.7). (These types of considerations can be found in the paper of S. Oharu [5]). Our basic result is the following:

<u>Theorem 4</u>. Suppose that $W: X \longrightarrow [0,\infty)$ is locally Lipschitz continuous and $z^* \in D$ is an equilibrium for (1.7). Suppose further that there are real constants α, ω and R such that

    (a) $W[T(t)x] \le e^{\omega t}W[x]$ for all $t \ge 0$, $x \in X$.

    (b) $W[x-z^* - h(Fx - Fz^*)] \ge (1-h\alpha)W[x-z^*]$ for all $x \in D$ with $W[x-z^*] \le R$.

    (c) $R > 0$ and $\alpha + \omega \le 0$.

Then for each $z \in E$ such that $W[z-z^*] < R$, the solution u_z to (1.7) satisfies

$$(1.19) \qquad W[u_z(t) - z^*] \le W[z - z^*]e^{(\omega+\alpha)t} \quad \text{for all } t \in [0,b_z).$$

In particular, if $\omega + \alpha < 0$ and there is a continuous, strictly increasing function $\Psi:[0,R] \longrightarrow [0,\infty)$ such that

(1.20) $W[x-z^*] \geq \Psi(|x-z^*|)$ for $x \in D$ with $W[x-z^*] \leq R$

then u_z exists on $[0,\infty)$ and $|u_z(t)-z^*| \longrightarrow 0$ as $t \longrightarrow \infty$.

Remark.   In comparing assumption (a) in Theorems 2 and 4, it is important to note that $M = 1$ in Theorem 4. This allows for the more general condition (b). In particular, α may be negative in Theorem 4, but not in Theorem 2. Also, if W is subadditive and positively homogeneous, then $W[Fx - Fz^*] \leq \alpha W[x-z^*]$ implies that (b) in Theorem 4 holds.

Proof of Theorem 4.   Suppose that $z \in D$ with $W[z-z^*] < R$. As in the proof of Theorem 3 we see that if $0 < s < t$,

$$u_z(t) - z^* = T(t-s)(u_z(s)-z^*) + \int_s^t T(t-r)[Fu_z(r)-Fz^*]dr$$
$$= T(t-s)(u_z(s)-z^*) + (t-s)(Fu_z(t)-Fz^*) + o(t-s)$$

where $|t-s|^{-1}o(t-s) \longrightarrow o$ as $s \longrightarrow t-$. Setting $t-s=h>0$ and using (a) and the fact that W is locally Lipschitz shows that

$$W[u_z(t)-z^* - h(fu_z(t)-Fz^*)] = W[T(t)(u_z(t-h)-z^*)] + o(h)$$
$$\leq e^{\omega h}W[u_z(t-h)-z^*] + o(h).$$

Therefore, by assumption (b),

$$(1-h\alpha)W[u_z(t) - z^*] \leq e^{\omega h}W[u_z(t-h) - z^*] + o(h)$$

and it follows that if $p(s) \equiv W[u_z(s)-z^*]$ for all s, then

$$p(t) - p(t-h) \leq h\alpha p(t) + (e^{\omega h}-1)p(t-h) + o(h) .$$

Dividing each side of this equation by $h > 0$ and letting $h \longrightarrow 0+$ shows that

$$D_-p(t) \leq (\alpha+\omega)p(t)$$

where D_- is the lower left Dini derivative. Since p is continuous this differential inequality implies

$$W[u_z(t) - z^*] \leq W[z-x^*]e^{(\alpha+\omega)t}$$

so long as $W(u_z(t) - z^*] \leq R$. Since $W[z-x^*] < R$ and $\alpha + w \leq 0$ it follows that (1.19) is valid. If (1.20) holds then φ^{-1} exists and

$$|u_z(t) - z^*| \leq \varphi^{-1}(W[u_z(t)-z^*]) \leq \varphi^{-1}(R)$$

and so $|u_z(t) - z^*|$ remains bounded. This shows that u_z is defined on $[0,\infty)$ and since $W[u_z(t)-z^*] \rightarrow 0$ as $t \rightarrow \infty$ when $\alpha + \omega < 0$, it is easy to see from (1.20) that $|u_z(t) - z^*| \rightarrow 0$ as $t \rightarrow \infty$. This completes the proof of Theorem 4.

§2. <u>Examples of Reaction-Diffusion Systems</u>.

In this section we indicate how the theorems in the preceding section can be applied to study solutions to reaction-diffusion systems of the form (3). We assume here that the hypotheses and notations for equations (1) - (3) in the introduction are valid and use the Banach space $L^p \equiv L^p(\Omega,\mathbb{R}^m)$, where $1 \leq p < \infty$. For each $\varphi = (\varphi_i)_1^m \in L^p$ define

$$\| \varphi \|_p = [\int_\Omega \sum_{i=1}^m |\varphi_i(x)|^p dx]^{\frac{1}{p}}.$$

Consider the homogeneous system

$$\begin{array}{lll}
\text{(a)} & \partial_t w_i(x,t) = d_i \Delta w_i(x,t) & t>0, \; x\in\Omega, \; i = 1,\ldots,m \\
\text{(2.1)} \quad \text{(b)} & \alpha_i w_i(x,t) + (1-\alpha_i)\frac{\partial}{\partial\nu}w_i(x,t) = 0 & t>0, \; x\in\partial\Omega, \; i = 1,\ldots,m \\
\text{(c)} & w_i(x,0) = \varphi_i(x) & x\in\Omega, \; i = 1,\ldots,m,
\end{array}$$

where $\varphi = (\varphi_i)_1^m \in L^p$ and define

$$(2.2) \qquad [T(t)\varphi](x) = (w_i(x,t))_1^m \text{ for each } \varphi \in L^p, \; t \geq 0 \text{ and } x \in \Omega.$$

Then T is a C_o linear semigroup on L^p and there are numbers $\lambda \geq 0$, $M \geq 1$ such that

$$(2.3) \qquad \| T(t)\varphi \|_p \leq Me^{-\lambda t}\| \varphi \|_p \text{ for all } t \geq, \; \varphi \in L^p.$$

Also, if α_i is as in (H1), then $\lambda > 0$ if $\alpha_i > 0$ for each $i = 1,\ldots,m$ and $\lambda = 0$ if $\alpha_i = 0$ for some $i = 1,\ldots,m$.

Also, let $|\xi|_\infty = \max\{|\xi_i|: i = 1,\ldots,m\}$ for each $\xi = (\xi_i)_i^m \in \mathbb{R}^M$ and for each φ $\in L^p$ define

$$\| \varphi \|_\infty = \operatorname*{ess\ sup}_{x\in\Omega} |\varphi(x)|_\infty .$$

Then $\varphi \longrightarrow \| \varphi \|_\infty$ is a positively homogeneous, subadditive and lower semicontinuous functional on L^p. Also, for each $\rho > 0$ define

$$Q_\rho(\xi_i)_1^m = (\eta_i)_1^m \text{ where } \eta_i = \xi_i \text{ if } |\xi_i| \leq \rho$$
$$\text{and } \eta_i = \rho\xi_i/|\xi_i| \text{ if } |\xi_i| > \rho .$$

Now define f^ρ on $\mathbb{R}_+^m = \{(\xi_i)_1^m \in \mathbb{R}^m: \xi_i \geq 0 \text{ for } i = 1,\ldots,m\}$ by

(2.4)
$$f^\rho(\xi) = f(Q_\rho\xi) \text{ for each } \rho > 0 \text{ and } \xi \in \mathbb{R}_+^m$$

where $f = (f_i)_1^m$ satisfies (H3) and (H4) in the introduction. Finally, for each $\rho > 0$ define

(2.5)
$$\{F^\rho\varphi\}(x) = f^\rho(\varphi(x)) \text{ for all } x \in \Omega, \varphi \in D \text{ where}$$

$$D = \{\varphi \in L^p: \varphi(x) \in \mathbb{R}_+^m \text{ for almost all } x \in \Omega\}.$$

Since each f^ρ is Lipschitz continuous on $\mathbb{R}_+^m$ and quasipositive [see (H4)] it follows that

(2.6)
(a) $\| F^\rho\varphi - F^\rho\bar\varphi \|_p \leq L_\rho \| \varphi-\bar\varphi \|_p$ for all $\rho > 0$ and $\varphi, \bar\varphi \in D$

(b) $\lim_{h\to 0+} d(\varphi + hF^\rho\varphi;D)/h = 0$ for all $\varphi \in D$ and $\rho > 0$

(c) There is a $\mu_\rho > 0$ such that $\| F^\rho\varphi \|_\infty \leq \mu_\rho$ for all $\rho > 0$

and $\delta \in D$.

Since it is assumed in (H1) that $\beta_i = 0$ whenever $\alpha_i = 0$, it follows that there is a function $\aleph = (\aleph_i)_1^m$ from $\bar\Omega$ into $\mathbb{R}_+^m$ such that

$$d_i\Delta\aleph_i(x) = 0 \quad \text{for all } x \in \Omega, i = 1,\ldots,m$$

(2.7)
$$\alpha_i\aleph_i(x) + (1-\alpha_i)\frac{\partial}{\partial\nu}\aleph_i(x) = \beta_i, \ x\in\partial\Omega, \ i = 1,\ldots,m$$

where it is assumed that $\aleph_i \equiv 0$ if $\beta_i = 0$. Therefore, if v is the solution to (1) in the introduction, then

$$v(x,t) = [T(t)(\varphi - \aleph)](x) + \aleph(x) \text{ for all } t > 0, \ x \in \Omega$$

where T is as in (2.2) and $\aleph$ as in (2.7). Since $v(x,t) \in \mathbb{R}^m_+$ by the maximum principle we see that

$$T(t)(\varphi - \aleph) + \aleph \in D \text{ whenever } \varphi \in D \text{ and } t \geq 0.$$

By [3, Proposition 2.4] this along with (b) in (2.6) implies that

(2.8) $\quad \lim\limits_{h \to 0+} d(T(h)(\varphi - \aleph) + \aleph + hF^\rho \varphi; D)/h = 0 \text{ for all } \varphi \in D, \ \rho > 0.$

Therefore, by Theorem 1, for each $\rho > 0$ and $\varphi \in D$ there is a unique solution $u = u_\varphi^\rho$ to

(2.9) $\qquad u(t) = T(t)(\varphi - \aleph) + \aleph + \int_0^t T(t-s)F^\rho u(s)ds, \ t \geq 0$

such that $u \colon [0,\infty) \to D$ (u is defined on $[0,\infty)$ since F^ρ is globally Lipschitz on D). Since the maximum principle implies $\| T(t)\varphi \|_\infty \leq \| \varphi \|_\infty$ for all $\varphi \in L^\rho$, we see from (2.9) and (c) in (2.6) that if u is the solution to (2.9) then

(2.10) $\qquad \| u(t) - \aleph \|_\infty \leq \| \varphi - \aleph \|_\infty + t\mu_\rho \text{ for all } t \geq 0$

and solutions u to (2.9).

Assertion (2.10) is important, for if u is a solution to (2.9) and $\| u(r) \|_\infty \leq \rho$ on, say the interval [0,b] and $u(x,t) \equiv [u(t)](x)$, then

$$[F^\rho u(t)](x) = f^\rho(u(x,t)) = f(u(x,t)),$$

and hence the solution u to (2.9) is in fact a solution to the original problem (3) for $(x,t) \in \Omega \times [0,b]$.

As a typical example of the type of results that can be established with these techniques we have the following:

<u>Proposition 1.</u> Suppose that $\alpha_i > 0$ for each $i = 1,\ldots,m$ and that the function $\aleph = (\aleph_i)_1^m$ in (2.7) is constant, say, $\aleph(x) \equiv \varsigma \in \mathbb{R}^m_+$ for all $x \in \Omega$. Suppose further that

$$(2.11) \qquad \lim_{\substack{\xi \to \varsigma \\ \xi \in \mathbb{R}_+^m}} \frac{|f(\xi) - f(\varsigma)|_\infty}{|\xi - \varsigma|_\infty} = 0.$$

Then there are positive numbers σ, M and R such that, if $|\aleph(x) - \varsigma|_\infty \le R$ for all $x \in \Omega$, the solution u to (3) exists on $\Omega \times [0, \infty)$ and satisfies

$$(2.12) \qquad |u(x,t) - \varsigma|_\infty \le Me^{-\sigma t} \| \aleph - \varsigma \|_\infty \text{ for } (x,t) \in \Omega \times [0, \infty).$$

Remark   Note that (2.11) implies the jacobian matrix of f exists and is zero at the point ς.

Proof of Proposition 1.   Since $\alpha_i > 0$ for each $i = 1, \ldots, m$, the largest eigenvalue of the operator $\aleph \to d_i \Delta \aleph$ subject to the boundary condition $\alpha_i \aleph + (1-\alpha_i)\frac{\partial}{\partial \nu}\aleph = 0$ is strictly negative for each $i = 1, \ldots, m$. Hence there are numbers $\delta > 0$ and $M \ge 1$ such that

$$\| T(t)\aleph \|_\infty \le Me^{-\delta t} \| \aleph \|_\infty \text{ for all } \aleph \in L^p$$

(see, e.g., Rothe [8, Lemma 3, p. 25]). By (2.11) we have for all large $\rho > 0$ that

$$|f^\rho(\xi) - f^\rho(\varsigma)|_\infty \le \epsilon(R)|\xi - \varsigma|_\infty \text{ if } |\xi - \varsigma|_\infty \le R$$

where $\epsilon(R) \to 0$ as $R \to 0+$. Thus, if $\|\aleph - \varsigma\|_\infty \le R$,

$$\|F^\rho \aleph - F^\rho \varsigma \|_\infty = \text{ess} \sup_{x \in \Omega} |f^\rho(\aleph(x)) - f^\rho(\varsigma)|_\infty \le \epsilon(R)\|\aleph - \varsigma\|_\infty$$

and, since (2.10) implies that (c) in Theorem 2 holds, we see that this result is a consequence of Theorem 2 by choosing $R > 0$ such that $\delta = \sigma - M\epsilon(R) > 0$.

Theorem 3 also has several applications to equation (3). As an interesting case, suppose that the α_i's in the boundary conditions are independent of i, say $\alpha_i = \alpha$ for all $i = 1, \ldots, m$. Consider the eigenvalue problem

$$(2.13) \qquad \Delta \aleph = \lambda \aleph \text{ on } \Omega \text{ and } \alpha \aleph + (1-\alpha)\frac{\partial \aleph}{\partial \nu} = 0 \text{ on } \partial\Omega$$

and let $\lambda_1 \ge \lambda_2 \ge \lambda_3 \ge \cdots \ge \lambda_k \ge \cdots$ be the eigenvalues of (2.13). Suppose further that the function $\aleph$ in (2.7) is constant:

$$(2.14) \quad \aleph(x) \equiv \varsigma \in \mathbb{R}_+^m \text{ for all } x \in \Omega \text{ and } \alpha_i = \alpha \text{ for all } i = 1, \ldots, m.$$

The β_i's are not necessarily independent of i in (2.7).

Proposition 2. Suppose that $\alpha_i = \alpha$ for each $i = 1,\ldots,m$, that (2.14) is satisfied. Suppose further that f is continuously differentiable and that

$$\text{the eigenvalues of } f'(\varsigma) + \lambda_k D$$

(2.15)

$$\text{have negative real parts for each } k = 1,2,\ldots$$

where $D = \text{diag}(d_1,\ldots,d_m)$ and $f'(\varsigma)$ is the $m \times m$ jacobian matrix of f at ς. Then there are positive numbers σ, M and R such that if $\|\varphi-\varsigma\|_\infty \le R$, then the solution u to (3) exists on $\Omega \times [0,\infty)$ and satisfies

(2.16) $\quad |u(x,t) - \varsigma|_\infty \le Me^{-\sigma t}\|\varphi-\varsigma\|_\infty$ for all $(x,t) \in \Omega \times [0,\infty)$.

Proof. Let P be the $m \times m$ matrix $f'(\varsigma)$ and consider the system

$$
\begin{aligned}
& w_t(x,t) = D\Delta w(x,t) + Pw(x,t) && t > 0,\ x \in \Omega \\
(2.17) \quad & \alpha w(x,t) + (1 - \alpha)\frac{\partial w}{\partial \nu}(x,t) = 0 && t > 0,\ x \in \partial\Omega \\
& w(x,0) = \varphi(x)
\end{aligned}
$$

where $w = (w_i)_1^m$ and $\varphi = (\varphi_i)_1^m$. Separating variables, assume that $w^k(x,t) = \varphi_k(x)z(t)$ where φ_k is an eigenfunction of (2.13) corresponding to the eigenvalue λ_k and $z(t) = (z_i(t))_1^m$. Substituting into (2.17) and using that $\Delta\varphi_k = \lambda_k\varphi_k$ show that z should satisfy the system

$$z'(t) = \lambda_k Dz(t) + Pz(t),\ t \ge 0.$$

Since $\lambda_k \longrightarrow -\infty$ as $k \longrightarrow \infty$ and the eigenvalues of $\lambda_k D + P$ have negative real parts by (2.15), it follows that there are numbers $M_1 \ge 1$ and $\sigma > 0$ (independent of k) such that if $|\cdot|_2$ denotes the euclidean norm on $\mathbb{R}^m$,

$$|z(t)|_2 \le |z(0)|_2 M_1 e^{-\sigma t} \text{ for all } t \ge 0 .$$

Since the orthogonal sequence $\{\varphi_k x\}_1^\infty$ is dense in $L^2(R,\mathbb{R})$, it follows that if w is a solution to (2.17), $\|w(\cdot,t)\|_2 \le M_1 e^{-\sigma t}\|\varphi\|_2$. From this it follows that there is an M

≥ 1 such that

$$\|w(\cdot,t)\|_\infty \leq Me^{-\rho t}\|\varphi\|_\infty.$$

(See e.g., Rothe [8, Lemma 3, p. 25]). Writing the nonlinear term f^o for ρ large in the form

$$f^o(\xi) = f^o(\varsigma) + P(\xi-\varsigma) + [f^o(\xi) - f^o\phi\varsigma) - P(\xi-\varsigma)]$$

shows that this proposition follows from Theorem 3 with $V[\varphi] \equiv \|\varphi\|_\infty$ and $F = F^o$, $[B\varphi](x) \equiv P\varphi(x)$ for all $\varphi \in L^p$, $x \in \Omega$, and

$$N(\varphi-x) \equiv F^o(\varphi) - F^o(x) - B(\varphi-x).$$

For in this case, $[S(t)\varphi](x) \equiv w(x,t)$ where w is the solution to (2.17), and we see that Theorem 3 applies to this proposition.

An illustration of Proposition 2 is the Brusselator, a model of a chemical morphogenetic process due to Turing, which has the form

$$(2.18) \quad \begin{aligned} \partial_t u_1 &= d_1\Delta u_1 - u_1 u_2^2 + Bu_2 \\ \partial_t u_2 &= d_2\Delta u_2 + u_1 u_2^2 - (B+1)u_2 + A \\ u_1 &= B/A,\ u_2 = A \end{aligned} \qquad \begin{aligned} &x \in \Omega,\ t > 0 \\[1em] &x \in \partial\Omega,\ t > 0 \end{aligned}$$

where A, B, d_1 and d_2 are positive constants. (See [1], [7]). It is easy to check that $\aleph(x) \equiv (B/A, A)$ is a constant equilibrium solution to (2.18). According to Proposition 2, if $\lambda = \lambda_k$ is an eigenvalue for $\Delta\varphi = \lambda\varphi$ on Ω and $\varphi = 0$ on $\partial\Omega$, then the equilibrium solution $(B/A,A)$ is asymptotically stable provided each of the eigenvalues of the matrix

$$\begin{bmatrix} -A^2 + d_1\lambda_k & -B \\ A^2 & B-1 + d_2\lambda_k \end{bmatrix}$$

have negative real parts for $k = 1,2,\dots$. (This is always the case if, for example, $2B-1 + d_2\lambda_k < 0$ for all k.) For global results for solutions to (2.18) see [2].

A model occuring in the theory of gas-liquid reactions is the system

$$\partial_t u_1 = d_1\partial_{xx}u_1 - ku_1(b_o - u_2)$$

$$\partial_t u_2 = d_2 \partial_{xx} u_2 + ku_1(b_o - u_2) \qquad\qquad t > 0,\ 0 < x < 1$$

(2.19)
$$u_1(0,t) = a_o,\ u_1(1,t) = 0$$

$$\partial_x u_2(0,t) = 0,\ u_2(1,t) = 0 \qquad\qquad t > 0$$

where k, a_o and b_o are positive constants (see [4]). For this system we take $p = 1$, $\Omega = (0,1)$, and define the functional W on $L^1 = L^1((0,1);\mathbb{R}^2)$ by

$$W[(\varphi_1,\varphi_2)] = \int_0^1 \cos(\tfrac{\pi}{2}x)\Big[|\varphi_1(x)| + |\varphi_2(x)|\Big]dx.$$

Clearly W is positively homogeneous, subadditive, and Lipschitz continuous on L^1. Furthermore, if (v_1,v_2) satisfies

$$\partial_t v_1 = d_1 \partial_{xx} v_1 \text{ and } \partial_t v_2 = d_2 \partial_{xx} v_2,\ t > 0,\ 0 < x < 1$$

(2.20)
$$v_1(0,t) = \partial_x v_2(0,t) = 0 \text{ and } v_1(1,t) = v_2(1,t) = 0$$

for $t > 0$

where $v_1(x,0)$, $v_2(x,0) \geq 0$, then v_1, $v_2 \geq 0$ for all $t > 0$, $0 \leq x \leq 1$ and if $q(t) = W[v_1(\cdot,t), v_2(\cdot,t)]$, then setting $\mu(x) = \cos(\pi x/2)$ for $x \in [0,1]$ and suppressing the variables and using integration by parts,

$$q' = \int_0^1 \mu(\partial_t v_1 + \partial_t v_2)dx$$

$$= \int_0^1 \mu(d_1 \partial_{xx} v_1 + d_2 \partial_{xx} v_2)dx$$

$$= [\mu(d_1 \partial_x v_1 + d_2 \partial_x v_2)]_0^1 - \int_0^1 \mu'(d_1 \partial_x v_1 + d_2 \partial_x v_2)dx$$

$$= -\mu d_1 \partial_x v_1(0,t) - \int_0^1 \mu'(d_1 \partial_x v_1 + d_2 \partial_x v_2)dx$$

$$= -\mu d_1 \partial_x v_1(0,t) - [\mu'(d_1 v_1 + d_2 v_2)]_0^1 + \int_0^1 \mu''(d_1 v_1 + d_2 v_2)$$

$$= -\mu d_1 \partial_x v_1(0,t) - \frac{\pi^2}{4}\int_0^1 \mu(d_1 v_1 + d_2 v_2).$$

But v_1, $v_2 \geq 0$ and $v_1(0,t) = 0$, so $\partial_x v_1(0,t) \geq 0$ and it follows that

$$q' \leq -\frac{\pi^2}{4}\int_0^1 \mu(d_1 v_1 + d_2 v_2) \leq -\frac{\pi^2}{4}\min\{d_1,d_2\}q$$

and hence if $\omega = -\dfrac{\pi^2}{4}\min\{d_1,d_2\}$ then $q(t) \leq q(0)e^{\omega t}$ for $t \geq 0$. Thus, if

$$T(t)(\wp_1,\wp_2) \equiv (v_1(\cdot,t),\ v_2(\cdot,t))\ \text{for}\ t \geq 0,\ (\wp_1,\wp_2) \in L^1$$

where $(v_1,\ v_2)$ is the solution to (2.20) that satisfies $v_1(\cdot,0) = \wp_1$ and $v_2(\cdot,0) = \wp_2$, then

$$W[T(t)\wp] \leq W[\wp]e^{\omega t}$$

whenever $t \geq 0$ and $\wp = (\wp_1,\ \wp_2) \in L^1$ with $\wp \geq 0$. If $\wp = (\wp_1,\ \wp_2) \in L^1$ and $\wp^{\pm} = (\wp_1^{\pm},\ \wp_2^{\pm})$ (where $\wp_i^+ = \max\{\wp_i,0\}$ and $\wp_i^- = -\min\{\wp_i,0\}$). Then

$$W[T(t)\wp] = W[T(t)\wp^+ - T(t)\wp^-] \leq W[T(t)\wp^+] + W[T(t)\wp^-]$$
$$\leq (W[\wp^+] + W[\wp^-])e^{\omega t} = W[\wp^+ + \wp^-]e^{\omega t} = W[\wp]e^{\omega t}$$

and we see that (a) in Theorem 4 holds with $\omega = -\pi^2\min\{d_1,d_2\}/4$. Set

$$D = \{\wp = (\wp_1,\wp_2) \in L^1;\ 0 \leq \wp_1 \lfloor a_o\ \text{and}\ 0 \leq \wp_2 \leq b_o\ \text{a.e. on}\ (0,1)\}$$

and define

$$F(\wp_1,\ \wp_2) = (-k\wp_1(b_o - \wp_2),\ k\wp_1(b_o - \wp_2))$$

for all $(\wp_1,\wp_2) \in D$. In this case one can check that

(2.21) $$W[\wp - \bar{\wp} - h(F\wp - F\bar{\wp})] \geq W[\wp - \bar{\wp}]$$

for all $\wp,\ \bar{\wp} \in D$ and $h > 0$. One can also show, using the maximum principle, that any solution $(u_1,\ u_2)$ to (2.19) such that $(u_1(\cdot,0),\ u_2(\cdot,0)) \in D$, satisfies $(u_1(\cdot,t),\ u_2(\cdot,t)) \in D$ for all $t \geq 0$, and hence (1.9) in Theorem 1 holds (one can also verify this directly). Now consider the time dependent version of (2.19):

$$d_1\aleph_1'' - k\aleph_1(b_o - \aleph_2) = 0 \qquad\qquad 0 < x < 1$$
(2.22) $$d_2\aleph_2'' + k\aleph_1(b_o - \aleph_2) = 0$$
$$\aleph_1(0) = a_o,\ \aleph_2'(0) = 0,\ \aleph_1(1) = 0,\ \aleph_2(1) = 0.$$

The invariance of D for (2.19) can be used to show that (2.22) has a solution $\aleph =$

$(\aleph_1, \aleph_2) \in D$, and taking $\overline{P} = \aleph$ in (2.21) and applying Theorem 4 shows that if $(u_1(\cdot, 0), u_2(\cdot, 0)) \in D$,

$$W[(u_1(\cdot, t), u_2(\cdot, t)) - \aleph] \le W[(u_1(\cdot, 0), u_2(\cdot, 0)) - \aleph]e^{\omega t}$$

for all $t \ge 0$, where $\omega < 0$. This shows that (2.22) can have at most one solution in D, and hence (2.19) has a unique equilibrium solution $\aleph = (\aleph_1, \aleph_2) \in D$. Combining these comments gives the following result concerning the behavior of the solutions to (2.19). The system (2.19) has a unique equilibrium solution $\aleph = (\aleph_1, \aleph_2)$ such that $0 \le \aleph_1 \le a_o$, $\le \aleph_2 \le b_o$. Furthermore, if $0 \le u_1(\cdot, 0) \le a_o$, $0 \le u_2(\cdot, 0) \le a_2$, then the solution (u_1, u_2) to (2.19) exists on $[0,1] \times [0, \infty)$, $0 \le u_1(x, t) \le a_o$, $0 \le u_2(x, t) \le b_o$ for all $(x, t) \in [0,1] \times [0, \infty)$ and

$$\int_0^1 \cos(\tfrac{\pi}{2}x)[u_1(x, t) - \aleph_1(x) + u_2(x, t) - \aleph_2(x)]dx$$

$$\le e^{-\delta t}\int_0^1 \cos(\tfrac{\pi}{2}x)[u_1(x, 0) - \aleph_1(x) + u_2(x, 0) - \aleph_2(x)]dx$$

for all $t > 0$ where $\delta = \pi^2 \min\{d_1, d_2\}/4$.

REFERENCES

1. J.F.G. Auchmuty and G. Nicolis, *Bifurcation Analysis of Nonlinear Reaction-Diffusion Equations - I. Evolution Equations and Steady State Solutions*, Bulletin of Math. Biology 37(1975), 323-365.

2. S. L. Hollis, R. H. Martin, Jr. and M. Pierre, *Global Existence and Boundedness in Reaction-Diffusion Systems* (to appear)

3. R. H. Martin, Jr., *Nonlinear Perturbation of Linear Evolution Systems*, J. Math. Soc. Japan 29(1977), 233-252.

4. R. H. Martin, Jr., *Mathematical Models in Gas-Liquid Reactions*, JNA-TMA 4(1980), 509-527.

5. S. Oharu, *On the Characterization of Nonlinear Semigroups Associated with Semilinear Evolution Equations* (to appear).

6. N. Pavel, *Invariant Sets for a Class Semilinear Equations of Evolution*, JNA-TMA 1(1977), 187-196.

7. E. Prigogine and G. Nicolis, *Biological Order, Structure and Instabilities*, Quart. Reviews of Biophysics 4(1971), 107-148.

8. F. Rothe, *Global Soluitons of Reaction-Diffusion Systems*, Lecture Notes in Math, 1072, Springer-Verlag, Berlin (1984).

ULTRASINGULARITIES IN NONLINEAR WAVES

Jeffrey Rauch[1]
Department of Mathematics
University of Michigan
Ann Arbor, MI 48109

Michael C. Reed[2]
Department of Mathematics
Duke University
Durham, NC 27706

The propagation of singularities in a linear strictly hyperbolic system with smooth coefficients in one space dimension,

$$(\partial_t + A\partial_x)u + Bu = 0, \qquad u|_{t=0} = u^{\circ}$$

is easy to describe. If S is the singular support of u° and $\{\lambda_i\}$ the eigenvalues of A, denote by S_i the flow out of S for t>0 under the vector field $\partial_t + \lambda_i\partial_x$. Then the maximal singular set for u for t>0 is the set US_i. Whether the entire set US_i consists of singularities depends on more detailed properties of the initial data.

In the semilinear case,

$$(1) \quad (\partial_t + A\partial_x)u + Bu = f(t,x,u), \qquad u|_{t=0} = u^{\circ},$$

the following example,

$$
\begin{aligned}
(\partial_t + \partial_x)v &= 0 & v(0,x) &= 1 - H(x + 1) \\
(2) \quad (\partial_t - \partial_x)w &= 0 & w(0,x) &= H(x-1) \\
\partial_t z &= vw & z(0,x) &= 0,
\end{aligned}
$$

(where H denotes the Heaviside function) shows that there are new phenomena. A simple calculation [7] shows that the solution u=(v,w,z) is singular on the rightward characteristic from (-1,0), on the leftward characteristic from (1,0), and on the forward t characteristic from (1,0), the dashed line in Figure 1.

---

(1) Partially supported by NSF Grant #MCS-8301061

(2) Partially supported by NSF Grant #DMS-8401590

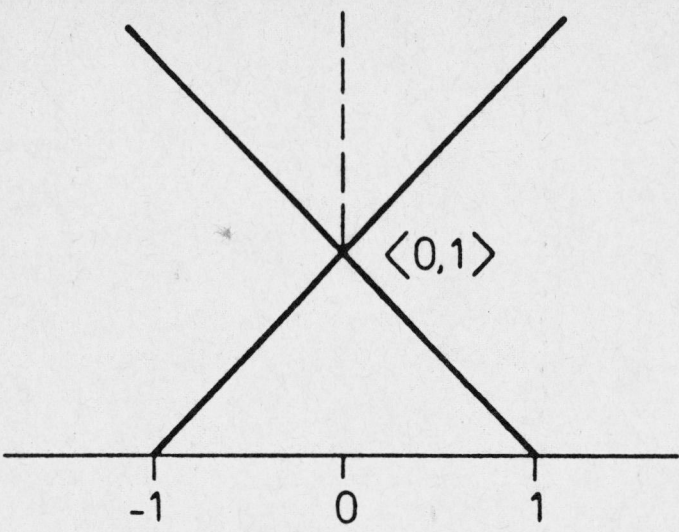

Figure 1

The dashed line is not a forward characteristic from one of the two initial
singular points so if the righthand sides were linear, the solution would be
smooth across the dashed line. In fact, in this example z is continuous
but $\partial_x z$ jumps across the dashed line. The interaction of the jump
discontinuities starting from (-1,0) and (1,0) at the point of collision,
(0,1), produces a new singularity which travels forward from (0,1) along
the dashed line. This phenomenon is well understood in one space dimension
[7], [9]; See [6] for boundary value problems, [5] for non-strictly
hyperbolic problems, and [4] for the quasi-linear analogue. In higher
space dimensions there are new phenomena, and many unsolved problems [1],
[2], [3], [10].

In one space dimension, the interaction of singularities in strictly
hyperbolic semilinear problems is governed by a simple sum law. If we work
in the piece-wise smooth category, then we say the solution has regularity
n across the characteristic if n derivatives are continuous but the (n+1)st
jumps. The sum law then reads as follows: When a characteristic of
regularity n_1 collides with a characteristic of regularity n_2 at p, then
the solution will (in general) have regularity $n_1 + n_2 + 2$ on the other
forward characteristics emerging from p. We say "in general" because the
solution might be more regular across the outgoing characteristics than n_1
$+ n_2 + 2$ because of special circumstances or cancellations. For example,

if the right-hand sides are linear, then the solution is smooth on the other
outgoing characteristics. In the example (2) above, the incoming
singularities to the point $(0,1)$ both have strength $-1 = n_1 = n_2$ (since one
must integrate once to make the jump continuous), so the sum law says that
the outgoing regularity across the z characteristic should be $(-1) + (-1)$
$+ 2 = 0$. This is exactly what happens, the solution is continuous but the
first derivative jumps. Notice that as long as $n_1 \geq -1$ and $n_2 \geq -1$, then
the new singularities produced by interaction will always be weaker than
the interacting singularities.

Several years ago, Cathleen Morawetz asked us what was the meaning of
these singularities for applied mathematics: how do sharp peaks or sharp
oscillations interact? Independently, Russ Caflisch and John Sylvester
asked us whether the sum law, $n_1 + n_2 + 2$, is true for more negative n's.
Both of these questions lead us to study (1) for highly singular initial
data. In the linear theory, one idealizes a sharp peak by the delta
function, a sharp oscillation by the derivative of the delta function.
Using linearity and the adjoint relation, the usual existence and
propagation of singularities theory can be extended to distribution initial
data. What about the semilinear case?

To see that there are at least some highly singular solutions, consider
the example (2) again. This time we will take for initial data:
$$v(o,x) = \delta(x+1), \quad w(0,x) = \delta(x-1), \quad z(0,x) = 0.$$
Then the solution is:
$$v(t,x) = \delta(x+1-t)$$
$$w(t,x) = \delta(x-1+t)$$
$$z(t,x) = \frac{1}{2} H(t-1) \, \delta(x).$$

In general, of course, it doesn't make sense to multiply distributions
but the products
$$\delta(x+1-t) \, \delta(x-1+t),$$
$$\frac{1}{2} H(t-1) \, \delta(x)$$
do make sense as distributions in the plane. Each of the terms in (2)
makes sense as a distribution and the three equations hold. Furthermore,
it is clear that the solutions v,w,z take on their initial values in a
natural way. The result is that the delta function propagating from $(-1,0)$
interacts with the delta function propagating from $(1,0)$ at the point
$(0,1)$. The interaction produces a new delta function which travels along
the forward z characteristic from $(0,1)$. Notice, that in our
classification of regularity a delta function is $n = -2$. Thus, the sum law

would predict that two interacting delta functions would produce new singularities of regularity $n_1+n_2+2=(-2)+(-2)+2= -2$, i.e., a delta function, and that is exactly what we see in this example. If one trys the same example with $v(o,x)$ and $w(0,x)$ equal to the derivative of the delta function (an idealization of a single sharp oscillation) then, since $(-3)+(-3)+2= -4$, the sum law predicts that the singularity emerging from the interaction will be the second derivative of the delta function, and one can compute that this is in fact the case.

It is clear that example (2) is very special in one respect. Only particular kinds of non-linearities are allowed. One could not permit terms involving v^2 or w^2, for example, on the right-hand side of the third equation. This raises the question of whether there are general classes of nonlinearities which permit highly singular data. There are several results of this kind in [11]; here we will present a special case of one of them.

Let us consider a strictly hyperbolic system in canonical form:

$$(3) \quad (\partial_t + \underline{\varLambda} \, \partial x)u = f(u),$$
$$u|_{t=0}=g+\gamma.$$

$\underline{\varLambda}$ is a diagonal matrix with smooth entries $\{\lambda_i(t,x)\}$, and f is assumed uniformly Lipschitz and bounded. The initial data has a classical part $g \in L^1$, and a singular part γ. We assume that v is a distribution whose support has Lebesque measure zero such that there is a sequence of smooth functions $\{h^\epsilon\}$ satisfying:

(i) $h^\epsilon \longrightarrow 0$ in measure,

(ii) $h^\epsilon \longrightarrow \gamma$ in ϵ'.

For example, if γ is supported at finitely many points and j_ϵ is the usual mollifier, then $h^\epsilon = j_\epsilon \times \gamma$ satisfies these hypotheses.

We define u^ϵ to be the solution of the problem with regularized data:

$$(4) \quad (\partial_t + \underline{\varLambda}\partial_x)u^\epsilon = f(u^\epsilon),$$
$$u^\epsilon|_{t=0} = g+h^\epsilon.$$

We want to ask: What is the limiting behaviour of u^ϵ as $\epsilon \longrightarrow 0$. This is a natural way to give meaning to the problem (3). Let $\bar{u}$ be the solution of the "classical" problem:

(5) $\quad (\partial_t + \_A\_\partial_x)\bar{u} = f(\bar{u}),$

$\quad\quad \bar{u}|_{t=0} = g,$

and let σ^ϵ and σ be the solutions of the linear problems:

(6) $\quad (\partial_t + \_A\_\partial_x)\sigma^\epsilon = 0$

$\quad\quad \sigma|_{t=0} = h^\epsilon$

(7) $\quad (\partial_t + \_A\_\partial_x)\sigma = 0$

$\quad\quad \sigma|_{t=0} = \gamma.$

Then we have the following:

Theorem: Under the above hypotheses:

(a) $\quad u^\epsilon - \bar{u} - \sigma^\epsilon \longrightarrow 0$ in $C([o,T]: L^1(R))$,

(b) $\quad u^\epsilon \longrightarrow \bar{u} + \sigma$ in ε'.

This theorem expresses a striking nonlinear superposition principle. The singular part of the solution propagates linearly. The classical part propagates by the nonlinear equation. And, the limit of the nonlinear solution u^ϵ as the data become more and more singular is the sum of the two parts. The intuitive reason for this splitting is that the peaking parts of the solution which occur on small sets make less and less difference in the nonlinear term since f is bounded.

References

[1] Beals, M., "Self-spreading and strength of Singularities for Solutions to Semilinear Wave Equations", Annals of Math 118 (1983), 187-214.

[2] Bony, J.M., "Second Microlocalization and Propagation of Singularities for semi-linear hyperbolic equations", Orsay Preprint, 1985

[3] Melrose, R. and N. Ritter, "Interaction of Nonlinear Progressing Waves for Semilinear Wave Equations", Annals of Math 121 (1985), 187-213.

[4] Messer, T. "The Propragation and Creation of Singularities of Solutions of Quasilinear, strictly hyperbolic systems in one space dimension," Duke University Thesis, 1984.

[5] Micheli, L. "Propagation of Singularities for Non-strictly hyperbolic semi-linear systems in one space dimension", Trans. Amer. Math. Soc., to appear.

[6] Oberguggenberger, M., "Semilinear mixed hyperbolic in two variables", J. Diff. Eq., to appear.

[7] Rauch, J. and M. Reed, "Jump Discontinuities of Semilinear, Strictly hyperbolic systems in one space dimension: Creation and Propagation", Comm. Math. Phys. 81 (1981), 203-227.

[8] Rauch, J. and M. Reed, "Propagation of Singularities for Semilinear Hyperbolic Equations in One Space Variable", Annals of Mathematics. 111(1980), 531-552.

[9] Rauch, J. and M. Reed, "Nonlinear Microlocal Analysis of Semilinear Hyperbolic Systems in One Space Dimension", Duke Mathematical Journal. 49(2), 397-475.

[10] Rauch, J. and M. Reed, "Propagation of Singularities in Non-Strictly Hyperbolic Semilinear Systems: Examples", Communications on Pure and Applied Mathematics. $\underline{35}$, (1982), 555-565.

[11] Rauch, J. and M. Reed, "Nonlinear Superposition and Absorption of Delta Waves in One Space Dimension", Preprint, 1985.

A Reaction-Hyperbolic System in Physiology

M. C. Reed[1]
Department of Mathematics
Duke University
Durham, NC 27706

J. J. Blum
Department of Physiology
Duke University Medical Center
Durham, NC 27710

Nerve cells have long axons which carry the depolarizing action potential from the central part of the cell, called the soma, to a synapse where
the arrival of the action potential triggers a sequence of events which affect the next cell. Axons typically have diameters of $1 - 10^2$ microns
and lengths of the order $10^4 - 10^6$ microns, so they are extremely long
and narrow. As part of a living cell, the axon has a great deal of cellular machinery which wears out and must be replaced. All of the cellular
apparati for making these replacement parts are located in the soma. The
parts are then shipped down the axon by transport systems and used where
needed. For example, in an unmyelinated axon, sodium pumps are needed
along the whole length of the axon and an accumulation of vesicles containing acetycholine (a neurotransmitter) is needed at the synapse end. It is
clear that there are at least two systems, a fast transport system at about
2-24 cm/day and a slow transport system at about .05 - 1 cm/day, which
transport different material and which may operate using different mechanisms. There is also a retrograde transport system which carries material
back toward the soma.

There are two reasons why a great deal of work has been done recently
on axonal transport. First, in all living cells materials (e.g. proteins)
are manufactured in one place and typically transported to other places to
be used. Understanding these transport mechanisms, which are often two
orders of magnitude or more faster than diffusion, is one of the fundamental problems of cell biology. Axonal transport is merely an example of

[1]Research partially supported by NSF Grant #DMS-8401590.

this kind of transport, but a spectacular one since the distances are so great. The great distances involved make experimental and theoretical study easier since the problem is essentially one space dimensional. The second reason for studying axonal transport is that in many neuropathies (e.g., Alzheimer's disease) the nerves die from the synapse end back. This suggests that a dysfunction in the axonal transport system may be involved.

One of the common ways of studying axonal transport during the past 20 years has been to inject radioactive amino acids into the soma. The amino acids are incorporated into proteins, some of which occur in vesicles or other organelles, that move via the fast transport system. Thus, one can follow the time course of the concentration of radioactivity as it propagates down the axon. Figure 1 shows typical results of such experiments.

Figure 1a: Taken from Ochs, 1972

Figure 1b: Taken from Gross and Beidler, 1975

Although the peak of the profile drops considerably because of the deposition of material along the axon, the wave front stays sharp and it moves with constant velocity.

It has long been thought that fast transport is associated in some way to microtubules, long polymer chains which run parallel to the length of the axon, and recently evidence from computer enhanced light microscopy studies (Allen et al., 1985; Miller and Lasek, 1985) shows that when vesicles are unattached to microtubules they don't move, but when they are attached via an intermediary molecule, termed kinesin, they are translocated down the axon.

One can model this as follows. Take the soma to be at $x = 0$ and let the positive x-axis denote distance down the axon. Let $P(x,t)$ denote the concentration of free vesicles, $E(x,t)$ the concentration of free kinesins, and $T(x,t)$ denote the concentration of free positions for translocation on the microtubules. The local chemistry can then be described:

$$P + nE \underset{k_2}{\overset{k_1}{\rightleftharpoons}} P \cdot nE \tag{1}$$

$$P \cdot nE + mT \underset{k_4}{\overset{k_3}{\rightleftharpoons}} P \cdot nE \cdot mT \tag{2}$$

$$E + T \underset{k_6}{\overset{k_5}{\rightleftharpoons}} E \cdot T \tag{3}$$

$$P + m(E \cdot T) \underset{k_8}{\overset{k_7}{\rightleftharpoons}} P \cdot m(E \cdot T) \tag{4}$$

$$E_0 = E + nP \cdot nE + nP \cdot nE \cdot mT + mP \cdot m(ET) + E \cdot T \tag{5}$$

$$T_0 = T + mP \cdot nE \cdot mT + E \cdot T + mP \cdot m(ET). \tag{6}$$

Equation (1) states that an organelle may interact with n kinesins to form an organelle-kinesin complex, $P \cdot nE \equiv C$. The organelle-kinesin complex may then interact with m free binding sites on a microtubule to form the moving organelle, $C \cdot mT$, according to equation (2). An alternative sequence of events has the kinesins first interacting with their binding sites on the microtubules (eq. 3), forming the complex $E \cdot T \equiv S$. An organelle may then interact with available sites on the cross-bridges attached to the microtubule to form a moving particle, $P \cdot mS$ (eq. 4). Equations (5) and (6) merely state that the total concentration at each x of kinesins, E_0, and microtubules, T_0, remain constant during the time required for an experiment. We shall assume for simplicity that $n = m$, (i.e., that during transport all the engines on an organelle are interacting with sites on the microtubule). Thus, there is only one kind of bound complex that is translocated, $P \cdot nE \cdot nT \equiv Q$. The changes in concentration of P, C, S, and Q with time at any point x, along the axon, may then be written as:

$$\frac{\partial P}{\partial t} = -k_1 PE^n + k_2 C - k_7 PS^n + k_8 Q \tag{7}$$

$$\frac{\partial C}{\partial t} = k_1 PE^n - k_2 C - k_3 CT^n + k_4 Q \tag{8}$$

$$\frac{\partial S}{\partial t} = k_5 ET - k_6 S - k_7 PS^n + k_8 Q \tag{9}$$

$$\frac{\partial Q}{\partial t} + \frac{\partial (vQ)}{\partial x} = k_3 CT^n - k_4 Q + k_7 PS^n - k_8 Q, \tag{10}$$

where

$$E = E_0 - nC - nQ - S \tag{11}$$

and

$$T = T_0 - S - nQ . \tag{12}$$

Equation (10) expresses the fact that only organelles that are attached to the microtubules are transported, and that they move at a velocity v. Finally, we need to specify how much is coming into the axon at $x = 0$,

$$Q(0) = Q_0 . \tag{13}$$

If we start with initial conditions $P(x,0) = Q(x,0) = C(x,0) = S(x,0) = 0$, then this set of coupled equations describes the propagation of material into an "empty" axon; that is, into an axon which is initially devoid of vesicles. This situation can be created experimentally. In cold block experiments a small region of the axon is cooled. This causes transport to stop in the cooled region. Upstream, material will pile up at the cold block while downstream, material will continue to be transported, leaving a virtually empty region distal to the cold block. When the axon is rewarmed, the piled up material is transported into the empty region.

However, the usual and more physiological situation is when radioactive material is propagating into a homogeneous axon which is in the steady state. Let $P(x,t)$ denote the concentration of radioactively labeled free vesciles, $p(x,t)$ the concentration of unlabeled free vesicles, with Q,q, and C,c defined similarly. In the steady state at each x:

$$P(x,t) + p(x,t) = P_e$$

$$Q(x,t) + q(x,t) = Q_e$$

$$C(x,t) + c(x,t) = C_e,$$

where P_e, Q_e, C_e (and S_e) are equilibrium values for the reactions specified by equations (1)-(4). Since the axon is homogeneous, P_e, Q_e, C_e, S_e
do not depend on x. By the steady state assumption, we must have Q_e =
Q_0; thus, the boundary condition (13) uniquely determines the equilibrium
values P_e, Q_e, C_e, S_e. Since the kinesins, microtubules, and kinesin-
microtubule complexes are not labeled, their concentrations are given by
the constants E_e, T_e, S_e. The values E_e and T_e are determined from
equations (11) and (12) by using $C = C_e$, $Q = Q_e$, $S = S_e$. Under these circumstances, equation (9) vanishes and equations (7), (8), and (10) become:

$$\frac{\partial P}{\partial t} = -(k_1 E_e^{\ n} + k_7 S_e^{\ n})P + k_2 C + k_8 Q \tag{14}$$

$$\frac{\partial C}{\partial t} = +k_1 E_e^{\ n}P - (k_2 + k_3 T_e^{\ n})C + k_4 Q \tag{15}$$

$$\frac{\partial Q}{\partial t} + V_0 \frac{\partial Q}{\partial x} = k_7 S_e^{\ n}P + k_3 T_e^{\ n}C - (k_4 + k_8)Q, \tag{16}$$

the boundary condition (13) remains unchanged and the initial conditions
are:

$$P(x,0) = Q(x,0) = C(x,0) = 0 . \tag{17}$$

For the homogeneous axon the radioactivity profile will be the solution of
the _linear_ hyperbolic mixed problem (13)-(17). To get a feel for how the
solution behaves, imagine solving by using the Trotter product formula,
alternately time stepping in the following two simple problems:

(1) set the right hand sides equal to zero; then the time step just trans-
     lates Q,

(2) set $V_0 = 0$; then the time step solves the system of ordinary differ-
     ential equations specifying the chemistry.

So, in the beginning Q_0 gets translated into the first part of the axon.
Then the chemistry makes some of these Q's hop off the microtubules, turn-

ing them into P's and C's. Then we translate again, adjust the chemistry again, and so forth. We will develop profiles for Q, P, C, which are monotone decreasing, which go to zero as $x \to \infty$, and which go to Q_e, P_e, C_e as $x \to 0$. See, for example, the profiles in Figure 2. In each of the four simulations, the total normalized radioactivity profile (P + Q + C) is plotted at 4, 8, 12, 16, 20, 24 hours; V_0 was taken to be 1.

This system (13)-(17) is linear and does not have mathematical travelling wave solutions except in the trivial case where $k_4 = 0 = k_8$ (then P and C remain zero and Q translates). Nevertheless, the physiological problem has approximate travelling waves, as does the mathematical problem (13)-(17) as shown by Figure 2a. So, the question is: when, i.e. for what values of the parameters, does (13)-(17) have a solution which is essentially a travelling wave on the space and time scale of the experiment? The rate constants $k_1, \ldots, k_8$, and E_0, T_0, and Q_0 determine the equilibrium values C_e, P_e, $Q_e = Q_0$. These in turn determine the "speed" of the wave. Behind the wave front, where the labelled vesicles have completely displaced the unlabelled ones, a typical vesicle spends $Q_e/(Q_e + P_e + C_e)$ of its time attached to the track moving at velocity V_0 and $(P_e + C_e)/(Q_e + P_e + C_e)$ off the track not moving at all. Thus, its average speed will be

$$V_0 \left(\frac{Q_e}{(Q_e + P_e + C_e)} \right) .$$

One can see from Figure 2 that the speeds of the wave are less than V_0 (which equals one there). Will the wave keep its shape as it moves? This is a much more subtle question since the answer depends on the details of the chemistry as well as the values of Q_0, E_0, T_0. For example, in the four simulations in Figure 2, all the parameters were the same except that Q_0 was very small in Figure 2a, and was increased successively in Figures 2b, 2c, and 2d. Figures 2a and 2b give quite nice approximate travelling waves, while Figures 2c and 2d are not approximate travelling waves. One can understand this qualitatively by noticing that raising Q_0 means that fewer free kinesins, E, and free track places, T, will be available at equilibrium and this causes a slow, less efficient exchange between the labelled and unlabelled vesicles.

Figure 2:   Taken from Blum and Reed, 1985

It is not difficult to enlarge the model discussed here to include deposition of material along the axon, degradation of deposited material, return to the soma via the retrograde transport system, and leakage to the environment. For appropriate parameter ranges, our simulations have most of the qualitative features of the experimental data. We think that the model will be useful for understanding fast axonal transport, and for the planning and interpretation of experiments.

REFERENCES

[1] Allen, R.D., Weiss, D.G., Hayden, J.H., Brown, D.T., Fujiwake, H., and
     Simpson, M., "Gliding movement of and bidirectional transport along
     single native microtubules from squid axoplasm: evidence for an
     active role of microtubules in cytoplasmic transport," J. Cell Biol.
     100 (1985), 1736-1752.

[2] Blum, J.J. and M. C. Reed, "A Model for Fast Axonal Transport," Cell
     Motility **5** (1985), 507-527.

[3] Blum, J.J. and M. C. Reed, "Effect of Deposition and Turnover on
     Radioactivity Profiles During Fast Axonal Transport," in preparation.

[4] Gross, G.W. and Beidler, L. M., "A quantitative analysis of isotope
     concentration profiles and rapid transport velocities in the C-fibers
     of the garfish olfactory nerve," J. Neurobiol. **6** (1975), 213-232.

[5] Miller, R. H. and R. J. Lasek, "Structural comparison of the cross
     bridges that mediate vesicle transport in axons," J. Protozool. **101**
     (1985), 388a.

[6] Ochs, S. "Rate of fast axoplasmic transport in mammalian nerve
     fibers," J. Physiol. **227** (1972), 627-645.

[7] Odell, G., "Theories of Axoplasmic Transport," in Lectures on
     Mathematics in the Life Sciences **9** (1977), Amer. Math. Soc.,
     Providence, p. 141-186.

[8] Rubinow, S.I. and Blum, J. J., "A theoretical approach to the analysis
     of axonal transport," Biophys. J. **30** (1980), 137-148.

COMPACT PERTURBATIONS OF LINEAR M-DISSIPATIVE OPERATORS WHICH LACK GIHMAN'S PROPERTY

Eric Schechter, Mathematics Department
Vanderbilt University, Box 21, Station B
Nashville, Tennessee 37235

Abstract: Some questions about abstract methods for initial value problems lead us to a study of the equation (*) $u'(t) = (A + B)u(t)$, where A is m-dissipative and B is compact. Does a solution to (*) necessarily exist? Earlier studies of this question, reviewed and then continued here, depend on an analysis of the related quasiautonomous equation (**) $u'(t) = Au(t) + f(t)$. We say A has *Gihman's property* if the mapping $f \mapsto u$ is continuous from $\mathcal{L}^1_w([0,T], K)$ into $\mathcal{C}([0,T]; X)$ for every compact $K \subset X$; this condition is closely related to the Lie-Trotter-Kato product formula. If A has this property, then (*) is known to have a solution. In this paper, we consider linear, m-dissipative operators A which lack Gihman's property. We obtain partial results regarding the existence of solutions of (*); but in general, the existence question remains open. Our method applies the variation of parameters formula to (**), but this requires a weakened topology when Range$(f) \not\subset \overline{D(A)}$. Two examples are studied: one in ℓ_∞, the other in the space of bounded continuous functions.

0. Introduction: two open problems. We begin with two fairly specific problems; later we shall relate these to more general questions.

Let BC be the Banach space of bounded, continuous functions from $\mathbf{R}$ into $\mathbf{C}$, with the usual supremum norm. Let $\beta : BC \to \mathbf{R}$ be a continuous function, and let $T > 0$ be given. Does a continuous function $u : [0, T] \to BC$ satisfying

$$u(t, \theta) = \int_0^t \exp\left[i\beta(u(s, \cdot)) + i(t - s + \theta)^2\right] ds \qquad (t \in [0, T], \theta \in \mathbf{R})$$

necessarily exist? Yes, if β is locally Lipschitz, or if β has the property that $\beta(x_n) \to \beta(x)$ whenever $x_n(\theta) \to x(\theta)$ uniformly for bounded θ. But in general, the answer is not known.

Alternatively, let ℓ_∞ be the Banach space of bounded sequences of complex numbers, with the supremum norm. Let β be a continuous function from ℓ_∞ into $\mathbf{R}$. Let T be a positive number. Does a continuous function $u = (u_1, u_2, u_3, \ldots) : [0, T] \to \ell_\infty$ satisfying

$$u_k(t) = \int_0^t \exp\left[ik(t - s) + i\beta(u(s))\right] ds \qquad (t \in [0, T]; \ k = 1, 2, 3, \ldots)$$

necessarily exist? Yes, if β is locally Lipschitz, or if $\beta(x_n) \to \beta(x)$ whenever $x_n \to x$ componentwise. But in general, the answer is not known.

1. Compact operators and dissipative operators. Consider initial value problems of the form

$$(1.1) \qquad \begin{cases} u'(t) \in G(u(t)) & (0 \le t \le T), \\ u(0) = z = \text{given}. \end{cases}$$

The operator G may be set-valued, may take values in finite- or infinite-dimensional vector spaces, and may be nonlinear and discontinuous. Hence the class of problems considered includes ordinary and partial differential equations, as well as functional differential equations, integral equations, population models, and other initial value problems. Here G and z are given, and $u(t)$ is unknown. A fundamental question is whether any solution at all exists. More precisely, what hypotheses on G and z guarantee that the initial value problem (1.1) will have at least one solution $u(t)$ for some $T > 0$?

The problem is not simple. We are far from knowing necessary and sufficient conditions for existence of solutions. The different examples of nonexistence are still relatively few in the literature [12, 14, 30]. It is not even a simple matter to choose an appropriate definition of "solution" — some useful definitions have permitted $u(t)$ to be non-differentiable (see §3, below), or even discontinuous [28].

Two of the main approaches to existence are via hypotheses of dissipativeness or compactness (or some variant thereof — see note at the end of this section). A large part of existence theory uses one or the other of these two hypotheses, although they go unmentioned in much research — for instance, compactness is implicit in finite-dimensional problems. (An introduction to the dissipative approach, and references introducing the compactness approach, are given later.)

The theory of dissipative operators and the theory of compact operators developed separately, for the most part, and they use different methods and tools. It is not yet known whether a simpler notion of "generativeness" can be formulated, to include both dissipativeness and compactness as special cases. An effort in that direction was made by Martin [26], who used a one-sided derivative involving measures of noncompactness. However, Martin assumed that his operators were bounded and uniformly continuous.

A number of other papers have also approached a notion of "generativeness" through studying the *dissipative plus compact* problem

$$(1.2) \qquad \begin{cases} u'(t) \in (A+B)u(t) & (0 \le t \le T), \\ u(0) = z = \text{given} \end{cases}$$

where A is an m-dissipative operator (possibly nonlinear and discontinuous) in a Banach space $(X, \| \ \|)$, and B is a compact operator (possibly nonlinear) in X. Does such an initial value problem necessarily have a solution? That question is still open; some partial answers will be surveyed in this paper.

The operator $A + B$ need not be either dissipative or compact, although it includes both of those possibilities as special cases (since A or B could be 0). Thus, $A+B$ may only possess the more general and less understood "generative" property alluded to earlier; and we can hope that a clearer understanding of (1.2) will lead to deeper insights into (1.1). However, the most successful studies of the dissipative plus compact problem (1.2) have made separate uses of the dissipativeness of A and the compactness of B, and have not investigated any directly verifiable "generative" property of the combined operator $A+B$. The arguments used in those papers are sketched in the next section.

We note that many variants on this problem are possible. The choice of A could be made more general: A could be quasi-dissipative [20] or Φ-dissipative [1], or it could satisfy a local dissipativeness condition [15, 32], or A could generate a semigroup in a locally convex space [11, 38]. The choice of B could also be made more general; for instance, we could use measures of noncompactness, as in [2, 37], or let B be set-valued, as in [6].

For definiteness, in this paper we shall assume B is a nonlinear compact mapping in a Banach space $(X, \| \ \|)$ — that is, $B : X \to X$ is continuous and B takes bounded sets to relatively compact sets. (For an introduction to such mappings, see [27].) We shall assume A is an m-dissipative operator in the Banach space X; an introduction to such operators will be given in §3. Starting in §4, we shall also assume A is linear but not densely defined.

2. Truncation, fixed points, and Gihman's property.

We are chiefly concerned with existence of solutions locally in time; the question of global continuability can be studied separately by other methods. In our problem (1.2), the value of T may be quite small, and it may depend on the initial value z. However, for small t, we can obtain *a priori* bounds on the solution's norm, $\|u(t)\|$ (see Theorem 2 in [34], or Theorem 2.3 in [16]). Hence, by a truncation argument replacing B by its composition with a radial retraction, we can assume that B actually has relatively compact range.

This truncation does not affect the basic nature of the problem, but it does simplify the form of the problem. We can now choose T in advance, independently of z. Hence we may work with the space $C([0,T];X)$ of continuous functions from $[0,T]$ into X, and the space $\mathcal{L}^1([0,T];X)$ of integrable functions, in a fixed point argument outlined below. We shall write $\mathcal{L}^1([0,T];K)$, to indicate those functions which have range contained in a given set $K \subseteq X$; and we shall use a subscript w, as in $\mathcal{L}^1_w([0,T];X)$, to indicate a weak topology. Weaker topologies have more compact sets; such sets are useful in fixed point arguments. In particular, if K is a compact convex subset of X, it can be shown [17, 31] that $\mathcal{L}^1_w([0,T];K)$ is compact and convex.

Associated with each m-dissipative operator A is the *quasi-autonomous* problem

$$\begin{cases} u'(t) \in Au(t) + f(t) & (0 \le t \le T), \\ u(0) = z = \text{given}. \end{cases}$$

This problem (discussed in greater detail in §3) is known to have a unique "limit solution" $u \in C([0,T];X)$, for each "forcing term" $f \in \mathcal{L}^1([0,T];X)$ and each initial value $z \in \overline{D(A)}$. Usually we shall keep A and z fixed, but permit f to vary; let us write $u = \mathcal{U}f$ to display the solution's dependence on f. Let us also denote $\mathcal{B}u = B \circ u$. Then solving the dissipative plus compact problem (1.2) amounts to finding a fixed point for the composition $\mathcal{U} \circ \mathcal{B}$ or the composition $\mathcal{B} \circ \mathcal{U}$. For this purpose we apply the Schauder-Tychonoff Fixed Point Theorem.

Let K be the closed convex hull of the range of B. Clearly, $\mathcal{B}$ is a continuous mapping from $C([0,T];X)$ into the compact convex set $\mathcal{L}^1_w([0,T];K)$. Hence, (1.2) has a solution, if $\mathcal{U}$ is continuous from $\mathcal{L}^1_w([0,T];K)$ into $C([0,T];X)$. When does $\mathcal{U}$ have such a property?

To investigate such continuous dependence, it will be useful to have a metric on $\mathcal{L}^1_w([0,T];K)$. Define

$$|||f||| = \max_{0 \le t \le T} \left\| \int_0^t f(s)\, ds \right\|, \qquad \text{for} \quad f \in \mathcal{L}^1([0,T];X),$$

where $\| \ \|$ is the norm of the Banach space X. It can be shown [31] that $||| \ |||$ is a norm on $\mathcal{L}^1([0,T];X)$. This norm is weaker than the usual one: it permits rapid oscillations to cancel out. For instance, $\sin(nt) \to 0$ as $n \to \infty$, in the topology of $|||\ \ |||$ but not in the usual norm topology of $\mathcal{L}^1([0, T]; \mathbf{R})$. It can be shown [17, 31] that the topology induced by $|||\ \ |||$ and the weak topology coincide on the set $\mathcal{L}^1([0, T]; K)$, if $K \subset X$ is compact.

We shall say that an m-dissipative operator A has *Gihman's property* if, for every compact set $K \subset X$, the solution operator $\mathcal{U}$ is continuous from $\mathcal{L}_w^1([0, T]; K)$ into $\mathcal{C}([0, T]; X)$. Equivalently, A has Gihman's property if and only if $\mathcal{U}f_n \to \mathcal{U}f_\infty$ whenever $\{f_n\}$ is a sequence of integrable functions with ranges all contained in a single compact set, and whose integrals satisfy $\int_0^t f_n(s)\, ds \to \int_0^t f(s)\, ds$ for all t.

This criterion is a special case of a more general continuous dependence principle, apparently first noted by I. I. Gihman [13]. For many classes of operators F — not necessarily continuous or linear — it can be shown that the solutions of $u'(t) = F_n(t, u(t))$ converge to the solution of $u'(t) = F_\infty(t, u(t))$ whenever the indefinite integrals $\int F_n(t, \cdot)\, dt$ converge in an appropriate sense to $\int F_\infty(t, \cdot)\, dt$. This principle has been extended to many classes of F's, not all continuous or in finite dimensions, in [10, 16, 21, 22, 24, 29, 31, 33-36] and other papers; see especially the bibliography of [35]. Some of these papers also give converses.

An interesting specialization of this principle is the (nonlinear) Lie-Trotter-Kato product formula: Fix some $T > 0$, and suppose

$$F_n(t, x) = \begin{cases} P(x) & \text{when } [\![2nt/T]\!] \text{ is even, and} \\ Q(x) & \text{when } [\![2nt/T]\!] \text{ is odd,} \end{cases}$$

where $[\![\ ]\!]$ is the greatest integer function. Then $F_\infty(t, x) \equiv \frac{1}{2}P(x) + \frac{1}{2}Q(x)$, and so the convergence of solutions $u_n(T) \to u_\infty(T)$ takes this form:

$$\left(\exp(\frac{T}{2n}Q) \exp(\frac{T}{2n}P) \right)^n z \quad \to \quad \exp\left(\frac{T}{2}(P + Q) \right) z \quad \text{when} \quad n \to \infty,$$

where e^{tA} is the semigroup generated by A (discussed further in §3). This formula has been studied intensively by many mathematicians — see [7, 23, 25], and other references cited therein. A particularly simple example of a failure of the Trotter product formula will be given in §6.

Returning to the quasi-autonomous problem, we summarize some known results. It is known that an m-dissipative operator A has Gihman's property if any of the following hypotheses are satisfied:

- A is continuous and $D(A)$ is closed [31].

- A is *semilinear*. More precisely, $A = A_1 + A_2$, where A_1 and A_2 are both m-dissipative, A_1 is linear and densely defined, and A_2 is continuous with $D(A_2) = X$ [34]. (We emphasize that A_1 must be densely defined in X; we shall discuss that point further.)

- X is uniformly smooth, and one of the following conditions also holds [16]:

    - A is locally bounded; i.e., each $x \in \overline{D(A)}$ has a neighborhood N such that $\bigcup\{Ay : y \in N \cap D(A)\}$ is bounded.

- A is homogeneous of some degree $\alpha > 0$; that is, $rD(A) \subseteq D(A)$ and $A(rx) = r^\alpha A(x)$ for all $r \geq 0$ and $x \in D(A)$.

- X is a Hilbert space, and one of the following conditions also holds [17, 33]:

  - $-A = \partial\varphi$, where $\varphi : X \to \mathbf{R}$ is some lower semicontinuous, proper, convex function.

  - $D(A)$ has nonempty interior.

  - X is finite-dimensional.

As we have noted, if A is an m-dissipative operator with Gihman's property, and B is any compact operator, then the dissipative-plus-compact problem (1.2) has a solution. But (1.2) is not so well understood if A lacks Gihman's property. It is to that case that we shall devote the rest of this paper.

Because this theory is still in its infancy, it needs as many concrete examples as possible. Therefore we shall concentrate on examples which are amenable to explicit computations. For the most part, that means linear examples. But if A is linear, m-dissipative, and densely defined, then A is known to have Gihman's property. Therefore we shall narrow our scope further, to operators A which are linear, m-dissipative, and not densely defined.

Very little of the literature is devoted specifically to such operators, and for good reason. If A is a linear, m-dissipative operator in a Banach space $(X, \| \ \|)$, then $(\overline{D(A)}, \| \ \|)$ is a Banach space in which A is linear, m-dissipative, and densely defined. For most purposes, $\overline{D(A)}$ is a better space than X for the study of A. For our purposes, however, non-densely-defined linear operators are of great interest. They provide us with a convenient (albeit contrived and unnatural) vehicle with which to explore some consequences and limitations of the abstract theory of m-dissipative operators.

The literature on linear, m-dissipative operators assumes a dense domain. The nonlinear (i.e., not necessarily linear) theory has weaker results, but makes no such assumption. Therefore, we turn to the nonlinear theory for our basic tools, even though we shall apply it to linear operators.

3. Review of nonlinear semigroups and dissipative operators.

This section is intended for newcomers to nonlinear semigroup theory; it contains "standard" results taken from [3-5, 8, 9, 27], and from other papers cited therein. Our notation follows that of [3]; minor adjustments must be made if other notations are used.

A mapping in a Banach space is *nonexpansive* (or a *contraction*) if it is Lipschitzian with Lipschitz constant less than or equal to 1.

A *semigroup* on a set C is a family of mappings $S(t) : C \to C$ $(t \geq 0)$ such that

$$S(0)z = z \text{ for all } z \in C, \text{ and } S(t) \circ S(s) = S(t + s) \text{ for all } t, s \geq 0.$$

Semigroups arise naturally in the study of initial value problems. If the initial value problem (1.1) has a unique solution $u_z : \mathbf{R}_+ \to C$ for each z in some set C, then (with most notions of "solution") $S(t)z \equiv u_z(t)$ defines a semigroup on C. This is made more precise in applications below.

A semigroup S on a topological space C is *strongly continuous* if

(3.1) $\qquad t \mapsto S(t)x$ is continuous from $\mathbf{R}_+$ into C, for each $x \in C$.

It is *jointly continuous* if

(3.2) $(t, x) \mapsto S(t)x$ is continuous from $\mathbf{R}_+ \times C$ into C.

The semigroup is *nonexpansive* if C is a subset of a Banach space $(X, \| \ \|)$ and

(3.3) $\|S(t)x - S(t)y\| \leq \|x - y\|$ for all $x, y \in C$ and all $t \geq 0$.

Clearly, any strongly continuous, nonexpansive semigroup is jointly continuous.

Let X be a Banach space. Let A be a set-valued mapping from some domain $D(A) \subseteq X$, into the set of all subsets of X. We define its

$$\text{resolvent } J_\lambda \equiv (I - \lambda A)^{-1} \quad \text{and } \textit{Yosida approximant } A_\lambda \equiv \lambda^{-1}(J_\lambda - I).$$

For motivation note that if X is a function space and A is a partial differential operator, then A may be discontinuous in X; but J_λ and A_λ may be integral operators, and thus may be much better behaved.

The operator A is *dissipative* (or, equivalently, $-A$ is *accretive*) if, for each $\lambda > 0$, the operator J_λ is single-valued and nonexpansive. Most of the theory of dissipative operators generalizes without substantial difficulty to operators A such that $A - cI$ is dissipative, where c is some constant, or even where c is a variable which is locally bounded **[15, 32]**. But the computations and notation take their simplest form when $c = 0$, and we shall follow a common practice of considering only that case. If A is dissipative, then A_λ is also dissipative, as well as Lipschitzian with Lipschitz constant $2/\lambda$.

If A is dissipative and $D(J_\lambda) \equiv R(I - \lambda A)$ contains $\overline{D(A)}$ for all $\lambda > 0$, then (Crandall-Liggett Theorem) A is the *generator* of a strongly continuous, nonexpansive semigroup S on $\overline{D(A)}$, in the sense that $\lim_{n \to \infty} J_{t/n}^n x = S(t)x$ for each $x \in \overline{D(A)}$. To display its dependence on A, the semigroup $S(t)$ generated by A will be denoted by e^{tA}. This is consistent with the classical definition $e^{tA} = \sum_{k=1}^{\infty}(tA)^k/k!$ when A is continuous and linear; then the two definitions give the same semigroup.

An operator A is *m-dissipative* in X if it is dissipative and for $\lambda > 0$, the resolvent J_λ has domain $D(J_\lambda) \equiv R(I - \lambda A)$ equal to all of X. Note then A_λ is also defined on all of X. Hence, by the classical contraction argument of Banach or Picard, for any $f \in \mathcal{L}^1([0,T]; X)$ and any initial value $z \in X$, the *λth approximate problem*

(3.4) $\begin{cases} u_\lambda'(t) = A_\lambda u_\lambda(t) + f(t) & (0 \leq t \leq T), \\ u_\lambda(0) = z \end{cases}$

has a unique strong solution u_λ — i.e., a function which is absolutely continuous on $[0, T]$ and which satisfies the differential equation almost everywhere on $[0, T]$. It can be shown that if $z \in \overline{D(A)}$, then the functions u_λ converge to a limit function $u(t)$ uniformly on $[0, T]$ as $\lambda \downarrow 0$. We define this function to be the *limit solution* (also known as *mild solution* or *integral solution*) of the quasi-autonomous problem

(3.5) $\begin{cases} u'(t) \in Au(t) + f(t) & (0 \leq t \leq T), \\ u(0) = z = \text{given}. \end{cases}$

If f is identically 0, it can be shown that this limit $u(t)$ is just $e^{tA}z$.

Newcomers to this subject may be surprised to learn that *the "limit solution" of* (3.5) *does not necessarily satisfy* (3.5) *in any classical sense.* The function $u(t)$ must be continuous but need not be differentiable; and it takes values in $\overline{D(A)}$ but not necessarily in $D(A)$. If (3.5) has a strong solution, then that function can be shown to coincide with the limit solution. If (3.5) does not have a strong solution, however, various motivating arguments can still be given for selecting the limit solution $u(t) \equiv \lim_{\lambda \downarrow 0} u_\lambda(t)$ as a natural "weak" or "generalized" solution of (3.5). Thus the theory of limit solutions gives us information even concerning initial value problems which only have solutions in a weak sense.

Several other, equally complicated constructions of this same limit u, with different motivating heuristics, can be found in the literature. Also, Bénilan [4, 5] has given an integral inequality which characterizes the solution independently of the method of construction; but his condition, too, is complicated. For our purposes, the method of Yosida approximants will suffice.

The limit solution is a continuous function of both the forcing term and the initial value. In fact, the mapping $(u(0), f) \mapsto u(\cdot)$ is nonexpansive from $\overline{D(A)} \times \mathcal{L}^1([0,T];X)$ into $\mathcal{C}([0,T];X)$; that is,

$$(3.6) \qquad \sup_{0 \leq t \leq T} \|u_1(t) - u_2(t)\| \leq \|u_1(0) - u_2(0)\| + \int_0^T \|f_1(s) - f_2(s)\|\, ds$$

if u_1, u_2 are limit solutions on $[0,T]$ for forcing terms f_1, f_2 respectively, with the same m-dissipative operator A.

If the m-dissipative operator A is also linear, then the limit solution u is given by the classical *variation of parameters* formula

$$u(t) \;\; = \;\; e^{tA} z + \int_0^t e^{sA} f(t-s)\, ds \;\; = \;\; e^{tA} z + \int_0^t e^{(t-s)A} f(s)\, ds,$$

whenever f has range contained in $\overline{D(A)}$. This restriction on f is necessary, since e^{tA} is only defined on $\overline{D(A)}$. If A is densely defined, then the condition on f is automatically satisfied for all $f \in \mathcal{L}^1([0,T];X)$. In the next section, we shall consider how to extend this formula to certain A's which are not densely defined.

4. Explicit formulas for solutions. Let A be linear and m-dissipative but not densely defined. Then A generates a linear, strongly continuous, nonexpansive semigroup e^{tA} on $\overline{D(A)}$, which is a proper subset of X. In §6 and §7 we shall consider examples in which e^{tA} extends naturally to a linear, nonexpansive semigroup $S(t)$ on all of X. Thus we are led to extend the variation of parameters formula; we expect the solution of the quasiautonomous problem (3.5) to be given by

$$(4.1) \qquad u(t) \;\; = \;\; S(t)z + \int_0^t S(s)f(t-s)\, ds \;\; = \;\; S(t)z + \int_0^t S(t-s)f(s)\, ds.$$

Some care must be exercised here, however. The semigroup S is not strongly continuous on all of $(X, \|\ \|)$ (see Corollary 4.8 below), and so the integrals in (4.1) may not make sense. This difficulty can be overcome by using a weaker topology, in which the semigroup S is strongly continuous. Typically, X might have a norm associated with uniform convergence;

the weaker topology might be that of, say, pointwise convergence, a natural setting in which to develop explicit formulas and compute examples. We remark that the arguments presented below can be extended to slightly weaker (albeit more complicated) hypotheses; but the arguments below are sufficiently general for our examples in §6 and §7.

Throughout the remainder of this section, we assume that

(4.2) $(X, \| \ \|)$ is a Banach space, and ξ is a weaker, locally convex, Hausdorff topology on the set X. This topology has the property that for each number $r > 0$, the set $\{x \in X : \|x\| \leq r\}$ is complete in X_ξ.

Here X_ξ denotes the set X endowed with the topology ξ. This topology need not be (and in our examples, it will not be) the usual weak topology generated on X by the topological dual X^* of the Banach space $(X, \| \ \|)$. Limits in the space X_ξ will be denoted by ξ-lim.

The completeness assumption implies easily that if $J \subseteq \mathbf{R}$ is an interval, and $u : J \to X_\xi$ and $\gamma : J \to \mathbf{R}_+$ are continuous functions satisfying $\|u(t)\| \leq \gamma(t)$ and $\int_J \gamma(t)\, dt < \infty$, then $\int_J u(t)\, dt$ exists as a Riemann integral in X_ξ, and $\| \int_J u(t)\, dt \| \leq \int_J \gamma(t)\, dt$. Note that $\| \ \|$ is a seminorm on X_ξ which is lower semicontinuous, but not necessarily continuous. Since X_ξ is weaker than X,

(4.3) for each continuous seminorm ρ on X_ξ there exists a constant $c = c(\rho)$ such that $\rho(x) \leq c\|x\|$ for all $x \in X$.

For simplicity we shall start with the semigroup rather than with its generator. We assume hereafter that

(4.4) S is a jointly continuous semigroup on X_ξ, which is nonexpansive on $(X, \| \ \|)$.

That is, S satisfies (3.3) in X_ξ and (3.4) in $(X, \| \ \|)$, but not necessarily (3.2) in $(X, \| \ \|)$. Define

(4.5) $$ Ax \equiv \xi\text{-}\lim_{t \downarrow 0} \frac{S(t)x - x}{t}, \qquad \text{with } D(A) \equiv \{x \in X : \text{ the limit exists.}\} $$

4.6. THEOREM. *The operator A is m-dissipative in $(X, \| \ \|)$, and $D(A)$ is dense in X_ξ. Let A_λ be the Yosida approximant of A. Then*

(4.7) $$ \xi\text{-}\lim_{\lambda \downarrow 0} \exp(sA_\lambda)x = S(s)x \qquad \text{for each } x \in X. $$

Thus, the strongly continuous, nonexpansive semigroup e^{tA} generated in $(X, \| \ \|)$ by A is the restriction to $\overline{D(A)}$ of the semigroup $S(t)$.

Proof. The first part of the proof is just an easy modification of arguments which can be found in Yosida [38], and so we shall state several results without proof. However, note that our notation differs slightly from Yosida's: we write J_λ where he would write $J_{1/\lambda}$. Also, we use conditions (3.4) and (4.3) in place of Yosida's assumption that the mappings $S(t)$ are equicontinuous.

For each $x \in D(A)$, the mapping $t \mapsto S(t)x$ is differentiable in X_ξ, with derivative equal to $AS(t)x = S(t)Ax$. It follows that $I - \lambda A$ is injective; so $(I - \lambda A)^{-1}$ is single-valued. For each $x \in X$ and $\lambda > 0$, the Riemann integral

$$J_\lambda x \equiv \frac{1}{\lambda} \int_0^\infty e^{-t/\lambda} S(t)x \, dt$$

exists in X_ξ. It satisfies $(I - \lambda A)J_\lambda x = x$ for all $x \in X$; hence $J_\lambda = (I - \lambda A)^{-1}$. Also, $\|J_\lambda x\| \leq \|x\|$, so A is m-dissipative in $(X, \|\ \ \|)$; and $\xi\text{-}\lim_{\lambda \downarrow 0} J_\lambda x = x$, so $D(A)$ is dense in X_ξ (but not necessarily in X).

It remains for us to demonstrate (4.7); but here our conditions (3.4) and (4.3) no longer suffice as a replacement for Yosida's assumption of equicontinuity. Instead, for a proof of (4.7) we shall modify an argument of Hille-Phillips, Theorem 6.3.3 in [18].

Although we define $J_\lambda x$ separately for each x by a Riemann integral in X_ξ, the linear operator J_λ is not necessarily continuous from X_ξ into X_ξ. However, it is continuous from $(X, \|\ \ \|)$ into $(X, \|\ \ \|)$, and therefore so is $A_\lambda \equiv \lambda^{-1}(J_\lambda - I)$. Hence the series

$$\exp(sA_\lambda) = e^{-s/\lambda} \exp\left(\frac{s}{\lambda} J_\lambda\right) = e^{-s/\lambda} \sum_{n=0}^\infty \frac{1}{n!} \left(\frac{s}{n}\right)^n J_\lambda^n$$

converges in the topology of the operator norm, and in any weaker topology as well.

We evaluate the nth term in this series; by induction on n we shall show that

$$J_\lambda^n x = \frac{1}{(n-1)!\,\lambda^n} \int_0^\infty t^{n-1} e^{-t/\lambda} S(t)x \, dt \qquad (x \in X;\ \lambda > 0;\ n = 1, 2, 3, \ldots).$$

Again, the integral is understood to be a Riemann integral in X_ξ. The calculus here is straightforward, but we must exercise some care with the topologies. Let $x \in X$, and let Λ be any continuous linear functional on X_ξ. In the computation below, equation (a) follows from the induction hypothesis. For fixed s, the map $S(s) : X_\xi \to X_\xi$ is continuous and linear, so it commutes with Riemann integration in X_ξ; this justifies (b). Equalities (c) and (e) follow from the continuity of the map $\Lambda : X_\xi \to \{\text{scalars}\}$. Finally, (d) is just calculus with scalar-valued functions. Thus we compute:

$$\Lambda\left[J_\lambda^n x\right] = \Lambda\left[J_\lambda J_\lambda^{n-1} x\right]$$

$$\overset{(a)}{=} \Lambda\left[\frac{1}{(n-2)!\,\lambda^n} \int_0^\infty e^{-s/\lambda} S(s) \left\{\int_0^\infty t^{n-2} e^{-t/\lambda} S(t)x \, dt\right\} ds\right]$$

$$\overset{(b)}{=} \Lambda\left[\frac{1}{(n-2)!\,\lambda^n} \int_0^\infty \left\{\int_0^\infty t^{n-2} e^{-(t+s)/\lambda} S(t+s)x \, dt\right\} ds\right]$$

$$\overset{(c)}{=} \frac{1}{(n-2)!\,\lambda^n} \int_0^\infty \int_0^\infty t^{n-2} e^{-(t+s)/\lambda} \Lambda\left[S(t+s)x\right] dt \, ds$$

$$\overset{(d)}{=} \frac{1}{(n-1)!\,\lambda^{n-1}} \int_0^\infty t^{n-1} e^{-t/\lambda} \Lambda\left[S(t)x\right] dt$$

$$\overset{(e)}{=} \Lambda \left[\frac{1}{(n-1)!\,\lambda^{n-1}} \int_0^\infty t^{n-1} e^{-t/\lambda} S(t)x\,dt \right].$$

Since X_ξ is a Hausdorff locally convex space, its continuous linear functionals Λ separate its points. This completes the induction proof of our formula for J_λ^n.

Hence

$$\exp(sA_\lambda)x = e^{-s/\lambda}x + \sum_{n=1}^\infty \int_0^\infty \frac{s^n t^{n-1}}{\lambda^{2n}\,(n-1)!\,n!} e^{-(t+s)/\lambda} S(t)x\,dt$$

$$= e^{-s/\lambda}x + \int_0^\infty \sum_{n=1}^\infty \frac{s^n t^{n-1}}{\lambda^{2n}\,(n-1)!\,n!} e^{-(t+s)/\lambda} S(t)x\,dt.$$

Again, to justify the above exchange of limits $\sum \int = \int \sum$, we apply an arbitrary continuous linear functional Λ to both sides of this equation, and take limits in X_ξ. The interchange of limits of scalars $\sum \int \Lambda(\ ) = \int \sum \Lambda(\ )$ follows from Fubini's theorem.

Following the notation of [18], let

$$K(s,t,\omega) = e^{-\omega(s+t)} \sum_{n=0}^\infty \frac{(\omega^2 s)^{n+1} t^n}{n!\,(n+1)!} = \omega\sqrt{s/t}\, e^{-\omega(s+t)} I_1(2\omega\sqrt{st}),$$

where I_1 is the modified Bessel function of first kind of order 1. Then

$$K(s,t;\omega) \geq 0, \qquad \int_0^\infty K(s,t;\omega)\,dt = 1 - e^{-\omega s},$$

and since $0 \leq I_1(\nu) \leq \frac{1}{2}\nu e^\nu$, we have $K(s,t;\omega) \leq \omega^2 s \exp[-\omega(\sqrt{s} - \sqrt{t})^2]$. Hence

$$\lim_{\omega\to\infty} \int_0^{s-\delta} K(s,t;\omega)\,dt = 0, \qquad \lim_{\omega\to\infty} \int_{s+\delta}^\infty K(s,t;\omega)\,dt = 0.$$

Therefore, $\xi\text{-}\lim_{\omega\to\infty} \int_0^\infty K(s,t;\omega)u(t)\,dt = u(s)$ for any continuous function $u: \mathbf{R}_+ \to X_\xi$ which is $\|\ \|$-bounded. In particular, taking $u(t) = S(t)x$ and $\omega = 1/\lambda$, we obtain (4.7). ∎

4.8. COROLLARY. *Define A as in (4.5). If S is strongly continuous on $(X, \|\ \|)$, then $D(A)$ is dense in $(X, \|\ \|)$.*

Proof. Suppose S is strongly continuous on $(X, \|\ \|)$. Define its infinitesimal generator in the norm topology:

$$A_1 x \equiv \|\ \|\text{-}\lim_{t\downarrow 0} \frac{S(t)x - x}{t}, \qquad \text{with domain } D(A_1) = \{x \in X: \text{the limit exists}\}.$$

By the classical Hille-Yosida Theorem, A_1 is densely defined and m-dissipative. The operator A is an extension of A_1, since the ξ topology is weaker than the norm topology. But $I - \lambda A_1$ is injective and $R(I - \lambda A_1) = X$ for $\lambda > 0$, since A_1 is m-dissipative. The operator A_1 cannot have a proper extension with these same properties. Thus $A_1 = A$. ∎

Remarks. We now turn to the variation of parameters formula (4.1) and the solution of the quasiautonomous problem (3.5). A subtle distinction must be made here: we are dealing with two different notions of "solution." The "limit solution" of (3.5), defined by $u = \lim u_\lambda$ in the previous section, is of interest to us because of general questions about the abstract theory of dissipative operators. The "classical solution", given by (4.1) with integration in a weak topology, is of interest because it gives us an explicit formula, with which we can do experimental calculations and computations. (For instance, later we shall use this formula to show that this m-dissipative operator A does not have Gihman's property.) We expect these two solutions, obtained by different methods, to coincide; otherwise the abstract theory is not of much use. Our next result, below, shows that the two solutions do indeed coincide. We remark that the variation of parameters formula has already been established in locally convex spaces **[11]**, but under different hypotheses than ours.

4.9. COROLLARY. *In addition to the preceding hypotheses, let f be an element of $\mathcal{L}^1([0,T];X)$. Also, suppose f is piecewise-continuous from $[0,T]$ into X_ξ — i.e., assume $f : [0,T] \to X_\xi$ has only finitely many discontinuities, and has left and right limits at those discontinuities. Then the limit solution of (3.5) is given by the extended variation of parameters formula (4.1), where the integrals are interpreted as Riemann integrals in X_ξ.*

Proof. For each $\lambda > 0$, let u_λ be the solution of (3.4). Since A_λ is a continuous linear operator on the Banach space $(X, \|\ \|)$, we have

$$u_\lambda(t) = \exp(tA_\lambda)z + \int_0^t \exp\left[(t-s)A_\lambda\right] f(s)\, ds,$$

with integration in $(X, \|\ \|)$. Our hypotheses on f imply that the Riemann integral also exists in X_ξ. As $\lambda \downarrow 0$, the integrand converges pointwise in X_ξ to $S(t-s)f(s)$. Hence, for any continuous seminorm ρ on X_ξ,

$$\rho\left(\int_0^t \exp\left[(t-s)A_\lambda\right] f(s)\, ds - \int_0^t S(t-s)f(s)\, ds\right)$$

$$\leq \int_0^t \rho\left(\exp\left[(t-s)A_\lambda\right] f(s) - S(t-s)f(s)\right)\, ds.$$

The right side converges to 0 when $\lambda \downarrow 0$, by the Dominated Convergence Theorem. Thus $\xi\text{-}\lim_{\lambda \downarrow 0} \int_0^t \exp\left[(t-s)A_\lambda\right] f(s)\, ds = \int_0^t S(t-s)f(s)\, ds$. Also $\xi\text{-}\lim \exp\left[tA_\lambda\right]z = S(t)z$ and $u_\lambda(t) \to u(t)$ in X, so $u(t)$ is given by (4.1). ∎

The hypotheses on f are satisfied, for instance, if f is a step-function, or if f is a continuous map from $[0,T]$ into $(X, \|\ \|)$. But the limit solution of (3.5) depends continuously on $f \in \mathcal{L}^1([0,T];X)$, as we see from (3.6). In some cases the explicit formula given by (4.1) may be extended to all $f \in \mathcal{L}^1([0,T];X)$; but the details of such an extension depend on the choice of X. Examples will be given in §6 and §7.

5. A weak Gihman's property; an existence theorem. The explicit solution (4.1) will be used in §6 and §7 to study two m-dissipative operators A, and the solutions to their respective quasi-autonomous problems. These two example operators were selected because they are very simple and linear — thus easy to work with — and because they do not have Gihman's property. We conjecture that with either of these choices of A, and some suitable choice of B, the dissipative-plus-compact problem does not have a solution. The author had originally hoped to demonstrate this conjecture by an explicit computational example, following the precedent of **[12, 14]**, *et al.*

But we shall see that both those operators satisfy a weak version of Gihman's property, developed below. This weak version is not enough to guarantee existence of solutions for the dissipative plus compact problem (1.2) for *all* choices of the compact operator B; but it is enough to guarantee existence in those cases where explicit computations are easily performed — i.e., in those cases where B can be analyzed componentwise or in some other weak topology. Thus, if there do exist choices of B for which (1.2) has no solution, those choices may be very hard to discover.

5.1. THEOREM. *In addition to the hypotheses of the preceding section, let K_ξ be a $\|\ \ \|$-bounded, ξ-compact subset of X_ξ. (This set will be given the topology induced by X_ξ.) Define*

$$(\mathcal{U}f)(t) \equiv S(t)z + \int_0^t S(t-s)f(s)\,ds.$$

Then the mapping $f \mapsto \mathcal{U}f$ is continuous from $C\left([0,T]; K_\xi\right)$ into $C\left([0,T]; X_\xi\right)$. This map takes $C\left([0,T]; K_\xi\right)$ into a subset of a compact convex set $\mathcal{G} \subset C\left([0,T]; X_\xi\right)$.

Proof. Throughout this argument, all topological notions — compactness, continuity, convergence, etc. — will be with respect to the topology of X_ξ, except where noted otherwise.

Let K_1 be the closed convex hull of $\{0\} \cup \{z\} \cup K_\xi$. Let $K_2 = \{S(t)x : (t,x) \in [0,T] \times K_1\}$. Let K_3 be the closed convex hull of K_2. Let r be any positive number. Then each of the sets K_1, K_2, K_3, rK_3 is compact, convex, and $\|\ \ \|$-bounded, by (4.2) and (4.4).

Let any $f \in C\left([0,T]; K_\xi\right)$ and $t \in [0,T]$ be given. The integrand $S(t-s)f(s)$ lies in K_2; hence the approximating Riemann sums lie in tK_3; hence so does the Riemann integral

$$(\mathcal{R}f)(t) = \int_0^t S(t-s)f(s)\,ds.$$

The function $t \mapsto S(t)z$ takes values in K_3; hence $\mathcal{U}f$ takes values in $(T+1)K_3$.

Next we shall show that for $f \in C\left([0,T]; K_\xi\right)$, the functions $\mathcal{U}f$ are uniformly equicontinuous on $[0,T]$. To see this, let $0 \le r \le t \le T$, and compute

$$(\mathcal{U}f)(t) - (\mathcal{U}f)(r)$$

$$= [S(t)z - S(r)z] + \int_r^t S(t-s)f(s)\,ds + \int_0^r \left[S(t-s) - S(r-s)\right]f(s)\,ds$$

$$= [S(t)z - S(r)z] + \int_r^t S(t-s)f(s)\,ds + \left[S(t-r) - I\right](\mathcal{R}f)(r).$$

Observe that $t \mapsto S(t)z$ is continuous, hence uniformly continuous on $[0, T]$. Next, $S(t-s)f(s)$ lies in K_2, hence in K_3; hence $\int_r^t S(t-s)f(s)\,ds$ lies in $(t-r)K_3$; therefore $\int_r^t S(t-s)f(s)\,ds$ converges to 0 as $(t-r) \to 0$, uniformly for all choices of t, r, f. Finally, $(\mathcal{R}f)(r)$ lies in the compact set TK_3, and $[S(t-r) - I] \to 0$ uniformly on compact sets as $t - r \to 0$, since S is jointly continuous. This proves uniform equicontinuity.

Let $\mathcal{N}$ be the family of all closed convex neighborhoods of 0 in X_ξ. Recall that $\mathcal{N}$ is a neighborhood base — i.e., every neighborhood of 0 contains an element of $\mathcal{N}$ — since X_ξ is a Hausdorff locally convex topological vector space. The condition of uniform equicontinuity can be restated as follows: for each $E \in \mathcal{N}$, there is some $\delta = \delta(E) > 0$ such that if $u = \mathcal{U}f$ for some $f \in C\left([0, T]; K_\xi\right)$, then

(5.2) $\qquad u(t) - u(r) \in E \qquad$ whenever $\qquad t, r \in [0, T]$ and $|t - r| \leq \delta(E)$.

Select some particular such function $\delta : \mathcal{N} \to (0, \infty)$. Now define

$$\mathcal{G} = \left\{ u \in C\left([0, T]; X_\xi\right) : \mathrm{Ran}(u) \subseteq (T+1)K_3 \text{ and } (5.2) \text{ holds for all } E \in \mathcal{N} \right\}.$$

Then $\mathcal{G}$ is easily seen to be closed and convex. By a sufficiently general version of the Arzela-Ascoli Theorem **[19]**, it is compact in $C\left([0, T]; X_\xi\right)$.

It remains only to show the continuity of $\mathcal{U}$. Suppose that some net $\{f_\alpha\}$ converges to a limit f in $C\left([0, T]; K_\xi\right)$; that is, $f_\alpha \to f$ uniformly on $[0, T]$. The f_α's and f all have range contained in the compact set K_ξ; and the semigroup S is continuous, hence uniformly continuous on the compact set $[0, T] \times K_1$. Hence $S(t-s)[f_\alpha(s) - f(s)] \to 0$ uniformly for $s, t \in [0, T]$. Fix any $t \in [0, T]$ and any closed neighborhood E of 0. For all α sufficiently large, we have $\{S(t-s)[f_\alpha(s) - f(s)] : 0 \leq s \leq t\} \subseteq E$. Then the Riemann sums for $[\mathcal{R}(f_\alpha - f)](t)$ also lie in E; hence so does that integral. Thus, $\mathcal{R}f_\alpha(t) - \mathcal{R}f(t)$ lies in any neighborhood of 0, for sufficiently large. Therefore $\mathcal{R}f_\alpha(t) \to \mathcal{R}f(t)$; hence also $\mathcal{U}f_\alpha(t) \to \mathcal{U}f(t)$. Finally, the convergence is uniform in t; that fact follows from the uniform equicontinuity of the functions $\mathcal{U}f_\alpha$. This proves continuity. ∎

5.3. COROLLARY. *In addition to the hypothese of the preceding section, suppose $B : X_\xi \to X_\xi$ is a continuous mapping, with range contained in a ξ-compact, $\| \quad \|$-bounded set. Then for any $T > 0$ and $z \in X$, there exists at least one continuous function $u : [0, T] \to X_\xi$ satisfying*

$$u(t) = S(t)z + \int_0^t S(t-s)B\left(u(s)\right)ds \qquad \text{for all } t \in [0, T].$$

Moreover, if the range of B is relatively compact in $(X, \| \quad \|)$, then u and $B \circ u$ are continuous from $[0, T]$ into $(X, \| \quad \|)$. Hence u is a limit solution of (3.5) with $f \equiv B \circ u$; so u is a solution of the dissipative plus compact problem (1.2) in the Banach space $(X, \| \quad \|)$.

Proof. Let K_ξ be the range of B. Define $K_1, K_2, K_3, \mathcal{G}$ as in the proof of the preceding theorem. Define $\mathcal{B} : \mathcal{G} \to C\left([0, T]; K_\xi\right)$ by taking $\mathcal{B}u \equiv B \circ u$. It is easy to see that this function is continuous: Suppose $u_\alpha \to u$ in $\mathcal{G}$. Then $u_\alpha(t) \to u(t)$ uniformly in t. All the u_α's and u have range in the compact set $(T+1)K_3$. The function B is continuous;

hence it is uniformly continuous on that compact set. It follows that $B(u_\alpha(t)) \to B(u(t))$ uniformly in t, proving our claim.

Thus, the composition $\mathcal{U} \circ B$ is a continuous self-mapping of the compact convex set $\mathcal{G}$. By the Schauder-Tychonoff Theorem, this mapping has at least one fixed point $u \in \mathcal{G}$. Then u satisfies the integral equation stated in the theorem.

Finally, suppose that the range of B is actually a relatively compact subset of $(X, \| \ \|)$. Since $B : X_\xi \to X_\xi$ is continuous and X_ξ is a Hausdorff topology weaker than that of X, it follows easily that B is in fact continuous from X_ξ into $(X, \| \ \|)$. Hence $f \equiv B \circ u$ is continuous from $[0, T]$ into $(X, \| \ \|)$. Then Corollary 4.9 is applicable, and the limit solution of (3.5) is given by the variation of parameters formula (4.1). $\blacksquare$

6. Example in a sequence space.

Let X be the Banach space ℓ_∞ of bounded sequences of complex numbers, with the usual supremum norm denoted by $\| \ \|$. Define a nonexpansive semigroup S on X by taking

$$S(t)\{x_k\} = \{e^{ikt}x_k\} \qquad \text{for each } \{x_k\} \in X.$$

For fixed $x \in X$, the map $t \mapsto S(t)x$ is not necessarily continuous into ℓ_∞. Indeed, its range may not even be separable. For instance, let $x = (1, 1, 1, \ldots)$. Then $\|S(t)x - S(s)x\| \geq \sqrt{3}$ whenever $0 \leq s < t < 2\pi$. (*Proof:* Let $r \equiv t - s$; then $r \in (0, 2\pi)$. At least one of the angles kr $(k = 1, 2, 4, 8, \ldots)$ lies in the interval $[2\pi/3, 4\pi/3]$, modulo 2π, and hence $|e^{ikr} - 1| \geq \sqrt{3}$.)

Let ξ be the topology of componentwise convergence — that is, let X_ξ have the topology induced by the product topology on the product space $\mathbf{C}^{\mathbf{N}}$. We easily verify (4.2) and (4.4). Define the operator A by (4.5); we find that

$$A(\{x_k\}) \equiv \{ikx_k\}, \qquad \text{with domain} \ \ D(A) \equiv \{\{x_k\} : \{ikx_k\} \text{ is bounded}\}.$$

Then $\overline{D(A)} = c_0 = \{\text{sequences converging to } 0\}$. The restriction of $S(t)$ to c_0 is the strongly continuous, nonexpansive semigroup e^{tA} generated by the m-dissipative operator A.

By Corollary 4.9, we obtain an explicit solution of the quasi-autonomous problem (3.5), at least whenever $f \in \mathcal{L}^1([0, T]; X) \cap \mathcal{C}([0, T]; X_\xi)$ — in particular, whenever $f \in \mathcal{C}([0, T]; X)$. That solution is given by $\mathcal{U}f = u = (u_1, u_2, u_3, \ldots)$, where

$$(6.1) \qquad u_k(t) = e^{ikt}z_k + \int_0^t e^{iks} f_k(t - s)\, ds.$$

Since $\mathcal{C}([0, T]; X)$ is dense in $\mathcal{L}^1([0, T]; X)$, we may take limits; thus (6.1) is valid for all $f \in \mathcal{L}^1([0, T]; X)$.

6.2. PROPOSITION. *A does not have Gihman's property.*

This proposition was proved in [34]. A different proof will be given below.

6.3. PROPOSITION. *Let $\alpha = (1,1,1,\ldots)$. Define $P(x) = A(x) + \alpha$ and $Q(x) = A(x) - \alpha$; then P and Q are m-dissipative with $D(P) = D(Q) = D(A)$. Let m be an odd positive integer, and let $T = 2\pi m$. Then we have the following failure of the Trotter product formula:*

$$\left[\exp\left(\frac{T}{2n}Q\right) \exp\left(\frac{T}{2n}P\right) \right]^n (0) \quad \text{does not converge to} \quad [\exp(TA)](0) \quad \text{as } n \to \infty.$$

Proof of Propositions 6.2 and 6.3. For $0 \le t \le T$, let $f^{[n]}(t) = (-1)^{[\![2nt/T]\!]} \alpha$, where $[\![\ ]\!]$ is the greatest integer function. Take initial value $u(0) = 0$, and let

$$(\mathcal{U}f^{[n]})(t) = u^{[n]}(t) = \left(u_1^{[n]}(t), u_2^{[n]}(t), u_3^{[n]}(t), \ldots \right).$$

Also, let $f^{[\infty]}$ be identically 0; then $u^{[\infty]} \equiv \mathcal{U}f^{[\infty]}$ is identically 0.

The $f^{[n]}$'s all have range contained in $K = \{-\alpha, +\alpha\}$, which is certainly a compact subset of ℓ_∞. Also, $\lim_{n \to \infty} \| \int_0^t f^{[n]}(s)\,ds \| = 0$ uniformly for $t \in [0, T]$; so $f^{[n]} \to f^{[\infty]}$ in $\mathcal{L}_w^1([0,T]; K)$. Hence, if A has Gihman's property, then $\|u^{[n]}(t)\|$ must converge to 0 uniformly for $t \in [0,T]$. Also observe that

$$\left[\exp\left(\frac{T}{2n}Q\right) \exp\left(\frac{T}{2n}P\right) \right]^n (0) = u^{[n]}(T),$$

and $[e^{TA}](0) = 0$. So to prove both propositions, it suffices to show that $\|u^{[n]}(T)\|$ does not converge to 0 as $n \to \infty$.

The explicit formula (6.1) yields

$$u_k^{[n]}(T) = e^{ikT} \int_0^T e^{-iks}(-1)^{[\![2ns/T]\!]}\,ds$$

$$= e^{ikT} \sum_{j=0}^{n-1} \left\{ \int_{(2jT)/2n}^{(2j+1)T/2n} e^{-iks}\,ds - \int_{(2j+1)T/2n}^{(2j+2)T/2n} e^{-iks}\,ds \right\}$$

$$= e^{ikT} \left\{ \sum_{j=0}^{n-1} \exp\left(-ik \cdot \frac{2jt}{2n}\right) \right\} \left\{ 1 - \exp\left(-ik \cdot \frac{T}{2n}\right) \right\} \int_0^{T/2n} e^{-iks}\,ds.$$

In particular, at $k = n$, we obtain

$$u_n^{[n]}(T) = e^{inT} \left\{ \sum_{j=0}^{n-1} e^{-ijT} \right\} \left\{ \frac{1}{in} \right\} \left\{ 1 - e^{-iT/2} \right\}^2 = \frac{4}{i},$$

finally using the fact that $T = 2\pi m$ where m is an odd integer. Thus $\|u^{\lfloor n \rfloor}(T)\| \geq |u_n^{\lfloor n \rfloor}(T)| = 4$. ∎

Since this A does not have Gihman's property, it is not known whether the dissipative plus compact problem (1.2) has a solution for arbitrary choices of B. We might seek to investigate this question by a study of particular examples. It is fairly easy to construct particular examples of compact mappings $B = (b_1, b_2, b_3, \ldots) : \ell_\infty \to \ell_\infty$: it suffices to choose the maps $b_j : \ell_\infty \to \mathbf{C}$ equicontinuous, and in such a fashion that B has relatively compact range. One easy method of choosing the b_j's continuous is to let each b_j be a continuous function of only finitely many coordinates: $b_j = b_j(x_1, x_2, \ldots, x_{k_j})$. But then it follows that B is continuous from X_ξ into X_ξ, and so a solution of (1.2) is known to exist by Corollary 5.3. It will be harder to construct examples for which Corollary 5.3 is not applicable.

7. Example with continuous functions. For our second example, let X be the Banach space BC of bounded continuous functions from $\mathbf{R}$ into $\mathbf{C}$, with the supremum norm. Points $x \in BC$ will be identified with functions $x(\theta)$. For consistency of notation, we shall denote space variables by $\theta, \sigma \in \mathbf{R}$, and time variables by $r, s, t \in [0, T]$. For instance, a mapping $u : [0, T] \to BC$ may instead be viewed as a complex-valued function $u(t, \theta)$ with $t \in [0, T]$ and $\theta \in \mathbf{R}$. The two views of u are not really the same, since different topologies are involved, but the choice of topology will be made clear in each context.

We consider the translation semigroup,

$$[S(t)x](\theta) = x(t + \theta) \qquad (x \in X,\, \theta \in \mathbf{R},\, t \geq 0).$$

This is a nonexpansive semigroup on BC; but for fixed $x \in BC$, the mapping $t \mapsto S(t)x$ generally is not continuous. Indeed, it may not even have separable range. For instance, let $x(\theta) = \exp(i\theta^2)$. Then $\|S(t)x - S(s)x\| = 2$ whenever $s \neq t$. (*Proof:* Choose θ so that $2(t - s)(\theta + s) + (t - s)^2 \equiv \pi \pmod{2\pi}$. Then $|[S(t)x](\theta) - [S(s)x](\theta)| = 2$.)

Let ξ be the topology of uniform convergence on compact subsets of $\mathbf{R}$; that is, $\xi\text{-}\lim_{n\to\infty} x_n = x$ means that $x_n(\theta) \to x(\theta)$ uniformly for bounded θ. We easily verify conditions (4.2) and (4.4). Define A as in (4.5). We find that $A = d/d\theta$, with domain $D(A) = \{x \in BC : Ax \in BC\} = \{x : x \text{ is continuously differentiable and } x' \text{ is bounded}\}$. Then $\overline{D(A)} = BUC \equiv \{\text{bounded, uniformly continuous functions}\}$. (*Proof:* BUC is closed and contains $D(A)$. Also, any $x \in BUC$ is approximated uniformly in θ, as $\varepsilon \downarrow 0$, by the functions $x_\varepsilon(\theta) = \varepsilon^{-1} \int_0^\varepsilon x(\theta + \sigma)\, d\sigma = \varepsilon^{-1} \int_\theta^{\theta+\varepsilon} x(\sigma)\, d\sigma$, which lie in $D(A)$.)

Corollary 4.9 gives us an explicit formula for the solution of the quasi-autonomous problem (3.5), at least whenever $f \in \mathcal{C}([0, T]; BC)$:

$$(7.1) \qquad (\mathcal{U}f)(t, \theta) = z(t + \theta) + \int_0^t f(t - s, s + \theta)\, ds.$$

We may take limits, using the Dominated Convergence Theorem. Since $\mathcal{C}([0, T]; BC)$ is dense in $\mathcal{L}^1([0, T]; BC)$, it follows that the explicit formula (7.1) is valid whenever $f \in \mathcal{L}^1([0, T]; X)$.

7.2. PROPOSITION. *The operator A does not have Gihman's property.*

Proof. For $\alpha, \theta \in \mathbf{R}$, let $g(\alpha; \theta) = \exp(i\alpha + i\theta^2)$. Observe that $\partial g / \partial \alpha = ig$ has absolute value 1; hence $|g(\alpha; \theta) - g(\beta; \theta)| \leq |\alpha - \beta|$ for all $\alpha, \beta, \theta \in \mathbf{R}$; hence $\|g(\alpha; \cdot) - g(\beta; \cdot)\| \leq |\alpha - \beta|$. Thus the mapping $\alpha \mapsto g(\alpha; \cdot)$ is continuous from $\mathbf{R}$ into BC. Since that mapping is also periodic with period 2π, its range is a compact set $K \subset BC$. Now define

$$f_n(t, \theta) = g\left(-(t-n)^2; \theta\right) = \exp\left[i\left(\theta^2 - (t-n)^2\right)\right].$$

Then $f_n \in C([0,T]; K) \subset \mathcal{L}^1([0,T]; K)$.

We claim that $f_n \to 0$ in $\mathcal{L}^1_w([0,T]; K)$. To show this, it suffices to demonstrate that $\|\int_0^t f_n(s, \cdot)\,ds\| \to 0$ as $n \to \infty$, uniformly for $t \in [0,T]$. For $n > T$, we have

$$\int_0^t f_n(s, \theta)\,ds = \int_0^t \exp\left(i\theta^2 - i(s-n)^2\right)\,ds$$

$$= \frac{-1}{2i}\int_{s=0}^{s=t} \frac{1}{(s-n)}\,d\exp\left(i\theta^2 - i(s-n)^2\right)$$

and now integrating by parts

$$= \left[\frac{-1}{2i(s-n)}\exp\left(i\theta^2 - i(s-n)^2\right)\right]_{s=0}^{s=t} + \frac{1}{2i}\int_{s=0}^{s=t} \exp\left(i\theta^2 - i(s-n)^2\right)\,d(s-n)^{-1}$$

$$= \left[\frac{-1}{2i(s-n)}\exp\left(i\theta^2 - i(s-n)^2\right)\right]_{s=0}^{s=t} - \frac{1}{2i}\int_0^t (s-n)^{-2}\exp\left(i\theta^2 - i(s-n)^2\right)\,ds.$$

Now, the complex exponentials are uniformly bounded, but the terms $1/(s-n)$ tend to 0 uniformly for s in $[0,T]$ as $n \to \infty$. This proves our claim.

Now take initial value $z = 0$, and let u_n be the solution of the quasi-autonomous problem (3.5) with forcing term f_n. If A has Gihman's property, then we must have $u_n \to 0$ in $C([0,T]; BC)$ as $n \to \infty$. But we easily show that that is not so. Indeed, from the explicit formula (7.1) we obtain

$$u_n(t, \theta) = \int_0^t f_n(t-s, s+\theta) = \int_0^t \exp\left(i(s+\theta)^2 - i(t-s-n)^2\right)\,ds.$$

In particular, taking $\theta = n - t$, we have $u_n(t, n-t) = t$; hence $\|u_n(t, \cdot)\| \geq t$. ∎

Again, since this A does not have Gihman's property, it is not known whether the dissipative plus compact problem (1.2) has a solution for arbitrary choices of B. We might seek to investigate this question by a study of particular examples. Compact operators in spaces of continuous functions can easily be obtained in the form of integral operators; see [27] for instance. But simple hypotheses guaranteeing that such an operator will be continuous for the norm topology may also guarantee that the operator will be continuous for the topology of convergence on bounded sets — implying, by Corollary 5.3, that a solution of (1.2) exists. It will be harder to construct examples for which Corollary 5.3 is not applicable.

Acknowledgment. The author is grateful to Professors Glenn Webb and Michael Freedman, and others, for several helpful discussions.

References. For brevity, we omit some important but older references which are cited in more recent references listed below.

1. W. ARENDT, P. R. CHERNOFF, and T. KATO, A generalization of dissipativity and positive semigroups, *J. Oper. Th.* **8** (1982), 167-180.

2. J. BANAŚ, A. HAJNOSZ, and S. WĘDRYCHOWICZ, Some generalization of Szufla's theorem for ordinary differential equations in Banach space, *Bull. Acad. Polon. Sci. Ser. Math.* **29** (1981), 459-464.

3. V. BARBU, *Nonlinear Semigroups and Differential Equations in Banach Spaces*, Noordhoff, Leyden, 1976.

4. P. BÉNILAN, *Équations d'évolution dans un espace de Banach quelconque et applications*, thesis, Orsay, 1972.

5. P. BÉNILAN, M. G. CRANDALL, and A. PAZY, book in preparation.

6. A. BRESSAN, Solutions of lower semicontinuous differential inclusions on closed sets, *Rend. Sem. Mat. Univ. Padova* **69** (1983), 99-107.

7. A. J. CHORIN, T. J. HUGHES, M. F. MC CRACKEN, and J. E. MARSDEN, Product formulas and numerical algorithms, *Comm. Pure Appl. Math.* **31** (1978), 205-256.

8. M. G. CRANDALL, Nonlinear semigroups and evolution governed by accretive operators, *M.R.C. Technical Summary Report* **2724**. To appear in the proceedings of the Symposium on Nonlinear Functional Analysis and Applications, held in Berkeley in July, 1983.

9. M. G. CRANDALL and A. PAZY, An approximation of integrable functions by step functions with an application, *Proc. Amer. Math. Soc.* **76** (1979), 74-80.

10. M. DAWIDOWSKI, On some generalization of Bogoliubov averaging theorem, *Functiones et Approximatio* **7** (1979), 55-70.

11. B. DEMBART, Perturbations of semigroups on locally convex spaces, *Bull. Amer. Math. Soc.* **79** (1973), 986-991.

12. M. A. FREEDMAN, Product integrals of continuous resolvents: existence and nonexistence, *Israel J. Math.* **46** (1983), 145-160.

13. I. I. GIHMAN, Concerning a theorem of N. N. Bogolyubov, *Ukr. Math. J.* **4** (1952), 215-218 (in Russian). (For English summary see *Math. Reviews* **17**, p. 738.)

14. A. N. GODUNOV, Peano's theorem in Banach spaces, *Funct. Anal. Appl.* **9** (1975), 53-55.

15. J. A. GOLDSTEIN, Locally quasi-dissipative operators and the equation $\partial u/\partial t = \phi(x, \partial u/\partial x)\partial^2 u/\partial x^2 + g(u)$, in: *Evolution Equations and their Applications (proceedings of the Graz conference on nonlinear differential equations; F. Kappel and W. Schappacher, Ed.)*, Pitman Research Notes in Mathematics no. 68, Boston, 1982.

16. S. GUTMAN, Evolutions governed by m-accretive plus compact operators, *Nonlin. Anal. Theory Methods Appl.* **7** (1983), 707-715.

17. S. GUTMAN, Topological equivalence in the space of integrable vector-valued functions, *Proc. Amer. Math. Soc.* **93** (1985), 40-42.

18. E. HILLE and R. S. PHILLIPS, *Functional Analysis and Semigroups*, AMS Colloq. Publns. **31**, AMS, Providence R.I., Revised Edition, 1957.

19. J. L. KELLEY, *General Topology*, Van Nostrand, N.Y., 1955; reprinted by Springer, N.Y., 1975.

20. Y. KOBAYASHI, Difference approximation of Cauchy problems for quasi-dissipative operators and generation of nonlinear semigroups, *J. Math. Soc. Japan* **27** (1975), 640-665.

21. M. A. KRASNOSEL'SKII and S. G. KREIN, On the principle of averaging in nonlinear mechanics, *Usp. Mat. Nauk* **10** (1955), 147-152. Russian. (For English summary see *Math. Reviews* **17** #152.)

22. T. G. KURTZ, An abstract averaging theorem, *J. Funct. Anal.* **23** (1976), 135-144.

23. T. G. KURTZ and M. PIERRE, A counterexample for the Trotter product formula, *J. Diff. Eqns.* **52** (1984), 407-414.

24. J. KURZWEIL, Generalized ordinary differential equations and continuous dependence on a parameter, *Czech. Math. J.* **7** (1957), 418-449.

25. L. LAPIDUS, Generalization of the Trotter-Lie formula, *Integral Equations and Operator Theory* **4** (1981), 366-415.

26. R. H. MARTIN, Approximation and existence of solutions to ordinary differential equations in Banach spaces, *Funk. Ekvac.* **16** (1973), 195-211.

27. R. H. MARTIN, *Nonlinear Operators and Differential Equations in Banach Spaces*, Wiley, N.Y., 1976.

28. J. J. MOREAU, Evolution problem associated with a moving convex set in a Hilbert space, *J. Diff. Eqns.* **26** (1977), 347-374.

29. L. W. NEUSTADT; On the solutions of certain integral-like operator equations: existence, uniqueness, and dependence theorems; *Arch. Rat. Mech. Anal.* **38** (1970), 131-160.

30. C. OLECH, An existence theorem for solutions of orientor fields, in: *Dynamical Systems: An International Symposium, vol.* 2, L. Cesari, J. K. Hale, and J. P. La Salle (eds.), Academic Press, N.Y., 1976; pp. 63-66.

31. E. SCHECHTER, Evolution generated by continuous dissipative plus compact operators, *Bull. London Math. Soc.* **13** (1981), 303-308.

32. E. SCHECHTER, Interpolation of nonlinear partial differential operators and generation of differentiable evolutions, *J. Diff. Eqns.* **46** (1982), 78-102.

33. E. SCHECHTER, Perturbations of regularizing maximal monotone operators, *Israel J. Math.* **43** (1982), 49-61.

34. E. SCHECHTER, Evolution generated by semilinear dissipative plus compact operators, *Trans. Am. Math. Soc.* **275** (1983), 297-308.

35. E. SCHECHTER, Necessary and sufficient conditions for convergence of temporally

irregular evolutions, *Nonlin. Analysis Theory Methods Appl.* **8** (1984), 133-153.

36. E. SCHECHTER, Correction to "Perturbations of monotone operators" and a note on injectiveness, *Israel J. Math.* **47** (1984), 236-240.

37. S. SZUFLA, On the equation $x' = f(t, x)$ in Banach spaces, *Bull. Acad. Polon Ser. Sci. Math. Astron. Phys.* **26** (1978), 401-406.

38. K. YOSIDA, *Functional Analysis*, Springer, New York, 1964.

TWO COMPACTNESS LEMMAS

Thomas I. Seidman*
Department of Mathematics
University of Maryland Baltimore County
Catonsville, Maryland 21228, USA

1. INTRODUCTION

We consider here, in somewhat more abstract form, two compactness arguments which have already proved useful in particular applications.

The first argument was originally developed [4] in the context of demonstrating the existence of periodic solutions of a system of equations arising in semiconductor theory. At the particular point in the argument the initial value problem had already been studied and it had been established that for initial data ζ at time s one would have a (unique) solution z with a Lipschitz condition

$$|z(t;\zeta) - z(t;\hat{\zeta})| \leq M|\zeta-\hat{\zeta}|, \tag{1}$$

where M could be taken independently of (t,s) (ranging over a bounded set: say, a period) and of ζ (ranging over a bounded set). Since the equations of the system (for the components z_k of z) were parabolic:

$$\dot{z}_k = \nabla \cdot D_k z_k + (\ldots) \tag{2}$$

with $D_k \geq \delta > 0$, it proved possible — using the standard parabolic argument of multiplying by z_k, integrating over $(s,\cdot) \times \Omega$, and using the Gronwall Inequality — to bound $z(\cdot;\zeta)$ in $L^2(I \to H^1(\Omega))$; from the equation itself, this bounds $\dot{z}$ in $L^2(I \to H^{-1}(\Omega))$. We recognize that this suffices for applicability of the Aubin Compactness Theorem [1] so the standard estimates are enough to ensure that $\{z(\cdot;\zeta): \zeta \in B\}$ is precompact in $L^2(I \to L^2(\Omega))$. (All of this, of course, is predicated upon the restriction of ζ to a bounded set B and the principal effort of [4] went to a maximum principle argument showing that a certain set B, bounded in $L^2(\Omega)$, was indeed invariant.) Now consider the Poincaré map:

$$\zeta = z(t_0;\zeta) \longmapsto z(t_0+\tau;\zeta), \qquad \zeta \in B. \tag{3}$$

*This research has been partially supported under grant #AFOSR-82-0271 by the U.S. Air Force Office of Scientific Research.

At this point we know that this map is a well-defined continuous (indeed, Lipschitzian) self-map of B and that $\{z(\cdot;\zeta): \zeta \in B\}$ is precompact in $L^2((t_0,t_0+\tau) \rightarrow L^2(\Omega))$, but this compactness result ·is in the wrong place to show that (3) is a compact map and permit application of the Schauder Fixpoint Theorem. The point of the argument to be presented here is to fill the gap between use of the Aubin Theorem and use of the Schauder Theorem — to show that, using (1), the Aubin compactness result does, indeed, suffice to show compactness of (3).

Essentially the same argument is used in [2] to show existence of periodic solutions for a quasi-variational inequality of evolution arising as the Hamilton-Jacobi-Bellman 'equation' for a stochastic optimal control problem (again the context is parabolic so the standard estimate serves for Aubin's Theorem and a maximal principle argument permits restriction to a bounded set).

The second argument was originally developed [5] in the context of optimal control. The evolution of the system was there governed by an abstract integral equation

$$z(t) \;=\; F(z,v)(t) \;:=\; z_0(t) + \int_0^t \varphi(x,z(s),v(s)) \; ds. \qquad (4)$$

Here v is a control and we write $z(\cdot;v)$ for the solution of (4), taking values in a Banach space $\mathcal{Z}$. Assumed Lipschitz continuity of φ in z gives contractivity of $F(\cdot,v)$, so such solutions exist uniquely. Omitting details irrelevant here, we note only that to show existence of an optimal control it was desirable to show that $z(\mathcal{V}) := \{z(\cdot;v): v \in \mathcal{V}\}$ is precompact in $C([0,T] \rightarrow \mathcal{Z})$. Under the assumptions above, one sees that the map: $v \mapsto z(\cdot;v)$ will be continuous and, for bounded $\mathcal{V}$, $z(\mathcal{V})$ will be bounded and equicontinuous in t. Since, in general, $\mathcal{V}$ will not be compact we cannot obtain compactness of $z(\mathcal{V})$ as the continuous image of a compact set. If $\mathcal{V}$ is merely bounded and $\mathcal{Z}$ is finite dimensional, then it is standard to conclude by applying the Arzela-Ascoli Theorem that $z(\mathcal{V})$ is precompact, but for distributed parameter control theory (infinite dimensional $\mathcal{Z}$) that argument fails since a bound on $\|z(t,\mathcal{V})\|_{\mathcal{Z}}$ no longer gives precompactness of the range. If $F(z,v)$ depends compactly on v for given z — as was the case for the application [5] — then one can hope that the condition,

$$\{F(z,v): z \in \mathcal{S}, v \in \mathcal{V}\} \text{ is precompact if } \mathcal{S} \text{ is compact,} \qquad (5)$$

will hold and we will show that, given the uniform contractivity of $F(\cdot,v)$, this suffices to ensure the desired precompactness of $z(\mathcal{V}) := \{z: z = F(z,v) \text{ for some } v \in \mathcal{V}\}$.

2. THE FIRST LEMMA

Let X be a reflexive Banach space and suppose one has an evolution system on X, i.e., a family of maps

$$T(t,s): X \to X \qquad (t \geq s) \tag{6}$$
$$T(t,s) \circ T(s,r) = T(t,r).$$

There is no assumption that the maps are to be linear. Given an interval $[\sigma,\tau]$, one has an induced map

$$\hat{T}: X \to X^{[\sigma,\tau]}: \zeta \mapsto \{t \mapsto T(t,\sigma)\zeta: t \in [\sigma,\tau]\} \tag{7}$$

and we will take as our principal hypothesis that, for some $p \geq 1$,

$$\hat{T}: X \to L^p([\sigma,\tau] \to X) \quad \text{is continuous and compact.} \tag{8}$$

We also assume the Lipschitz condition:

$$\|T(\tau,s)\zeta - T(\tau,s)\hat{\zeta}\| \leq M(s)\|\zeta - \hat{\zeta}\| \tag{9}$$

with $M(\cdot)$ in $L^q_+(\sigma,\tau)$ $(1/p + 1/q = 1)$.

LEMMA 1: Let T, $\hat{T}$ be as in (5), (6) and satisfy the hypotheses (7), (8). Then $T(\tau,\sigma)$ is a continuous compact map from X to X.

Proof: We introduce an auxiliary map $A: L^p([\sigma,\tau] \to X) \to X$ defined by

$$Af = (\tau-\sigma)^{-1} \int_\sigma^\tau T(\tau,s) f(s) \, ds. \tag{10}$$

(Note that f(s) — and hence T(τ,s)f(s) — may only be defined a.e. on $[\sigma,\tau]$.) Note that (9) gives

$$\|Af - Ag\|_X \leq (\tau-\sigma)^{-1} \int_\sigma^\tau \|T(\tau,s)f(s) - T(\tau,s)g(s)\|_X \, ds$$
$$\leq (\tau-\sigma)^{-1} \int_\sigma^\tau M(s)\|f(s) - g(s)\|_X \, ds$$
$$\leq (\tau-\sigma)^{-1}\|M\|_{L^q}\|f-g\|_{L^p}.$$

Thus, A is (Lipschitz) continuous.

The proof will continue by showing the factorization

$$T(\tau,\sigma) = A \circ \hat{T}. \tag{11}$$

This will be sufficient to demonstrate compactness of $T(\tau,\sigma)$ since, for any bounded set $\mathcal{S} \subset X$, one has $\mathcal{K} := \hat{T}\mathcal{S}$ precompact in $L^p([\sigma,\tau] \to X)$ by (8) (so $A\mathcal{K}$ is the continuous image of a compact set and so is compact by a standard theorem of topology). Similarly, the composition of continuous maps is continuous.

To see the factorization, we consider the definition of $A\hat{T}\zeta$, using (7), (10) and (6). Thus,

$$A\hat{T}\zeta \; := \; (\tau-\sigma)^{-1} \int_\sigma^\tau T(\tau,t) [T(t,\sigma)\zeta] \; dt$$

$$= \; (\tau-\sigma)^{-1} \int_\sigma^\tau T(\tau,\sigma)\zeta \; dt$$

$$= \; [(t-\sigma)^{-1} \int_\sigma^\tau dt] T(\tau,\sigma)\zeta \; = \; T(\tau,\sigma)\zeta.$$

That this holds for arbitrary ζ gives (11) and so, as noted above, the desired result. □

Note that the logic of this argument rests essentially on the fact that the "causality identity" (6) couples with a general averaging process to give the factorization (11). The particular choice of hypotheses then match some available distributed compactness result with the necessity for being able to show A is continuous in the relevant context. The generality presented here is already greater than that of [4] but clearly does not exhaust the possibilities. In particular, we observe that, once one has obtained an invariant set K for $T(\tau,\sigma)$ and so identified $K_t := \{T(t,\sigma)\zeta: \zeta \in K\}$, an easy modification of the argument shows that (9) need only be verified for $\zeta, \hat{\zeta} \in K_s$ $(\sigma < s < \tau)$ and one need only verify precompactness in $L^p([\sigma,\tau] \to X)$ of $\hat{T}K$ to see that $T(\tau,\sigma)K$ is precompact in X.

2. THE SECOND LEMMA

We begin with the basic abstract lemma and then indicate contexts in which the hypotheses can be verified.

LEMMA 2: Let X be a complete metric space, A an index set, and suppose $F: X \times A \to X$ satisfies

$$d(F(x,\alpha), F(\hat{x},\alpha)) \; \leq \; \vartheta d(x,\hat{x}) \tag{12}$$

for $x, \hat{x} \in X$, $\alpha \in A$ and a fixed $\vartheta < 1$. Suppose, also, that

For any compact $K \subset X$ one has $F(K,A) := \{F(x,\alpha): x \in K, \alpha \in A\}$ precompact in X. (13)

Then the set of fixpoints $\Phi_F := \{x \in X : x = F(x,\alpha) \text{ for some } \alpha \in A\}$ is precompact in X.

Proof: By Banach's Contractive Mapping Principle there is a unique fixpoint $x(\alpha) \in \Phi_F$ such that $x(\alpha) = F(x(\alpha),\alpha)$; further, starting with $x_0(\alpha) := x_0$ (fixed), the recursively defined sequence

$$x_n(\alpha) := F(x_{n-1}(\alpha),\alpha)$$

converges to $x(\alpha)$ with

$$d(x_n(\alpha),x(\alpha)) \leq d(x_1(\alpha),x_0)\vartheta^n/(1-\vartheta). \tag{14}$$

We set $K_0 := \{x_0\}$ and recursively define

$$K_n := F(\overline{K}_{n-1},A) := \{F(x,\alpha): x \in \overline{K}_{n-1}, \alpha \in A\}.$$

By the assumption (13) we have K_n precompact, so $\overline{K}_n$ is compact by induction on n. Note that our definition gives $x_n(\alpha) \in K_n$ for $n = 1,2,\ldots$ and each $\alpha \in A$.

To show Φ_F is precompact we show it is totally bounded: for each $\varepsilon > 0$ there is a finite set $\{\xi_j^\varepsilon: j = 1,\ldots,J_\varepsilon\}$ in X such that $\Phi_F \subset \cup_j \{x \in X: d(x,\xi_j^\varepsilon) < \varepsilon\}$. To see this we proceed as follows: Since $\overline{K}_1$ is compact and each $x_1(\alpha) \in \overline{K}_1$, there is some M such that (14) gives $d(x_n(\alpha),x(\alpha)) \leq M\vartheta^n$ (uniformly in $\alpha \in A$) for $n = 1,2,\ldots$. Thus, given $\varepsilon \in 0$ one has, as $\vartheta < 1$, an $n = n(\varepsilon)$ such that

$$d(x(\alpha),\overline{K}_n) \leq d(x_n(\alpha),x(\alpha)) \leq M\vartheta^{n(\varepsilon)} < \varepsilon/2$$

for each $\alpha \in A$. Since $\overline{K}_n$ is compact it is itself totally bounded and so has a finite cover by balls of radius $\varepsilon/2$, say, with centers $\{\xi_j^\varepsilon\}$; i.e.,

$$\hat{x} \in \overline{K}_n \implies d(\hat{x},\xi_j^\varepsilon) \leq \varepsilon/2 \text{ for some } j = j(\hat{x}).$$

Then $d(x(\alpha),\xi_j^\varepsilon) \leq \varepsilon$ for $j = j(x_n(\alpha))$. $\square$

For the context of (4) we take the index set A to be the set of controls and X to be $C([0,T] \to \mathcal{Z})$ with the norm

$$\|z\|_X := \sup\{e^{-\rho t}\|z(t)\|_z: 0 \leq t \leq T\}$$

for some ρ to be determined. Assuming the Lipschitz condition

$$\|\varphi(s,\zeta,\nu) - \varphi(s,\hat{\zeta},\nu)\|_z \leq M(s)\|\zeta-\hat{\zeta}\|_z \tag{15}$$

(uniformly in ν — where ν is a possible value for $v(s)$), we see that if $M(\cdot) \in L^1(0,T)$ we have

$$\|e^{-\rho t}[F(x,\alpha) - F(\hat{x},\alpha)](t)\|_{\mathcal{J}}$$

$$\leq \ e^{-\rho t} \int_0^t \|\varphi(s,x(s),v(s)) - \varphi(s,\hat{x}(s),v(s))\|_{\mathcal{J}} \ ds$$

$$\leq \int_0^t M(s)e^{-\rho(t-s)} \ ds \ \|x-\hat{x}\|_X,$$

which gives (12) if

$$\vartheta \ = \ \vartheta_\rho \ := \ \sup\{\int_0^t M(s)e^{-\rho(t-s)} \ ds: \ 0 \leq t \leq T\} \ < \ 1.$$

Some elementary real analysis shows $\vartheta_\rho \to 0$ as $\rho \to \infty$, so (15) suffices to ensure that we can define the norm to have (12). Similar considerations apply if in (4) one had $\varphi = \varphi(t,s,y,v)$ and replaced (15) by a similar estimate with suitable $M(t,s)$.

We consider two situations, each of the form (4), in which one obtains (13):

(a) If $v(s) \in \mathcal{V} =$ compact for each s and $\varphi = \varphi(s,y(s),v(s))$ is continuous. Clearly $[y,s] \mapsto y(s)$ is continuous from $Y \times [0,T]$ with $Y = C([0,T] \to \mathcal{J})$ so $y(s)$ takes values in some (fixed — once K is given) compact set. Then $\varphi(s,y(s),v(s))$ takes values in a compact set whence the integral does. This, with the equicontinuity in t which one easily gets, shows $F(K,A)$ precompact by the (generalized) Arzela-Ascoli Theorem.

(b) If the admissible set for $v(\cdot)$ is a ball in $L^p([0,T] \to \mathcal{V})$ with $\mathcal{V}$ a Banach space, suppose one were to have an estimate of the form

$$\varphi(t,s,\eta,v) \in \psi_K(t-s)[c_0 + \|v\|_{\mathcal{V}}]C \tag{16}$$

where $\psi_K \in L_+^q(0,T)$ (depending on the compact set K in (13) — as earlier, we note that this gives $\eta = y(s)$ in a compact subset of $\mathcal{J}$) and C compact in $\mathcal{J}$. Then

$$\int_0^t \varphi(t,s,y(s)v(s)) \ ds \ \in \ \int_0^t \psi_K(t-s)[c_0 + \|v(s)\|_{\mathcal{V}}] \ dsC \subset \lambda C$$

for some fixed λ depending on the size of the ball of admissible $v(\cdot)$, etc. Again, one easily obtains equicontinuity in t and so (13).

This and some other related results were discussed in [3].

REFERENCES

[1] J. Aubin, Un theorème de compacité, C.R. Acad. Sci., Paris 265(1963), 5042-5044.

[2] S. Belbas and T.I. Seidman, Periodic solutions of certain quasi-linear variational inequalities, to appear.

[3] T.I. Seidman, Optimally controlled fixed points, UMBC Math. Research Report 82-13, (1982) 41pp.

[4] T.I. Seidman, Time-dependent solutions of a nonlinear system arising in semiconductor theory, II: boundedness and periodicity, Nonlinear Analysis TMA, to appear.

[5] T.I. Seidman, S.P. Sethi, N.A. Derzko, Dynamics and optimization of a sales-advertising model with population migration, J.O.T.A., to appear.

THE RICCATI EQUATION: WHEN NONLINEARITY REDUCES TO LINEARITY

Andrew Vogt
Department of Mathematics
Georgetown University
Washington, D. C. 20057

Some of us resist the wave of the future and remain true to the old creed that everything is linear. We grant that nonlinear structures are of interest, and that the chaotic behavior of some nonlinearities offers a new perspective on natural phenomena and on the possible limits of mathematical explanation. Some of us might even concede that if everything is linear everything is also nonlinear (as Jim Sandefur observed). Still, the old linear ways have much to recommend them and possess depths not yet plumbed.

One's first naive thought after dealing with the manifestly linear is to deal with the quadratic. Thus one must show that second degree equations are linear. Our ancestors - Riccati, the Bernoullis and company - have done this (at least in some cases).

Recall that the scalar Riccati equation is the (general!) quadratic differential equation in a single scalar variable x:

$$\begin{cases} dx/dt = a(t)x^2 + b(t)x + c(t) \\ \\ x(t_0) = x_0 \end{cases}$$

where we include initial data and take $a(t)$, $b(t)$, $c(t)$ to be continuous functions of t on some interval I containing t_0.

By means of the transformation

$$x = u/v ,$$

this can be transformed into the linear system:

$$\begin{cases} du/dt \ = \ k(t)u + c(t)v \\[2mm] dv/dt \ = \ -a(t)u + (k(t) - b(t))v \\[2mm] u(t_0) \ = \ x_0 \\[2mm] v(t_0) \ = \ 1 \\[2mm] v \text{ nonvanishing.} \end{cases}$$

Here $k(t)$ is an arbitrary continuous function. It can be shown [5, Chapter I] that the first system has a solution if and only if the second does. If (u,v) solves the second, $x = u/v$ solves the first. Singularities of x are related to zeros of v so that the linear system gives information about the original system even with regard to non-existence of solutions.

Naturally the ideas present in the scalar case have been generalized. There is a matrix Riccati equation (Reid [5]):

$$dX/dt \ = \ XA(t)X + B_1(t)X + XB_2(t) + C(t) \tag{0.1}$$
$$n \text{ by } m \quad m \text{ by } m \quad \ n \text{ by } n \quad m \text{ by } n$$

where X is an m by n matrix. This equation has applications in calculus of variations, oscillation theory, and optimal control. Riccati equations in Banach algebras have also been studied by Hille [2, pp. 479-491], Reid [6, section 8.6], and others. In connection with infinite-dimensional control theory, J. L. Lions [4, p. 157 and thereabouts] and others have considered a Riccati equation in $L(H)$, the space of bounded linear transformations of a Hilbert space H into itself.

In the present note we reverse the usual order and emphasis. Instead of passing to a linear system from a quadratic one of a special type, we begin with a linear system in variables u and v, apply the transformation $x = u/v$ or $x = v^{-1}(u)$, and determine when a dynamical system in the new variable x results.

Dynamical systems in x thus obtained, nonlinear though they may appear, are essentially linear. They can be solved by applying linear methods to the original linear systems in u and v. By arriving at systems in this way - familiar Riccati equations and other not so familiar quadratic systems, we not only obtain a better understanding of the Riccati method but we also extend the boundaries of what is linear.

Our treatment below runs as follows. We introduce basic definitions. Then we state and prove a fundamental theorem characterizing Riccati equations. This theorem makes use of algebraic structures which are developed more fully in [7]. Next we demonstrate that the matrix Riccati equations (0.1) fall within this characterization. Then we examine autonomous quadratic equations without linear or constant terms - Riccati equations of the form $dx/dt = a(x)(x)$. We make some general observations about such systems, and display a complete list of these systems for the first nontrivial case - when x is in R^2.

§1. DEFINITIONS

Our first task is to assign a meaning to the transformation

$$x = v^{-1}(u) \quad , \tag{1.1}$$

which links the linear variables u and v with the new variable x. We propose to take the operation between v^{-1} and u to be a bilinear multiplication and to take v^{-1} to be a member of an algebra. More precisely, let x and u be members of a Banach space X and let v be a member of a Banach algebra A contained in the Banach algebra $L(X)$. We assume that A is a closed subalgebra of $L(X)$ sharing the unit of $L(X)$. The transformation (1.1) thus consists in applying the inverse of an invertible element v of A to an element u in the Banach space X. The scalar field is fixed throughout our discussion and, except where otherwise noted, can be taken to be either the reals or the complexes.

Let I be an open interval of the real line R. Let $M : I \longrightarrow L(X \oplus A, X)$, $N : I \longrightarrow L(X \oplus A, A)$, and $F : I \times X \longrightarrow X$ be continuous functions. The topologies on the spaces of linear transformations are the uniform operator topologies associated with the operator norms. (Our requirements that M and N be continuous with respect to these topologies and that F be globally defined in X are rather strong hypotheses!). We use the abbreviation (M,N,F) for functions M, N, and F defined in this manner, and we call (M,N,F) a _triple_ associated with X,A.

A triple (M,N,F) is called a _Riccati_ _triple_ if and only if the following condition is fulfilled:

whenever $t \longmapsto (u(t),v(t))$ is a C^1 function from an open subinterval J of I into $X \oplus A$ such that for all t in J

$$\begin{cases} du/dt = M(t)(u(t),v(t)), \\ dv/dt = N(t)(u(t),v(t)), \text{ and} \\ v(t) \text{ is invertible in A;} \end{cases} \qquad (1.2)$$

then $t \longmapsto x(t) = v(t)^{-1}(u(t))$ is a C^1 function from J into X such that for all t in J

$$dx/dt = F(t,x(t)). \qquad (1.3)$$

If $F : I \times X \longrightarrow X$ is a continuous function and there exists a pair of functions M,N such that (M,N,F) is a Riccati triple, then F is called a _Riccati_ _function_ with respect to the algebra A and (1.3) is called a _Riccati_ _equation_ with respect to A.

§2. THE FUNDAMENTAL THEOREM

Let (M,N,F) be a Riccati triple. Suppose that t_0, x_0, and v_0 are given elements of I, X, A, with v_0 invertible in A. By a familiar existence theorem ([3], p. 67) there is an open interval J in I with t_0 in J and a solution $t \longmapsto (u(t),v(t))$ of (1.2) such that $(u(t_0),v(t_0)) = (v_0(x_0),v_0)$. The invertibility condition may be realized by contracting J if necessary so that $\|v(t) - v_0\| < \|v_0^{-1}\|^{-1}$ for t in J.

Differentiating the equation $x(t) = v(t)^{-1}(u(t))$ by the quotient rule, we obtain:

$$dx/dt = -(v(t)^{-1} \circ (dv/dt) \circ v(t)^{-1})(u(t)) + v(t)^{-1}(du/dt).$$

When we substitute in the right sides of (1.2) and (1.3), apply the operator v(t), and evaluate at t_0, this becomes:

$$v_0(F(t_0,x_0)) = -N(t_0)(v_0(x_0),v_0)(x_0) + M(t_0)(v_0(x_0),v_0).$$

Varying the initial data t_0,x_0,v_0 and the interval J, we arrive at:

$$v(F(t,x)) = -N(t)(v(x),v)(x) + M(t)(v(x),v) \qquad (2.1)$$

for any t in I, any x in X, and any invertible element v of A.

In particular, if $v = e =$ the unit of A and $L(X)$, (2.1) becomes:

$$F(t,x) = -N(t)(x,e)(x) + M(t)(x,e), \qquad (2.2)$$

which demonstrates the quadratic character of the Riccati function F.

Substitution of (2.2) into (2.1) yields

$$-v(N(t)(x,e)(x)) + v(M(t)(x,e)) =$$
$$-N(t)(v(x),v)(x) + M(t)(v(x),v) . \qquad (2.3)$$

Replacing x by λx in (2.3), where λ is a scalar variable, and using the linearity of $N(t)$ and $M(t)$, we may equate like coefficients of powers of λ to obtain:

$$v(N(t)(x,0)(x)) = N(t)(v(x),0)(x) \qquad (2.4i)$$

$$-v(N(t)(0,e)(x)) + v(M(t)(x,0)) = \qquad (2.4ii)$$
$$-N(t)(0,v)(x) + M(t)(v(x),0)$$

$$v(M(t)(0,e)) = M(t)(0,v) . \qquad (2.4iii)$$

Eqns. (2.4) hold for all t in I, all x in X, and all v in A. Here v need not be invertible. Indeed, one may replace v in (2.4) by $e + \lambda v$, where v is any element of A and λ is a varying scalar close enough to 0 so that $e + \lambda v$ is invertible. Then linearity of the three equations in (2.4) allows us to separate out and cancel the terms containing e alone (the terms obtained when $\lambda = 0$) and to cancel the scalar factor λ from the other terms. What remains is (2.4) for arbitrary v.

Now set $a(t) = -N(t)(\cdot,0)$, $b_1(t) = -N(t)(0,e)$, $b_2(t) = M(t)(\cdot,0)$, and $c(t) = M(t)(0,e)$. Evidently a, b_1, b_2, and c are continuous functions from I into L(X,A), A, L(X), and X, respectively. Equations (2.4) may be recast as:

$$v(a(t)(x)(x)) = a(t)(v(x))(x) \qquad (2.5i)$$
$$N(t)(0,v) = b_2(t) \circ v - v \circ b_2(t) - v \circ b_1(t). \qquad (2.5ii)$$
$$M(t)(0,v) = v(c(t)). \qquad (2.5iii)$$

To arrive at a characterization, we introduce two sets:

$$M(X,A) = \{a : a \text{ is in } L(X,A), \text{ and } (v \circ a(x))(x) =$$
$$a(v(x))(x) \text{ for all } x \text{ in } X \text{ and } v \text{ in } A\} \qquad (2.6i)$$

and

$$\iota(A) = \{b : b \text{ is in } L(X), \text{ and } (b \circ a - a \circ b) \text{ is in } A$$
$$\text{for all } a \text{ in } A\}. \qquad (2.6ii)$$

The sets $M(X,A)$ and $\iota(A)$, defined in terms of X and A, are closed linear subspaces of $L(X,A)$ and $L(X)$, respectively, with the inherited operator norm topologies. Additional properties of these sets are discussed in [7]. For present purposes we only need note that by (2.5i) $a(t)$ is in $M(X,A)$ for all t, and by (2.5ii), since $N(t)(0,v)$ and $b_1(t)$ are in A, $b_2(t)$ is in $\iota(A)$ for all t.

Theorem 2.1 (Fundamental Theorem): Let X be a Banach space, and A a closed subalgebra of L(X) containing the unit of L(X). Then a triple (M,N,F) associated with X,A is a Riccati triple if and only if there exist continuous functions
$t \longmapsto a(t), b_1(t), b_2(t), c(t)$ from I into $M(X,A)$, A, $\iota(A)$, and X respectively, such that

$$M(t)(u,v) = b_2(t)(u) + v(c(t)) \qquad (2.7i)$$

$$N(t)(u,v) = -a(t)(u) - v \circ b_1(t) + b_2(t) \circ v - v \circ b_2(t) \qquad (2.7ii)$$
and
$$F(t,x) = a(t)(x)(x) + b_1(t)(x) + b_2(t)(x) + c(t) \qquad (2.7iii)$$

for all t in I, u and x in X, and v in A.

Proof: If (M,N,F) is a Riccati triple, the earlier discussion shows that $a(t)$, $b_1(t)$, $b_2(t)$, and $c(t)$, as defined prior to Eqns. (2.5), have the domain, ranges, and continuity properties in the theorem. Substitution of these functions into (2.2) yields (2.7iii), and similar substitutions making use of (2.5) yield (2.7i) and (2.7ii).

For the converse, suppose that $a(t)$, $b_1(t)$, $b_2(t)$, and $c(t)$ are as in the theorem and that M,N,F are defined by (2.7). It follows easily that $M(t)$, $N(t)$, and $F(t,x)$ lie in $L(X \oplus A,X)$, $L(X \oplus A,A)$, and X, respectively, and are continuous functions of their arguments. If $u(t)$ and $v(t)$ satisfy (1.2) on a subinterval J of I for this M and N, then $t \longmapsto x(t) = v(t)^{-1}(u(t))$ is a C^1 function from J into X and

$$
\begin{aligned}
dx/dt &= d/dt(v^{-1}(u)) \\
&= -(v^{-1} \circ dv/dt \circ v^{-1})(u) + v^{-1}(du/dt) \\
&= -(v^{-1} \circ N(t)(u,v) \circ v^{-1})(u) + v^{-1}(M(t)(u,v)) \\
&= -v^{-1} \circ \{-a(t)(u) - v \circ b_1(t) + b_2(t) \circ v - v \circ b_2(t)\}(x) \\
&\quad + v^{-1}(b_2(t)(u) + v(c(t))) \\
&= (v^{-1} \circ (a(t)(u)))(x) + b_1(t)(x) + b_2(t)(x) + c(t).
\end{aligned}
$$

Since $a(t)(u)(x) = a(t)((v \circ v^{-1})(u))(x) = a(t)(v(x))(x) = (v \circ (a(t)(x)))(x)$, we conclude that $dx/dt = F(t,x(t))$ for all t in J. ∎

Remarks: The general Riccati equation associated with the pair X,A is of the form:

$$dx/dt = a(t)(x)(x) + b(t)(x) + c(t) \tag{2.8}$$

where $t \longmapsto a(t)$, $b(t)$, $c(t)$ are continuous functions from I into $M(X,A)$, $\mathcal{l}(A)$, and X, respectively. The difference between (2.7iii) and (2.8) is accounted for by the fact that $A + \mathcal{l}(A) \subseteq \mathcal{l}(A)$. Accordingly $b_1(t)$ and $b_2(t)$ in (2.7iii) can be compressed into a single element $b(t)$ of $\mathcal{l}(A)$ to yield (2.8).

Since the decomposition $b = b_1 + b_2$ is not unique, there may be distinct linear systems yielding the same Riccati equation. In fact, sometimes $a(t)$ is not uniquely determined from the second degree term $a(t)(x)(x)$ in F (even when A is given and $a(t)$ is assumed to be in $M(X,A)$). Another source of nonuniqueness is duality, mentioned below.

§3. THE STANDARD RICCATI EQUATIONS

The fundamental theorem above provides a method for generating

Riccati equations. We begin with an interval I, a Banach space X, and an algebra $A \subseteq L(X)$. Then we choose continuous functions a, b, and c from I into $M(X,A)$, $\mathcal{L}(A)$, and X. In [7] we obtain characterizations of $M(X,A)$ and $\mathcal{L}(A)$ for various familiar choices of A, thereby arriving at a, b, c, and F. Here we shall content ourselves with demonstrating that the standard Riccati equations (in particular, the finite-dimensional ones of [5]) can be generated in this manner.

Let Y and Z be Banach spaces, and let X be the Banach space $L(Y,Z)$. $L(Y)$ can be regarded as a subalgebra of $L(X)$ acting on X by composition on the right: for v in $L(Y)$ and u in X,

$$v(u) = u \circ v . \tag{3.1}$$

Using this action we identify elements of $L(Y)$ with elements of $L(X)$, and we denote the set of these elements by $L(Y)_r$. $L(Y)_r$ is a closed subalgebra of $L(X)$ sharing its unit. The multiplication on $L(Y)_r$ is opposite to that of $L(Y)$, and the identity map is an isometric anti-homomorphism from $L(Y)$ to $L(Y)_r$.

Proposition 3.1: Let I be an interval, let Y and Z be Banach spaces, and let $X = L(Y,Z)$. Let $F : I \times X \longrightarrow X$ be a continuous function. Then F is a Riccati function with respect to $L(Y)_r$ if there exist continuous functions $t \longmapsto a(t)$, $b_1(t)$, $b_2(t)$, $c(t)$ from I into $L(Z,Y)$, $L(Z)$, $L(Y)$, and $L(Y,Z)$, respectively, such that

$$F(t,x) = x \circ a(t) \circ x + b_1(t) \circ x + x \circ b_2(t) + c(t) \tag{3.2}$$

for all t in I and x in X.

Proof: Let

$$M(t)(u,v) = b_1(t) \circ u + c(t) \circ v$$
$$N(t)(u,v) = -a(t) \circ u - b_2(t) \circ v$$

for t in I, u in X, and v in $L(Y)$. Identifying v and $N(t)(u,v)$ with members of $L(Y)_r$, we see easily that (M,N,F), with F as in (3.2), is a triple associated with $X, L(Y)_r$. If $t \longmapsto (u(t),v(t))$ is a solution to (1.2) on some subinterval J of I, consider $x(t) = v(t)^{-1}(u(t))$ for t in J. Differentiating this function and taking (3.1) into account, we obtain:

$$dx/dt = -v^{-1}(dv/dt)v^{-1}(u) + v^{-1}(du/dt)$$
$$= -x \circ dv/dt \circ v^{-1} + du/dt \circ v^{-1}$$
$$= x \circ (a(t) \circ u + b_2(t) \circ v) \circ v^{-1} + (b_1(t) \circ u + c(t) \circ v) \circ v^{-1}$$
$$= x \circ a(t) \circ x + b_1(t) \circ x + x \circ b_2(t) + c(t) \ .$$

Thus F satisfies (1.3) and is a Riccati function with respect to $X, L(Y)_r$. ∎

This proof makes use only of the definitions in §1 and not of the fundamental theorem. With the aid of the fundamental theorem one may prove a converse (see [7]): all Riccati functions associated with the algebra $L(Y)_r$ are of the form (3.2). This form is precisely the one encountered in the matrix case, and its generalization to infinite dimensions offers no surprises.

Instead of allowing $L(Y)$ to act on $L(Y,Z)$ by composition on the right, we might with equal reason let $L(Z)$ act on $L(Y,Z)$ by composition on the left. It can be shown that functions of the form (3.2) are also Riccati functions with respect to the algebra $L(Z)_\ell$. The associated linear systems are dual to the ones obtained in the proof of Proposition 3.1. Functions of the form (3.2) are the only Riccati functions in finite dimensions or when Y and Z are reflexive Banach spaces. However, in some infinite-dimensional spaces there are additional Riccati functions for the algebra $L(Z)_\ell$ not shared by the algebra $L(Y)_r$. See [7] for a detailed discussion of these matters.

§4. THE EQUATION $dx/dt = a(x)(x)$

For the remainder of our discussion we restrict our attention to Riccati functions with no time dependence and with no linear or constant term in x.

By Theorem 2.1 these Riccati functions have the form $F(t,x) = F(x) = a(x)(x)$, where a is a member of the set $M(X,A)$ defined by (2.6i). Thus for some subalgebra A of $L(X)$, $a : X \longrightarrow A$ is a continuous linear transformation satisfying $(v \circ a(x))(x) = a(v(x))(x)$ for all x in X and v in A.

If we attempt to deemphasize the algebra A, we are led to consider continuous linear transformations a : X $\longrightarrow$ L(X) such that a(y)∘a(x)(x) = a(a(y)(x))(x) for all x, y in X. Going a step further, we suppress the final x on each side of this equation and obtain the following result.

Proposition 4.1: Let X be a Banach space, let e be the unit of L(X), and let a : X $\longrightarrow$ L(X) be a continuous linear transformation with the property that

$$a(a(x)(y)) = a(x)\circ a(y) \tag{4.1}$$

for all x and y in X. Then a(x)(x) is a Riccati function and the system

$$dx/dt = a(x)(x) , \quad x(0) = x_0 \tag{4.2}$$

has the solution

$$x(t) = (e - ta(x_0))^{-1}(x_0) \tag{4.3}$$

in any interval about t = 0 in which e - ta(x_0) is invertible in L(X).

Proof: Let A = {v: v is in L(X) and a(v(y)) = v∘a(y) for all y in X}. Then A is a closed subalgebra of L(X) containing the unit e. By hypothesis a is in M(X,A), and thus F(t,x) = a(x)(x) is a Riccati function with respect to X,A. The linear transformation e - ta(x_0) is in A and so is its inverse when the latter exists. Differentiating (4.3), we obtain:

$$\begin{aligned} dx/dt &= (e - ta(x_0))^{-1}\circ a(x_0)\circ (e - ta(x_0))^{-1}(x_0) \\ &= (e - ta(x_0))^{-1}\circ a(x_0)(x(t)) \\ &= a((e - t\,a(x_0))^{-1}(x_0))(x(t)) \\ &= a(x(t))(x(t)) \ . \end{aligned}$$

Since x(0) = x_0, (4.2) is established. ∎

Some generic examples:

1. Let X = L(Y,Z) and define a : X $\longrightarrow$ L(X) by a(x)(y) = x∘a_0∘y

for x and y in X, where a_0 is a fixed member of $L(Z,Y)$. Then for z in X, $a(a(x)(y))(z) - a(x) \circ a(y)(z) = a(x)(y) \circ a_0 \circ z - x \circ a_0 \circ (a(y)(z)) = (x \circ a_0 \circ y) \circ a_0 \circ z - x \circ a_0 \circ (y \circ a_0 \circ z) = 0$, and (4.1) is fulfilled. The equation $dx/dt = x \circ a_0 \circ x$ is of course a standard Riccati equation.

2. Let $X = L(Y,Z)$ again and define a : $X \longrightarrow L(X)$ by $a(x)(y) = y \circ a_0 \circ x$, reversing the last definition. This map a also satisfies (4.1) and yields the same Riccati equation as the last example.

3. Let X be an arbitrary Banach space and let f be in X^*. Define a : $X \longrightarrow L(X)$ by $a(x)(y) = f(x)y$. Then $a(a(x)(y))(z) - a(x) \circ a(y)(z) = f(x)f(y)z - f(x)f(y)z = 0$ for all x,y,z in X. So $dx/dt = f(x)x$ is a Riccati equation. In fact, this is essentially a special case of (3.2), obtained by taking Y to be the scalar field.

4. Let X be a Banach space. Let $\{\Pi_i\}$ be a finite family of continuous linear projections of X into X satisfying $\Pi_j \circ \Pi_k = \delta_{jk}\Pi_k$ and $\sum \Pi_i = e =$ the identity of $L(X)$. Let $\{f_i\}$ be a family of continuous linear functionals on X with the same indexing as the projections. Define a : $X \longrightarrow L(X)$ by $a(x) = \sum_i f_i \circ \Pi_i(x)\Pi_i$. Then it is easy to verify that this map a satisfies the conditions of the Proposition. The differential equation $dx/dt = a(x)(x)$ in this instance uncouples: it can be regarded as a system of equations of the form $dz_i/dt = f_i(z_i)z_i$, where $z_i = \Pi_i(x)$.

5. Let X be a Banach space, and let T : $X \longrightarrow X$ be a continuous linear transformation satisfying $T^2 = ae + bT$ for some scalars a,b. Let g be a member of X^*. Define a : $X \longrightarrow L(X)$ by $a(x) = g(x)T + (g \circ T(x) - bg(x))e$. That $a(a(x)(y)) = a(x) \circ a(y)$ follows from $a(T(x)) = T \circ a(x)$. For the latter, observe that $a(T(x)) - T \circ a(x) = g(T(x))T + (g \circ T^2(x) - bg(T(x)))e - g(x)T^2 - (g \circ T(x) - bg(x))T = \{g \circ (ae + bT)(x) - bg(T(x))\}e - ag(x)e - bg(x)T + bg(x)T = 0$. Hence, $dx/dt = g(x)T(x) + (g(T(x)) - bg(x))x$ is a Riccati equation.

Let us now specialize to two dimensions. Take the scalars to be the real numbers and X to be R^2. We thus interest ourselves in equations of the form:

$$\begin{cases} dx/dt = ax^2 + bxy + cy^2 , \\ dy/dt = dx^2 + exy + fy^2 , \end{cases} \tag{4.4}$$

where a,..., f are fixed real numbers and (x,y) or $[\begin{smallmatrix}x\\y\end{smallmatrix}]$ denotes a point in R^2. Which equations of this form are Riccati equations with respect to some subalgebra of $L(R^2)$?

In order for (4.4) to be a Riccati system, it must arise from a continuous linear transformation a : $R^2 \longrightarrow L(R^2)$, where

$$a\left(\begin{bmatrix}x\\y\end{bmatrix}\right) = \begin{bmatrix}ax + b_1y & b_2x + cy\\dx + e_1y & e_2x + fy\end{bmatrix} \text{ with } b_1 + b_2 = b, \; e_1 + e_2 = e.$$

The map a necessarily satisfies $a(a([\begin{smallmatrix}x'\\y'\end{smallmatrix}])([\begin{smallmatrix}x\\y\end{smallmatrix}]))([\begin{smallmatrix}x\\y\end{smallmatrix}]) = a([\begin{smallmatrix}x'\\y'\end{smallmatrix}])\circ a([\begin{smallmatrix}x\\y\end{smallmatrix}])([\begin{smallmatrix}x\\y\end{smallmatrix}])$ (cf., the discussion just prior to Proposition 4.1). Equating like coefficients in this equation as $[\begin{smallmatrix}x'\\y'\end{smallmatrix}]$ and $[\begin{smallmatrix}x\\y\end{smallmatrix}]$ vary, we obtain twelve scalar (nonlinear) equations:

$a^2 + b_1d = a^2 + b_2d$

$2ab_2 + b_1e_2 + cd = ab + b_2e$

$b_2{}^2 + ce_2 = ac + b_2f$

$ab_1 + b_1e_1 = b_1a + cd$

$ac + b_2b_1 + b_1f + ce_1 = b_1b + ce$

$b_2c + cf = b_1c + cf$

$da + e_1d = da + e_2d$

$db_2 + e_2a + e_1e_2 + fd = db + e_2e$

$e_2b_2 + fe_2 = dc + e_2f$

$db_1 + e_1{}^2 = e_1a + fd$

$dc + e_2b_1 + 2e_1f = e_1b + fe$

$e_2c + f^2 = e_1c + f^2$.

We apply brute force and solve these equations directly, obtaining the following cases:

Case	Hypotheses	Conclusions	Equations
1.	$c = d = 0$, $b_1 \neq 0 \neq e_2$	$e_1 = b_2 = 0$, $a = e_2$, $f = b_1$	$\begin{cases}dx/dt = (ax + fy)x\\dy/dt = (ax + fy)y\end{cases}$
2.	$c = d = e_2 = 0$, $b_1 \neq 0$, $b_2 \neq f$	$a = e_1 = b_2 = 0$, $f = b_1$	$\begin{cases}dx/dt = (fy)x\\dy/dt = (fy)y\end{cases}$
3.	$c = d = e_2 = 0$, $b_1 \neq 0$, $f = b_2$	$e_1 = 0$, $f = b_1$	$\begin{cases}dx/dt = ax^2 + 2fxy\\dy/dt = fy^2\end{cases}$
4.	$c = d = b_1 = 0$, $b_2 = 0 \neq f$, $0 \neq e_1$	$a = e_1 = e_2$	$\begin{cases}dx/dt = ax^2\\dy/dt = 2axy + fy^2\end{cases}$
5.	$c = d = b_1 = 0$, $b_2 = e_1 = 0 \neq f$	$e_2 = 0$	$\begin{cases}dx/dt = ax^2\\dy/dt = fy^2\end{cases}$

6. $c = d = b_1 = 0$, $a =$ nonzero one of $\begin{cases} dx/dt = (ax)x \\ dy/dt = (ax)y \end{cases}$
    $b_2 = f = 0$, e_1 or e_2
    exactly one of
    e_1 and e_2 nonzero

7. $c = d = b_1 = 0$, $\begin{cases} dx/dt = ax^2 \\ dy/dt = 0 \end{cases}$
    $b_2 = f = 0$, both
    of e_1 and e_2 zero

8. $c = d = b_1 = 0$, $a = e_1 = e_2$ $\begin{cases} dx/dt = ax^2 \\ dy/dt = 2axy \end{cases}$
    $b_2 = f = 0$,
    neither e_1 nor
    e_2 zero

9. $c = d = b_1 = 0$, $a = e$, $f = b_2$, $\begin{cases} dx/dt = (ax + fy)x \\ dy/dt = (ax + fy)y \end{cases}$
    $b_2 \neq 0$ $e_2 = 0$

Beyond these is one additional case, Case 10, when c and d are not both zero. Then $e_1 = e_2 = e/2$ and $b_1 = b_2 = b/2$, while the remaining equations reduce to:

$be = 4cd$
$2ae + 4df = 2bd + e^2$ or $(a - e/2)(e/2) + (f - b/2)d = 0$
$4ac + 2bf = 2ce + b^2$ or $(a - e/2)c + (f - b/2)(b/2) = 0.$

Since the vectors $(c,b/2)$ and $(e/2,d)$ are dependent but not both zero, there exists a vector (u,v) and scalars λ, σ, and τ such that:

$u^2 + v^2 = 1$
$(c,b/2) = \lambda(u,v)$
$(e/2,d) = \sigma(u,v)$
$(f - b/2, e/2 - a) = \tau(u,v).$

The equations for Case 10 resulting from these expressions are:

$$\begin{cases} dx/dt = (\sigma u - \tau v)x^2 + 2\lambda vxy + \lambda uy^2 \\ dy/dt = \sigma vx^2 + 2\sigma uxy + (\lambda v + \tau u)y^2 . \end{cases} \tag{4.5}$$

The argument so far shows that the equations corresponding to Cases 1 through 10 are the only candidates for Riccati equations in R^2. In fact, all of these equations are Riccati equations. This is apparent for Cases 1, 2, 5, 6, 7, and 9 since they are instances of generic examples 3 and 4.

The remaining cases are instances of generic example 5. For Case 3, let the map $a : R^2 \longrightarrow L(R^2)$ be given by:

$$a\left(\begin{bmatrix} x \\ y \end{bmatrix}\right) = (x) \begin{bmatrix} 0 & f \\ 0 & -a \end{bmatrix} + (ax + fy) \begin{bmatrix} 1 & 0 \\ 0 & 1 \end{bmatrix}.$$

Letting $g(x,y) = x$ and $T(x,y) = (fy, -ay)$, we see that $T^2 = -aT$, from which it follows that $(ax + fy) = g \circ T(x,y) - (-a)g(x,y)$. Thus the conditions in example 5 are fulfilled. Similar arguments apply to Cases 4 and 8. For Case 10, define

$$a\left(\begin{bmatrix} x \\ y \end{bmatrix}\right) = (vx + uy) \begin{bmatrix} 0 & \lambda \\ \sigma & \tau \end{bmatrix} + ((\sigma u - \tau v)x + \lambda vy) \begin{bmatrix} 1 & 0 \\ 0 & 1 \end{bmatrix}.$$

Letting $g(x,y) = vx + uy$ and $T(x,y) = (\lambda y, \sigma x + \tau y)$, we see that $T^2 = \lambda \sigma e + \tau T$, from which it follows that $(\sigma u - \tau v)x + \lambda vy = g \circ T(x,y) - \tau g(x,y)$. Hence (4.5) is a Riccati equation of the Example 5 type.

We thus have proved the following.

<u>Proposition 4.2</u>: The Riccati equations associated with the real Banach space R^2 are precisely the equations of Cases 1 through 9 above and equations (4.5).

Inspection of these equations shows that the system $dx/dt = y^2$, $dy/dt = x^2$ is not a Riccati system. Nor are the Volterra-Lotka equations $dx/dt = (a - by - cx)x$, $dy/dt = (dx - e - fy)y$, with a, b, d, and e positive and c and f nonnegative. Note that by the fundamental theorem if a quadratic equation without linear or constant terms is not a Riccati equation, it will not be a Riccati equation when such terms are included. See [1] or the more recent [8] for a qualitative survey of all quadratic systems in R^2, including ones with linear terms.

We conclude with a brief examination of the structure of the Riccati equations discovered above.

As already noted, Cases 1, 2, 5, 6, 7, and 9 have the straightforward structure described in generic examples 3 and 4. Consider Case 3, which is also representative of Cases 4 and 8:

$$dx/dt = ax^2 + 2fxy, \quad dy/dt = fy^2 \; . \tag{4.6}$$

If a is nonzero, we can make the transformation $x_1 = x + y(f/a)$, $y_1 = y/a$ and obtain:

$$dx_1/dt = ax_1^2, \quad dy_1/dt = afy_1^2 \; .$$

Thus the system uncouples. If a is zero, this does not happen. The transformation $x_1 = x$, $y_1 = fy$, with $a = 0 \neq f$, converts (4.6) to:

$$dx_1/dt = 2x_1y_1, \quad dy_1/dt = y_1^2 \; . \tag{4.7}$$

Here one variable is governed by a scalar Riccati equation, the other is coupled to it.

Eqns. (4.5) of Case 10 have a variable structure. If $(\tau/2)^2 + \sigma\lambda = \gamma^2 > 0$ and $0 \neq \lambda$, the transformation $x_1 = (1/2)\{x/\lambda + y/\gamma - \tau x/2\lambda\gamma\}$, $y_1 = (1/2)\{x/\lambda - y/\gamma + \tau x/2\lambda\gamma\}$ converts (4.5) into:

$$\begin{cases} dx_1/dt = x_1^2\{2(\tau^2/4 + \sigma\lambda)u + 2\gamma(\tau u/2 + \lambda v)\} \\ dy_1/dt = y_1^2\{2(\tau^2/4 + \sigma\lambda)u - 2\gamma(\tau u/2 + \lambda v)\} \; . \end{cases}$$

If $\lambda = 0 \neq \tau$, we get a different uncoupling. With $x_1 = \sigma x/\tau + y$ and $y_1 = -x/\tau$, (4.5) becomes:

$$dx_1/dt = \tau u x_1^2, \quad dy_1/dt = (\tau^2 v - \sigma\tau u)y_1^2 \; .$$

When $(\tau/2)^2 + \sigma\lambda = 0 \neq (\tau/2)^2 + \lambda^2 = \rho$, the substitution $x_1 = \lambda x/\rho + \tau y/2\rho$, $y_1 = -\tau x/2\rho + \lambda y/\rho$ converts (4.5) into:

$$\begin{cases} dx_1/dt = 2x_1y_1(\tau u/2 + \lambda v)(\lambda - \sigma) + y_1^2(-\tau v/2 + \lambda u)(\lambda - \sigma) \\ dy_1/dt = y_1^2(\tau u/2 + \lambda v)(\lambda - \sigma). \end{cases}$$

This system is of the form:

$$dx_1/dt = 2ax_1y_1 + fy_1^2, \quad dy_1/dt = ay_1^2 \; . \tag{4.8}$$

If a is nonzero, the substitution $x_2 = ax_1 + fy_1$, $y_2 = y_1$ yields:

$$dx_2/dt = 2ax_2y_2, \quad dy_2/dt = ay_2^2 \; ,$$

which is a variation of Case 3 treated earlier. If $a = 0 \neq f$ in (4.8), the substitution $x_2 = x_1$, $y_2 = fy_1$ yields:

$$dx_2/dt = y_2{}^2, \quad dy_2/dt = 0 \ . \tag{4.9}$$

This is a new type of equation: y_2 is static, but the evolution of x_2 depends on the value of y_2. If $\lambda = \tau = 0$ in (4.5), (4.5) has a form which is a variation of (4.8) and the considerations above also apply.

Finally, suppose that $(\tau/2)^2 + \sigma\lambda = -\gamma^2 < 0$ in (4.5). Then the substitution $x_1 = x/\lambda$, $y_1 = -\tau x/2\gamma\lambda + y/\gamma$ converts (4.5) into:

$$\begin{cases} dx_1/dt = -\gamma^2 ux_1{}^2 + 2\gamma(\tau u/2 + \lambda v)x_1 y_1 + \gamma^2 uy_1{}^2 \\ dy_1/dt = -\gamma(\tau u/2 + \lambda v)x_1{}^2 - 2\gamma^2 ux_1 y_1 + \gamma(\tau u/2 + \lambda v)y_1{}^2 \ . \end{cases}$$

These equations are of the form:

$$\begin{cases} dx_1/dt = -ax_1{}^2 + 2bx_1 y_1 + ay_1{}^2 \\ dy_1/dt = -bx_1{}^2 - 2ax_1 y_1 + by_1{}^2 \ . \end{cases} \tag{4.10}$$

If a is nonzero, the substitution $x_2 = ax_1$, $y_2 = ay_1$ leads to equations of the form:

$$\begin{cases} dx_2/dt = -x_2{}^2 + 2px_2 y_2 + y_2{}^2 \\ dy_2/dt = -px_2{}^2 - 2x_2 y_2 + py_2{}^2 \end{cases}$$

with $p = b/a$. Yet another substitution, this time $x_3 = py_2 - x_2$ and $y_3 = px_2 + y_2$, yields:

$$dx_3/dt = x_3{}^2 - y_3{}^2, \quad dy_3/dt = 2x_3 y_3 \ . \tag{4.11}$$

These equations can be recognized as the real and imaginary parts of the complex Riccati equation in one dimension: $dz/dt = z^2$ with $z = x_3 + iy_3$. If $a = 0 \neq b$ in (4.10), the substitution $x_3 = by_1$, $y_3 = bx_1$ converts (4.10) into (4.11) by a different route.

<u>Summary</u>: Riccati equations in R^2 of the form $dx/dt = a(x)(x)$ are thus of the following types: uncoupled scalar equations (relative to a suitable basis), equations based on a linear functional times the identity transformation (see generic example 3), two types of degen-

eracies represented by Eqns. (4.7) and (4.9), and the real and imaginary parts of a single complex Riccati equation.

References

[1] W. A. Coppel, A survey of quadratic systems, J. Diff. Eqns. $\underline{2}$ (1966), 293-304.

[2] E. Hille, Lectures on Ordinary Differential Equations, Addison-Wesley, Reading, Mass., 1969.

[3] E. Hille and R. S. Phillips, Functional Analysis and Semigroups, Amer. Math. Soc., Providence, R. I., 1957.

[4] J. L. Lions, Contrôle optimale de systèmes gouvernés par des équations aux dérivées partielles, Études Mathématiques, Dunod and Gauthier-Villars, Paris, 1968.

[5] W. T. Reid, Riccati Differential Equations, Academic Press, New York, 1972.

[6] W. T. Reid, Sturmian Theory for Ordinary Differential Equations, Springer, New York, 1980 (prepared for publication by J. Burns, T. Herdman, and C. Ahlbrandt).

[7] A. Vogt, A generalization of the Riccati equation (available from the author).

[8] Yan Qian Ye, Some problems in the qualitative theory of ordinary differential equations, J. Diff. Eqns. $\underline{46}$ (1982), 153-164.

Lecture Notes in Mathematics

Edited by A. Dold and B. Eckmann

918

Zbigniew Semadeni

Schauder Bases in Banach Spaces of Continuous Functions

Springer-Verlag
Berlin Heidelberg New York 1982

Author

Zbigniew Semadeni
Instytut Matematyczny, Polskiej Akademii Nauk
ul. Sniadeckich 8, skr. pocztowa 137, 00-950 Warszawa

AMS Subject Classifications (1980): 41 A 15, 46 B XX, 46 B 15, 46 B 25, 46 B 30, 46 E 15

ISBN 3-540-11481-5 Springer-Verlag Berlin Heidelberg New York
ISBN 0-387-11481-5 Springer-Verlag New York Heidelberg Berlin

Printing and binding: Beltz Offsetdruck, Hemsbach/Bergstr.
2141/3140-543210

Preface

Writing these lecture notes I had several goals in mind:

1) To present (in Chapters 1 and 4) theoretical aspects of Schauder bases in spaces $C(X)$ and $C_o(X)$, in the spirit of the isometric theory of Banach spaces, including a construction of a basis in a generic separable space $C_o(X)$.

2) To show (in Chapter 2) constructions of some classical bases of functions of a single variable. This includes the Faber-Schauder system in $C(I)$, where $I = [0,1]$, the Franklin system in $C(I)$, and the Haar system as a basis in $C(2^\omega)$.

3) To give an introduction to the work of Z.Ciesielski, S.Schoenefeld and others on spline bases in various spaces of differentiable functions and to the work of S.V.Bočkariëv on bases in certain Banach spaces of analytic functions (Section 3.9).

4) To present (in Section 4.8) results concerning nonexistence of Schauder bases with some special properties; typical here is the celebrated theorem of A.M.Olevskiĭ that no uniformly bounded orthonormal system can be a basis in $C(I)$.

5) To show (in Chapter 3) constructions of bases in spaces $C(I^d)$ where $1 < d < \infty$. For two (or perhaps even four) decades it has been known how to construct such bases, consisting of certain spline functions; in these notes we consider bases of splines of degree 1, that is, of functions continuous on I^d and affine on each simplex of some triangulation of I^d. In spite of the regularity of the construction of these bases and their nice properties, they have not yet attracted people working in numerical methods. An obvious reason is that not only these functions are not differentiable, but also the rate of convergence is worse than with higher-order splines. Yet, another reason may be that in the existing literature the descriptions of the basis functions of several variables are geometrical, outlined only, without explicit formulas for the functions. This is why the presentation of bases in $C(I^d)$ is so detailed here. Also, though arranging the basis functions into a single sequence is quite easy, it is technically not convenient to work with. Therefore, I prefer to use dyadic numbers rather than natural numbers as indices.

These lecture notes are written in a text-book style and are self-contained (with the exception that a few proofs, technical and easily available in the literature, are omitted). The proofs presented here are detailed and necessary auxiliary material is included. In the text there are several references to "Banach spaces of continuous functions", vol.1, by the author, marked with # (e.g. # 8.6.1).

Each section is accompanied by exercises intended to help the reader to understand the concepts and methods involved. The exercises include counter-examples, corollaries and sidelights on the preceding material; most of them are easy. They are followed by Notes with additional information and references. The bibliography includes also material related to bases in C(X) but not discussed in these lecture notes. The reader should be aware that the number of papers on general properties of bases in C(X) is relatively small whereas there is an abundance of papers dealing with expansions with respect to special, classical bases (like the Haar system). I have tried to select papers relevant to properties of bases in the supremum norm.

A part of these lecture notes were written in the summer 1981 at the Institut für Angewandte Mathematik der Universität Bonn. I particularly thankful to Prof.Diethard Pallaschke for his help there. I am also indebted to Professors Zbigniew Ciesielski, Tadeusz Figiel, Aleksander Pełczyński and Józef Siciak for providing me with a lot of information on the subject matter of these notes.

Zbigniew Semadeni

Warsaw, November 1981

CONTENTS

Preface

Chapter 1

GENERAL PROPERTIES OF BASES IN C(X)

The purpose of this chapter is to prove general theorems on bases in $C(X)$ for an arbitrary compact metric space X. More advanced results in this line will be shown in Chapter 4.

1.1. Preliminaries on bases in Banach spaces

1.1.1. Definitions. A <u>Schauder basis</u> in an infinite-dimensional Banach space F is a sequence (e_n) of elements of F satisfying the following condition: for every f in F there is a unique sequence (a_n) of scalars such that

$$(1) \qquad f = \sum_{n=1}^{\infty} a_n e_n, \quad \text{i.e.,} \quad \left\| f - \sum_{n=1}^{m} a_n e_n \right\| \longrightarrow 0 \quad \text{as} \quad m \longrightarrow \infty.$$

A <u>basic sequence</u> in F is a sequence which is a Schauder basis in its closed linear span. If F is finite-dimensional and $\dim F = m$, then a Schauder basis in F is just any vector-space basis and (1) is to be understood as $f = \sum_{n=1}^{m} a_n e_n$; if F is 0-dimensional, the empty set is a basis. In these lecture notes the term <u>basis</u> will always mean a Schauder basis.

It should be understood that a basis is a sequence $(e_n)_{n \in \mathbb{N}}$, with a prescribed order of terms, and not merely a set $\{e_n\}_{n \in \mathbb{N}}$. Yet, we shall often use indexes other than natural numbers; if it is written $(e_v)_{v \in V}$ or $\sum_{v \in V} a_v e_v$, say, it is to be meant $(e_{v_n})_{n \in \mathbb{N}}$ and $\sum_{n=1}^{\infty} a_{v_n} e_{v_n}$, the order in V being defined separately (sometimes it follows from the context).

The coefficients a_j depend on f; let $e_n^*(\sum_{j=1}^{\infty} a_j e_j) = a_n$. Thus, $f = \sum_{n=1}^{\infty} e_n^*(f) e_n$. From the uniqueness of (1) it follows that e_n^* is a linear functional on F. Moreover,

$$(2) \qquad e_n^*(e_m) = \delta_{nm}.$$

The partial sums of a given f will be denoted by $s_n = \sum_{j=1}^{n} a_j e_j$. If f is a variable, the partial sum is a linear operator

(3) $S_n : F \longrightarrow F$,   where $S_n(f) = \sum\limits_{j=1}^{n} e_j^*(f)e_j.$

Obviously the range of S_n is n-dimensional and $S_nS_m = S_mS_n = S_n$ if $m \geq n$.
A basis (e_n) in F is <u>normalized</u> iff $\| e_n \| = 1$ for each n, and is <u>seminormalized</u> iff

(4) $0 < \inf \| e_n \| \leq \sup \| e_n \| < \infty.$

Clearly, if (e_n) is any basis, then $e_n / \| e_n \|$ is a normalized basis.

   1.1.2. <u>Lemma</u>. Let (e_n) be a basis in F. Then for every f in F of the form (1) the number

$$\||\, f \,\|| = \sup_{n} \| \sum_{j=1}^{n} a_j e_j \|$$

is finite and $\|| \cdot \||$ is a norm on F equivalent to the initial norm on F, i.e., there is a constant $c > 0$ such that $\| f \| \leq \||\, f \,\|| \leq c \| f \|$ for f in F.

   Proof. We shall first show that $(F, \||\cdot\||)$ is complete.
Let $f_m \in F$ and $\||\, f_p - f_q \,\|| \longrightarrow 0$ as $p,q \longrightarrow \infty$. Choose $\varepsilon > 0$. Then there is an n_o such that

(5) $\| S_n(f_p) - S_n(f_q) \| < \varepsilon$ for $p,q \geq n_o$ and for $n = 1,2,\ldots$

Hence $\| (e_n^*(f_p) - e_n^*(f_q))e_n \| < 2\varepsilon$ and $e_n^*(f_p)$, for each fixed n, is a Cauchy sequence of numbers, convergent to some number b_n as $p \longrightarrow \infty$. Consequently,

$$S_n(f_p) \longrightarrow \sum_{j=1}^{n} b_j e_j, \quad \text{as} \quad p \longrightarrow \infty.$$

Consider the inequality (5) for fixed ε, p, n, letting $q \longrightarrow \infty$. Then

(6) $\| S_n(f_p) - \sum\limits_{j=1}^{n} b_j e_j \| \leq \varepsilon$ for $p \geq n_o$, $n = 1,2,\ldots$.

Let $n < m$. Then

$$\| S_m(f_p) - S_n(f_p) - \sum_{j=n+1}^{m} b_j e_j \| \leq 2\varepsilon \quad \text{for} \quad p \geq n_o.$$

Since (e_n) is a basis, for each p there is an index m_p such that

$m > n > m_p$ implies $\| S_m(f_p) - S_n(f_p) \| < \varepsilon.$

Hence $\| \Sigma_{j=n+1}^{m} b_j e_j \| < 3\varepsilon$ for $m > n > m_p$. Thus, $\Sigma_{j=1}^{\infty} b_j e_j$ is convergent to some f in F. Inequality (6) means now that

$$\| S_n(f_p) - S_n(f) \| \le \varepsilon \quad \text{for} \quad p \ge n_o \text{ and for } n = 1,2,\ldots .$$

Thus $\||f_p - f\|| \le \varepsilon$ for $p \ge n_o$. We have shown that $(F, \||\cdot\||)$ is complete.

Consequently, by Banach's theorem, the identity map $(F, \||\cdot\||) \longrightarrow (F, \||\cdot\||)$ has a continuous inverse. $\Box$

1.1.3. <u>Theorem</u>. Let (e_n) be a sequence in F. Then the following conditions are equivalent:

(i) (e_n) is a basic sequence,

(ii) $e_n \ne 0$ for all n and there exists a constant c such that for each sequence (a_n) of scalars and each $n < m$ we have

(7) $\| \sum_{j=1}^{n} a_j e_j \| \le c \| \sum_{j=1}^{m} a_j e_j \|$,

(iii) $e_n \ne 0$ and $\underset{\gamma>0}{\exists} \ \underset{n,k}{\forall} \ \underset{(a_j)}{\forall} \ \| \sum_{j=1}^{n} a_j e_j \| = 1 \Rightarrow \| \sum_{j=1}^{n+k} a_j e_j \| \ge \gamma.$

Proof. (i) $\Rightarrow$ (ii) follows from 1.1.2. Indeed, if (e_n) is a basis in its span H, then for each h in H we have

$$\| S_n(h) \| \le \||S_n(h)\|| \le \||S_m(h)\|| \le c\| S_m(h) \| .$$

Now, assume (ii). Let H_1 be the set of all elements of the form $\Sigma_{n=1}^{\infty} a_n e_n$. Such an expansion must be unique, because if $\Sigma_{n=1}^{\infty} a_n e_n = 0$, then $\|\Sigma_{j=1}^{n} a_j e_j \| \le c\| \Sigma_{j=1}^{m} a_j e_j \|$ and letting $m \longrightarrow \infty$ we have $\Sigma_{j=1}^{n} a_j e_j = 0$ for $n = 1,2,\ldots$ and hence, by induction, $0 = a_1 = a_2 = \ldots$ The proof that H_1 is closed is similar to that of 1.1.2. Thus, we get (i).

The equivalence (ii) $\Leftrightarrow$ (iii) with $\gamma = c^{-1}$ is obvious. $\Box$

1.1.4. <u>Proposition</u>. Let (e_n) be a basis in F. Then (3) is a bounded linear projection operator onto $\text{span}(e_1,\ldots,e_n)$ and

(8) $\sup_n \| S_n \| < \infty.$

Moreover, each $e_n(\cdot)$ is a bounded linear functional. If (e_n) is seminormalized, then

$$(9) \qquad \sup_n \| e_n^* \| < \infty.$$

Proof. Continuity of each S_n follows from 1.1.2. Inequality (8) is a consequence of the Banach-Steinhaus theorem. Finally, (8) implies (9), as

$$S_n(f) - S_{n-1}(f) = e_n^*(f) e_n \quad \text{and} \quad \| e_n^* \| \le 2 \| e_n \|^{-1} \sup \| S_n \| . \quad \square$$

1.1.5. <u>Definitions</u>. The number $\sup \| S_n \|$ is the <u>basis constant</u> of (e_n). A basic sequence (e_n) is <u>monotone</u> iff for any scalars $a_1, a_2, \ldots$

$$(10) \qquad \| \sum_{j=1}^{n} a_j e_j \| \le \| \sum_{j=1}^{n+1} a_j e_j \| \qquad \text{for} \quad n = 1, 2, \ldots .$$

Equivalently, (e_n) is monotone iff the basis constant (with respect to its closed span) is 1; also, (e_n) is monotone iff $\| S_n \| = 1$ for $n = 1, 2, \ldots,$ i.e., $\| S_n(f) \| \le \| f \|$ for each f in F.

1.1.6. <u>Proposition</u>. Let (e_n) be a basis in F. Then (e_n^*) is a basic sequence in F^*. Moreover, the basis constants of (e_n) and of (e_n^*) are equal.

Proof. Let $S_n^* : F^* \longrightarrow F^*$ be the operator adjoint to $S_n : F \longrightarrow F$. Clearly, $S_n^* S_m^* = S_m^* S_n^* = S_n^*$ for $n \le m$ and

$$S_n^* (\sum_{j=1}^{m} b_j e_j^*) = \sum_{j=1}^{n} b_j e_j^* \qquad \text{for} \quad n \le m.$$

Hence $\| \sum_{j=1}^{n} b_j e_j^* \| \le c \| \sum_{j=1}^{m} b_j e_j^* \|$, where $c = \sup \| S_n^* \| = \sup \| S_n \|$. By 1.1.3, (e_n^*) is a basic sequence. $\square$

1.1.7. <u>Proposition</u>. A Banach space F has a monotone basis if and only if there exists a sequence of projections $S_n : F \longrightarrow E_n$ such that

(α) $\| S_n \| = 1$,

(β) $\dim E_n = n$, $E_n = S_n(F)$,

(γ) $E_n \subset E_{n+1}$ for $n = 1, 2, \ldots$

(δ) UE_n is dense in F.

Proof. Suppose that projections $S_n : F \longrightarrow E_n$ satisfy $(\alpha)-(\delta)$. Let $e_1 \in E_1$, $\| e_1 \| = 1$. Since the restriction $\tilde{S}_n$ of S_n to E_{n+1} $(n \geq 1)$ is a projection $\tilde{S}_n : E_{n+1} \longrightarrow E_n$, there exists an e_{n+1} in E_{n+1} such that $\| e_{n+1} \| = 1$ and $S_n(e_{n+1}) = 0$. Hence, for any numbers $a_1, \ldots, a_{n+1}$,

$$\| \sum_{k=1}^{n} a_k e_k \| = \| S_n (\sum_{k=1}^{n+1} a_k e_k) \| \leq \| \sum_{k=1}^{n+1} a_k e_k \| . \quad \square$$

1.1.8. <u>Theorem</u> (Krein-Milman-Rutman). Let (e_n) be a basis in a Banach space F, with a basis constant c. Let (h_n) be another sequence in F such that

$$\gamma = \sum_{n=1}^{\infty} \| e_n^* \| \, \| e_n - h_n \| < 1.$$

Then (h_n) is also a basis in F with a basis constant $c' \leq c(1 + \gamma)(1 - \gamma)^{-1}$.

Proof. If $f \in F$, then $\sum_{n=1}^{\infty} | e_n^*(f) | \, \| e_n - h_n \| \leq \gamma \| f \| < \infty$. Consequently, the series

$$\Phi f = \sum_{n=1}^{\infty} e_n^*(f)(e_n - h_n)$$

is convergent and $\Phi f \in F$. This defines a linear operator $\Phi : F \longrightarrow F$. Moreover, $\| \Phi \| \leq \gamma < 1$. Consequently, the operator $\Psi = I - \Phi$ is invertible, Ψ^{-1} maps F onto F and $\| \Psi \| \cdot \| \Psi^{-1} \| \leq (1 + \gamma)(1 - \gamma)^{-1}$. Since

$$\Phi(e_m) = \sum_{n=1}^{\infty} e_n^*(e_m)(e_n - h_n) = e_m - h_m,$$

we get $\Psi(e_n) = h_n$. Thus, Ψ is a linear homeomorphism which maps e_n onto h_n; therefore (h_n) must also be a basis and its biorthogonal sequence of functionals h_n^* is given by $h_n^* = (\Psi^{-1})^*(e_n^*)$. $\square$

1.1.9. <u>Corollary</u>. Let F be a Banach space with a basis (e_n) having a basis constant c, and let H be a dense linear subset of F. Then for every $\varepsilon > 0$ there exists a basis (h_n) in F such that $h_n \in H$ and the basis constant of (h_n) is less than $c + \varepsilon$. $\square$

1.1.10. <u>Definitions</u>. Let F be a Banach space, $e_n \in F$, $e_n^* \in F^*$. The pair (e_n), (e_n^*) of sequences is a <u>biorthogonal</u> system iff it satisfies (2). We shall also say that the sequence (e_n^*) is biorthogonal to (e_n).

For any biorthogonal sequences (e_n), (e_n^*) we have two sequences of operators: operators $S_n : F \longrightarrow F$ defined by formula (3) and operators

$$(11) \quad T_n : F^* \longrightarrow F^* \quad \text{where} \quad T_n(\xi) = \sum_{k=1}^{n} \xi(e_k) e_k^* \quad \text{for} \quad \xi \quad \text{in} \quad F^*.$$

1.1.11. <u>Proposition</u>. Let (e_n), (e_n^*) be a biorthogonal system. Then:

(a) $S_n f \longrightarrow f$ for each f in $E = \text{span}\{e_k\}_{k \in \mathbb{N}}$.

(b) If $\sup \| S_n \| < \infty$, then (e_n) is a basic sequence.

(c) If $(e_n^*)_{n \in \mathbb{N}}$ separates points of F and $\| g - S_n f \| \longrightarrow 0$ for some $f, g \in F$, then $f = g$.

(d) Operators T_n are adjoint to S_n; hence $\| T_n \| = \| S_n \|$.

(e) If $\sup_n \| T_n(\xi) \| < \infty$ for ξ in F^*, then (e_n) is a basic sequence.

Proof. (a), (b) and (c) are obvious; (d) follows from the identity

$$(T_n \xi) f = \sum_{k=1}^{n} \xi(e_k) e_k^*(f) = \xi(S_n f).$$

Finally, (e) follows from the Banach-Steinhaus theorem. ☐

1.1.12. <u>Definition</u>. Let F, G be Banach spaces and $f_n \in F$, $g_n \in G$ $(n = 1, 2, \ldots)$. The sequences (f_n) and (g_n) are said to be <u>equivalent</u> iff there exists a constant $c > 0$ such that for each n and each finite sequence of scalars $a_1, \ldots, a_n$

$$(12) \quad c^{-1} \left\| \sum_{k=1}^{n} a_k f_k \right\|_F \leq \left\| \sum_{k=1}^{n} a_k g_k \right\|_G \leq c \left\| \sum_{k=1}^{n} a_k f_k \right\|_F.$$

1.1.13. <u>Exercises</u>. (A) Show that every basis is monotone with respect to the norm $||| \cdot |||$.

(B) Show that the best constant c in 1.1.2 is equal to $\sup \| S_n \|$ in 1.1.4 and is equal to the best constant c in 1.1.3.

(C) Show that the functions $e_n(t) = t^n$, $n = 0, 1, 2, \ldots$ in $C[a,b]$ do not satisfy 1.1.3 (ii).

(D) Let (e_n) be a seminormalized basis. If $g_n = e_1 + \ldots + e_n$ are bounded, then (g_n) is a basis.

(E) Let (e_n) be a basis and let η_n be a fixed sequence, $\eta_n = \pm 1$. Then $(\eta_n e_n)_{n \in \mathbb{N}}$ is also a basis (cf. 1.5.2).

(F) Let $E_1 \subset E_2 \subset \ldots$ be a sequence of subspaces of F such that $\dim E_n = n$ and there exist projections $S_n : E_{n+1} \longrightarrow E_n$ with $\| S_n \| = 1$, $n = 1, 2, \ldots$ Show that there exist S_n satisfying the assumptions of 1.1.7.

(G) Let F be a Banach space non-isomorphic to $C_o(\mathbb{N})$ and let (e_n) be a normalized basis in F. (a) Show that there exists a sequence (a_n) of numbers tending to O such that $\Sigma a_n e_n$ diverges. (b) The biorthogonal sequence (e_n^*) is not equivalent (in the sense of 1.1.12) to the unit-vector basis of l^1.

1.1.14. <u>Notes</u>. The above exposition is based on Banach ['32], Day ['73] and Lindenstrauss and Tzafiri ['77], ['80]. A very detailed presentation with bibliographical comments is given by I.Singer ['70], ['81]. 1.1.7 was observed by S.Mazur (see C.Bessaga ['57], E.Michael and A.Pełczyński ['66]).

1.2. A survey of bases in C(X)

1.2.1. C(X) will denote the space of real-valued continuous functions on a compact space X with the norm $\| f \| = \sup\{ |f(x)| : x \in X \}$. Since a space C(X) with a basis must be separable, we shall always assume that X is metrizable (#7.6.2).

If (e_n) is a basis in C(X), then the same real-valued functions (e_n) form a basis in the space $C(X, \mathbb{C})$ of complex-valued continuous functions on X.

We list below examples of compact spaces and bases.

X	basis in C(X)	reference
finite	unit vectors $e_j = (\delta_{jk})_{k=1, \ldots, n}$	
one-point compactification of countable discrete	unit vectors augmented with the constant function 1	1.2.2
2^ω (Cantor set)	Haar system	2.2
general compact O-dimensional (including countable X)		2.5
I = [0,1]	Schauder hat-function system, Franklin system	2.3 2.6

I^d $2 \leq d \leq \omega$	regular pyramidal basis,	3.3
	squew pyramidal basis,	3.4
	products of Schauder (or Franklin) functions,	1.5
	cube-multiaffine system	3.8
I^ω	prism-pyramidal basis	4.1
general closed subset of I^ω		4.4

1.2.2. If X is a locally compact space, γX will denote its one-point compactification. The space $C(\gamma X)$ will be identified with the space of continuous functions on X having limit at infinity; $C_o(X)$ will denote the subspace of functions vanishing at infinity.

If $(e_n)_{n=1,2,...}$ is any basis in $C_o(X)$, then $e_\infty, e_1, e_2, ...$ where $e_\infty = 1_X$, is a basis in $C(\gamma X)$. Specifically, if $f \in C(\gamma X)$, then

$$f = a_\infty(f) e_\infty + \sum_{n=1}^{\infty} a_n(f) e_n \quad \text{where} \quad a_\infty(f) = \lim_{x \to \infty} f(x).$$

In particular, if $X = \mathbb{N}$, $C_o(\mathbb{N})$ is the space of sequences tending to 0. The unit vectors $e_n = (\underbrace{0,...,0}_{n-1},1,...)$, $n \in \mathbb{N}$ form a basis in $C_o(\mathbb{N})$. Hence $e_\infty, e_1, ...,$ where $e_\infty = (1,1,...)$, is a basis in the space $C(\gamma \mathbb{N})$ of all convergent sequences. If $b = (b_1, b_2, ...)$ is a sequence convergent to b_∞, then

$$b = b_\infty e_\infty + \sum_{n=1}^{\infty} (b_n - b_\infty) e_n.$$

1.2.3. <u>Notes</u>. The notion of a Schauder basis was introduced by J.P.Schauder ['27] who gave a construction of a basis in $C(I)$ (cf. 2.4.10); see also Banach ['32]. The possibility of constructing a basis in $C(I^d)$ was in folklor in 1950's and was probably known by 1932; anyway, in Banach's book a basis in $C(I^2)$ is not mentioned, though there are explicit problems concerning the existence of bases in the space $C^{(1)}(I^2)$ of functions differentiable on I^2 and in the space A of functions continuous on the disc $\{z : |z| \leq 1\}$ and analytic inside of it; these two problems were solved only in 1970's (see 3.9).

A description of a basis in $C(X_1 \times ... \times X_n)$, where $X_k \subset I$ for $k = 1,...,n$, is in L.A.Gurevič ['54] (see 1.4.10 and 2.5.4); this includes a construction of a basis in $C(X)$ for any countable compact X. F.S.Vaher ['55] announced a construction of a basis in C(X) for an arbitrary compact metric space X, with an outline of proof (see 4.3.3). However, her proof is valid only when X is uncountable and in this case the existence of a basis in C(X) follows trivially from a theorem of Miliutin (proved in his thesis in 1952 but published only in ['66]) that if X is any uncountable compact metric space then C(X) is isomorphic to C(I). See also 4.2.6.

1.3. Interpolating bases

1.3.1. <u>Definition</u>. Let X be a locally compact space such that γX is metrizable and let $t_n \in X$, $n = 1,2,\ldots$ A basis (e_n) in $C_o(X)$ is called <u>interpolating with nodes</u> (t_n) iff for each f in $C_o(X)$ and each n we have

$$(1) \qquad s_n(t_m) = f(t_m) \qquad \text{for} \quad m = 1,\ldots,n,$$

where $s_n = \Sigma_{k=1}^n e_k^*(f) e_k = S_n(f)$. Thus, the n-th partial sum s_n interpolates f at n points $t_1,\ldots,t_n$.

E.g., the basis of unit vectors $e_1, e_2, \ldots$ in $C_o(\mathbb{N})$ and the basis $e_\infty, e_1, e_2, \ldots$ in $C(\gamma \mathbb{N})$ are interpolating. We shall see that many important bases (e.g., the Faber-Schauder system in C(I)) are also interpolating.

1.3.2. <u>Proposition</u>. Let (e_n) be a sequence in $C_o(X)$. Suppose that for each f in $C_o(X)$ there exists at least one sequence (a_n) such that

$$(2) \qquad f = \sum_{n=1}^{\infty} a_n e_n.$$

Let (t_n) be a sequence in X, $t_n \neq t_m$ for $n \neq m$. Then the following conditions are equivalent:

(i) (e_n) is an interpolating basis with nodes (t_n),

(ii) $e_n(t_n) \neq 0$ and $e_n(t_m) = 0$ for $n > m$.

Proof. (ii) $\Rightarrow$ (i). We have to show that the expansion (2) is unique. From (ii) is follows that

$$(3) \qquad f(t_1) = a_1 e_1(t_1) \qquad \text{and} \qquad e_1(t_1) \neq 0,$$

$$(4) \qquad f(t_m) = a_m e_m(t_m) + \sum_{k=1}^{m-1} a_k e_k(t_m) \qquad \text{and} \qquad e_m(t_m) \neq 0.$$

By induction we prove that $a_1, a_2, \ldots$ are uniquely determined by $f(t_1), f(t_2), \ldots$ Hence (e_n) is a basis in $C_o(X)$. Moreover, (e_n) is interpolating. Indeed, for each f of the form (2) we have, by (ii),

$$f(t_m) = \sum_{k=1}^{\infty} a_k e_k(t_m) = \sum_{k=1}^{n} a_k e_k(t_m) = s_n(t_m) \quad \text{if } n \geq m.$$

(i) $\Rightarrow$ (ii) If a function (2) satisfies $f(t_m) = s_m(t_m)$, then

$$\sum_{n=1}^{\infty} a_n e_n(t_m) = \sum_{n=1}^{m} a_n e_n(t_m)$$

and hence $\sum_{n=m+1}^{\infty} a_n e_n(t_m) = 0$. Since a_n are arbitrary, we get $e_n(t_m) = 0$ for $n > m$. Furthermore for each n in $\mathbb{N}$ there is a $g_n = \sum_{k=1}^{\infty} b_{nk} e_k$ in $C_o(X)$ such that $g_n(t_n) = 1$ and $g_n(t_m) = 0$ for $m < n$. By (i), $1 = g_1(t_1) = b_{11} e_1(t_1)$ and hence $e_1(t_1) \neq 0$. Suppose $e_1(t_1), \ldots, e_{n-1}(t_{n-1})$ are different from 0, $n > 1$. Then $0 = g_n(t_1) = b_{n1} e_1(t_1)$ and hence $b_{n1} = 0$. Suppose $b_{n1}, \ldots, b_{nk}$ are equal to 0 where $1 \leq k < n-1$. Then

$$0 = g_n(t_{k+1}) = \sum_{j=1}^{k+1} b_{nj} e_j(t_{k+1}) = b_{n,k+1} e_{k+1}(t_{k+1});$$

therefore $b_{n,k+1} = 0$. Consequently, $b_{n,j} = 0$ for $j < n$ and hence

$$g_n = \sum_{k=n}^{\infty} b_{nk} e_k.$$

Thus, $1 = g_n(t_n) = \sum_{k=n}^{\infty} b_{nk} e_k(t_n) = b_{nn} e_n(t_n)$ and $e_n(t_n) \neq 0$. $\square$

1.3.3. Proposition. Let (e_n) be an interpolating basis in $C_o(X)$ with nodes (t_n) and let $e_n(t_n) = 1$ for $n = 1, 2, \ldots$ Then for each f in $C_o(X)$ the equations (3), (4) have a solution

(5) $\quad a_1 = f(t_1),$

(6) $\quad a_n = f(t_n) + \sum_{k=1}^{n-1} f(t_k) \sum_{p=1}^{n-k} \sum_{(m_j)}^{(k,n)} (-1)^p \prod_{j=1}^{p} e_{m_{2j-1}}(t_{m_{2j}}), \quad n = 2, 3, \ldots,$

where $\sum_{(m_j)}^{(k,n)}$ is extended over all sequences $(m_1, \ldots, m_{2p})$ such that

$$k = m_1 < m_2 \leq m_3 < m_4 \leq \cdots \leq m_{2p-1} < m_{2p} = n.$$

Proof. Assume (6) to be true for numbers $1, \ldots, n$. Then, from (4) we get

$$a_{n+1} = f(t_{n+1}) - \sum_{k=1}^{n} a_k e_k(t_{n+1}) = f(t_{n+1}) + \sum_{k=1}^{n} f(t_k)[-e_k(t_{n+1})] +$$

$$+ \sum_{k=1}^{n} \sum_{q=1}^{k-1} f(t_q) \sum_{(m_j)}^{k-q} \sum_{p=1}^{(q,n)} (-1)^{p+1} e_k(t_{n+1}) \prod_{j=1}^{p} e_{m_{2j-1}}(t_{m_{2j}}).$$

We can now verify that the sums on the right-hand side consist of exactly these terms which appear in the sum in (6) for $n+1$. $\square$

1.3.4. <u>Corollary</u>. Let (e_n) be an interpolating basis in $C_o(X)$ with nodes (t_n) and let (μ_n) be the biorthogonal sequence of Radon measures on X. Then the carrier of each μ_n is contained in $\{t_1, \ldots, t_n\}$. Moreover, if $e_n(t_n) = 1$, then $\mu_n = \mu_{t_n} - \mu_n'$ where $\operatorname{car}(\mu_n')$ is contained in $\{t_1, \ldots, t_{n-1}\}$. $\square$

1.3.5. <u>Corollary</u>. Let (e_n) be an interpolating basis in $C_o(X)$ with nodes (t_n). Then the closure of $\operatorname{span}(\mu_n)$ in $M(X)$ is the set of all atomic measures of the form $\sum a_n \delta_{t_n}$ with $\sum |a_n| < \infty$. $\square$

1.3.6. <u>Proposition</u>. Let (e_n) be an interpolating basis with nodes (t_n) and let $e_n(t_n) = 1$ for $n = 1, 2, \ldots$ Let $\mu_n = \delta_{t_n} - \mu_n'$ be as in 1.3.4. Then (e_n) is a monotone basis if and only if $\|\mu_n'\| \leq 1$ for n in $\mathbb{N}$.

Proof. Suppose that $\|\mu_n'\| \leq 1$. Then for any $a_1, \ldots, a_n \in \mathbb{R}$

$$\| \sum_{k=1}^{n} a_k \mu_k \| = \| \sum_{k=1}^{n-1} a_k \mu_k - a_n \mu_n' \| + |a_n| \geq \| \sum_{k=1}^{n-1} a_k \mu_k \|.$$

Thus, the basis constant of (μ_n) is equal to 1; hence (by 1.1.6) the basis constant of (e_n) is also equal to 1.

Suppose now that $\|\mu_n'\| \leq 1$ and $\mu_n(t_k) = b_k$ for $k = 1, \ldots, n-1$. By backward induction we construct numbers $a_{n-1}, a_{n-2}, \ldots, a_1$ such that $\sum_{j=1}^{n-1} a_j \mu_j(t_k) = b_k$ for $k = 1, \ldots, n$. Hence $\sum_{j=1}^{n-1} a_j \mu_j = \mu_n'$ and

$$\| \sum_{j=1}^{n-1} a_j \mu_j + \mu_n \| = \| \sum_{j=1}^{n-1} a_j \mu_j + \delta_{t_n} - \mu_n' \| = 1 < \|\mu_n'\| = \| \sum_{j=1}^{n-1} a_j \mu_j \|.$$

Consequently, (μ_n) is not a monotone basic sequence; apply 1.1.6. $\square$

1.3.7. <u>Proposition</u>. Let (e_n) be an interpolating basis in $C_o(X)$ with nodes (t_n). Then the set $T = \{t_n\}_{n \in \mathbb{N}}$ is dense in X.

Proof. We may assume that $e_n(t_n) = 1$. If T were not dense in X, there would exist an f in $C_o(X)$ such that $\|f\| = 1$ and

$f(t_n) = 0$ for $n = 1, 2, \ldots$ Since (e_n) is an interpolating basis, f would be of the form (2) and conditions (3), (4) would hold. Consequently, $0 = a_1 = a_2 = \ldots$ and $f = 0$. ▯

1.3.8. <u>Proposition</u>. Let $t_n \in X$ and let (e_n) be a sequence in $C(X)$ satisfying (ii). Suppose that there is a constant $c > 0$ such that for each k and each $f \in \text{span}\{e_1, \ldots, e_k\}$

(7) $$\sup_{i=1,\ldots,k} |f(t_i)| \geq \gamma \|f\| .$$

Then (e_n) is a basic sequence in $C(X)$ with basis constant $c = 1/\gamma$.

Proof. Let $s_m = \Sigma_{j=1}^m a_j e_j$, $m = 1, 2, \ldots$ If $k < n$, then by (ii),

$$\|s_k\| \leq \frac{1}{\gamma} \sup_{i=1,\ldots,k} |s_k(t_i)| = c \sup_{i=1,\ldots,k} |s_n(t_i)| \leq c\|s_n\| .$$

Thus, from 1.1.3 it follows that (e_n) is a basic sequence. ▯

1.3.9. <u>Proposition</u>. Let $t_1, \ldots, t_n \in X$ and let F be an n-dimensional subspace of $C(X)$ such that (7) holds for each f in F. Then the Banach-Mazur distance between F and l_n^∞ does not exceed c; here l_n^∞ denotes the space of all $a = (a_1, \ldots, a_n)$ with $\|a\| = \max |a_i|$.

Proof. Consider the operator $U : F \longrightarrow l_n^\infty$ defined by

$$U(f) = (f(t_1), \ldots, f(t_n)).$$

Obviously, $\|U\| \leq 1$. If $f \in F$ and $\|f\| = 1$, then, by (7), $\|U(f)\| \geq \gamma \|f\| = \gamma > 0$. Thus, U is one-to-one and hence it is onto, and $\|U^{-1}\| \leq 1/\gamma$. Therefore $\|U\| \cdot \|U^{-1}\| \leq 1/\gamma = c$. ▯

1.3.10. <u>Exercises</u>. (A) Let $X = \{0, 1\}$. Construct an interpolating basis (e_1, e_2) in $C(X)'$ with nodes $t_1 = 0$, $t_2 = 1$, such that $\|e_1\| = \|e_2\| = 1$ and $e_1(0) < 1$.

(B) Generalize formulas (5), (6) to the case where (e_n) is any interpolating basis (replacing assumption $e_n(t_n) = 1$ by $e_n(t_n) \neq 0$).

(C) Construct a non-monotone interpolating basis in $C(I)$ with nodes (t_n) and such that $\|e_n\| = e_n(t_n) = 1$, $n = 1, 2, \ldots$ (Hint: take the Faber-Schauder system (φ_n) (see 2.4.1) and consider a new basis: $\varphi_0, \varphi_1, \varphi_2, \varphi_3, \varphi_4 + \frac{1}{2}\varphi_5, \varphi_5, \varphi_6, \ldots$).

(D) Find a basic sequence (e_n) in $C(I)$ satisfying condition 1.3.2(ii) for some sequence (t_n) dense in I and such that $\text{span}(e_1, \ldots)$ is not dense in $C(I)$. (Hint: hat functions (see 2.3.3) with lengths of supports quickly tending to O.)

(E) Show that if a basis (e_n) in $C(X)$ is orthogonal with respect to some atomless measure λ on X (e.g. (e_n) may be the Haar system in $C(2^\omega)$ or the Franklin system in $C(I)$, see 2.2 and 2.6), then the basis is not interpolating.

(F) Let (e_n) be an interpolating basis in $C(I)$. Show that (e_n) cannot be a basis in $L^1(I)$.

(G) Let (e_n), (μ_n) be as in 1.3.4 and let $e_n(t_n) = 1$. Then $\mu_n = \delta_{t_n} - \mu_n'$, where μ_n' is determined by conditions $\mu_n'(e_k) = e_k(t_n)$ for $k = 1, \ldots, n-1$ and $\mu_n'(e_m) = 0$ for $m \geq n$.

1.3.11. <u>Notes</u>. The definition of an interpolating basis was explicitly formulated in Semadeni ['63b]. 1.3.2 combines results of that paper and Ryll ['73].

1.3.7 and 1.3.9 are from Gurariĭ ['66a]; 1.3.8 is implicitly contained in the proof of Theorem 1 in that paper.

1.4. Cartesian products, c_0-products and tensor products of bases.

1.4.1. <u>Proposition</u>. Let $(e_n^{(k)})$ be a basis in $C_0(X_k)$, $k = 1, \ldots, m$. Let $\bar{e}_n^{(k)}$ be the canonical extension of $e_n^{(k)}$ to the disjoint sum $X_1 + \ldots + X_m$. Then the sequence

$$\bar{e}_1^{(1)}, \ldots, \bar{e}_1^{(m)}, \bar{e}_2^{(1)}, \ldots, \bar{e}_2^{(m)}, \ldots, \bar{e}_n^{(1)}, \ldots, \bar{e}_n^{(m)}, \ldots$$

is a basis in $C_0(X_1 + \ldots + X_m) \cong C_0(X_1) \times \ldots \times C_0(X_m)$. $\quad\square$

1.4.2. <u>Proposition</u>. Let $(e_n^{(k)})_{n \in \mathbb{N}}$ be a basis in $C_0(X_k)$ with a basis constant c_k, $k = 1, 2, \ldots$ Let $\bar{e}_n^{(k)}$ be the canonical extension of $e_n^{(k)}$ to the disjoint sum $S_{k=1}^\infty X_k$. Arrange the double sequence $\bar{e}_n^{(k)}$ into a single sequence using the diagonal method:

(1) $\qquad \bar{e}_1^{(1)}, \bar{e}_1^{(2)}, \bar{e}_2^{(1)}, \bar{e}_1^{(3)}, \bar{e}_2^{(2)}, \bar{e}_3^{(1)}, \bar{e}_1^{(4)}, \ldots$

If $c = \sup c_k < \infty$, then this new sequence is a basis in the space $C_0(S_{k=1}^\infty X_k)$, which is canonically isometric to the space

(2) $\qquad (C_0(X_1) \times C_0(X_2) \times \ldots)_{c_0}$

of all sequences $f = (f_k)_{k \in \mathbb{N}}$ such that $f_k \in C_0(X_k)$ and $\|f_k\| \longrightarrow 0$ as $k \longrightarrow \infty$.

Proof. It is clear that the sequence (1) satisfies the condition (ii) of 1.1.3. Moreover, the linear span of (1) is dense in $C_o(S_{m=1}^{\infty} X_k)$. ☐

1.4.3. <u>Theorem</u> (B.Gelbaum and J.Gil de Lamadrid). Let F,G be Banach spaces and let $\|\ \|$ be a uniform cross-norm on $F \otimes G$, i.e., a norm such that $\| f \otimes g \| = \| f \| \cdot \| g \|$ and $\| U \otimes T \| \leq \| U \| \cdot \| T \|$ for any operators $U : F \longrightarrow F$, $T : G \longrightarrow G$. Let (φ_m) and (ψ_m) be bases in F and G, respectively, with biorthogonal sequences (μ_m) in F^* and (ν_n) in G^*. Arrange the tensor products $\varphi_m \otimes \psi_n$ into a single sequence by the rectangular method: $\varphi_1 \otimes \psi_1$, $\varphi_2 \otimes \psi_1$, $\varphi_2 \otimes \psi_2$, $\varphi_1 \otimes \psi_2$, $\varphi_3 \otimes \psi_1$, $\varphi_3 \otimes \psi_2$, $\varphi_3 \otimes \psi_3$, $\varphi_2 \otimes \psi_3$, $\varphi_1 \otimes \psi_3$, $\varphi_4 \otimes \psi_1, \ldots$ Then this sequence is a basis in the completion of $(F \otimes G, \|\cdot\|)$ with the biorthogonal sequence $\mu_m \otimes \nu_n$ ordered analogously.

Proof. Let $P_m : F \longrightarrow F$, $Q_n : G \longrightarrow G$ and $S_k : F \otimes G \longrightarrow F \otimes G$ be the operators of partial sums with respect to (φ_m), (ψ_n), $(\varphi_m \otimes \psi_n)$, respectively. Thus,

$$P_m(f) = \sum_{i=1}^{m} \mu_i(f)\varphi_i, \quad Q_n(g) = \sum_{j=1}^{n} \nu_j(g)\psi_j.$$

Let $h = \sum_{q=1}^{r} f_q \otimes g_q \in F \otimes G$. The partial sum $S_k(h)$ has the simplest form when $k = n^2$, i.e., when the summation is extended over a square in the matrix with entries $\varphi_i \otimes \psi_j$, $i,j = 1,2,\ldots$, namely

$$S_{n^2}(h) = \sum_{i=1}^{n} \sum_{j=1}^{n} (\mu_i \otimes \nu_j)(h)\varphi_i \otimes \psi_j = \sum_{q=1}^{r} P_n(f_q) \otimes Q_n(g_q).$$

Hence $S_{n^2} = P_n \otimes Q_n$ and $\| S_{n^2} \| \leq \| P_n \| \cdot \| Q_n \|$. Moreover,

$$\| h - S_{n^2} \| = \| \sum_{q=1}^{r} [f_q \otimes (g_q - Q_n g_q) + (f_q - P_n f_q) \otimes Q_n g_q] \|$$

$$\leq \sum_{q=1}^{r} [\| f_q \| \| g_q - Q_n g_q \| + \| f_q - P_n f_q \| \| Q_n \| \| g_q \|].$$

Since r, f_q, g_q are fixed and $\| Q_n \|$ is bounded, $\| h - S_{n^2} h \| \longrightarrow 0$ as $n \longrightarrow \infty$. Thus the partial sums $S_{n^2} h$ tend to h and the norms $\| S_{n^2} \|$ are bounded. Yet, convergence of a subsequence is not enough to guarantee that $S_n h$ tend to h (cf. example with Walsh functions in 2.2.8 and 2.2.10). If $1 \leq k \leq n+1$, then

$$\| S_{n^2+k}(h) - S_{n^2}(h) \| = \| \sum_{j=1}^{k} \sum_{q=1}^{r} \mu_{n+1}(f_q) \nu_j(g_q) \varphi_{n+1} \otimes \psi_j \|$$

$$= \| \sum_{q=1}^{r} (P_{n+1} - P_n)(f_q) \otimes Q_k(g_q) \|$$

$$\leq \sum_{q=1}^{r} \| (P_{n+1} - P_n)(f_q) \| \, \| Q_k \| \, \| g_q \| \, .$$

Thus, $\| S_{n^2+k}(h) - S_{n^2}(h) \| \longrightarrow 0$ as $n \longrightarrow \infty$ and

$$\| S_{n^2+k} - S_{n^2} \| \leq \| P_{n+1} - P_n \| \, \| Q_k \| \leq \text{const.}$$

Similar estimations are valid for $n + 1 < k \leq 2n$. ☐

1.4.4. <u>Corollary</u>. Let $(\varphi_n^{(j)})_{n \in N}$ be a basis in a Banach space F_j, $j = 1,\ldots,m$. Then the products

$$(3) \qquad \varphi_{n_1,\ldots,n_m} = \varphi_{n_1}^{(1)} \otimes \ldots \otimes \varphi_{n_m}^{(m)},$$

suitably ordered, form a basis in $F_1 \hat{\otimes} \ldots \hat{\otimes} F_m$, where $\hat{\otimes}$ denotes the projective tensor product (with the greatest cross-norm). The products (3) also form a basis in the weak tensor product $F_1 \check{\otimes} \ldots \check{\otimes} F_m$.

Proof. This follows the fact that the norm in $F_1 \hat{\otimes} F_2$ and the norm in $F_1 \check{\otimes} F_2$ are uniform cross-norms (cf. Schatten ['50]). ☐

1.4.5. <u>Corollary</u>. Let $(\varphi_n^{(j)})_{n \in N}$ be a basis in $C_o(X_j)$, $j = 1,\ldots,m$. Then the functions

$$(4) \qquad \varphi_{n_1,\ldots,n_m}(x_1,\ldots,x_m) = \varphi_{n_1}^{(1)}(x_1) \varphi_{n_2}^{(2)}(x_2) \cdots \varphi_{n_m}^{(m)}(x_m),$$

suitably ordered, form a basis in $C_o(X_1 \times \ldots \times X_m)$. ☐

1.4.6. <u>Corollary</u>. Let (φ_n) be a basis in $C_o(X)$ and let (e_n) be a basis in a Banach space F. Then the functions $h_{m,n} : X \longrightarrow F$ defined as

$$h_{m,n}(x) = \varphi_m(x) e_n \qquad \text{for} \quad x \in X, \quad m,n \in N,$$

form a basis in $C_o(X,F)$. ☐

1.4.7. <u>Definition</u>. The basis (4) is called the <u>tensor product of bases</u> $(\varphi_n^{(j)})_{n \in N}$, $j = 1,\ldots,m$.

A significant example of tensor product of bases will be presented in 2.2.10. Specifically, we shall show that Walsh functions may be regarded as tensor products of Rademacher functions.

It turns out that even if each basis $(\varphi_n^{(j)})_{n\in\mathbb{N}}$ is monotone, the basis constant of the sequence (3) may tend to ∞ as $m \longrightarrow \infty$. Consequently, it may happen that $X_1, X_2, \ldots$ are compact, $\varphi_1^{(j)} = 1_{X_j}$ for each j, the functions (4) are well defined on $X = \mathbb{P}_{j=1}^{\infty} X_j$, and the linear combinations of the functions (4) are dense in $C(X)$, and yet the functions (4) do not form a basis in $C(X)$.

1.4.8. <u>Proposition</u>. If each basis $(\varphi_n^{(j)})_{n\in\mathbb{N}}$ is interpolating in $C_o(X_j)$ with nodes (t_n), then the tensor product basis is also interpolating.

Proof. This follows easily from 1.3.2. ☐

1.4.9. <u>Exercises</u>. (A) Prove 1.4.5 directly (for $m = 2$) without using tensor products.

(B) Suppose that the basis constant c_k of $(e_n^{(k)})_{n\in\mathbb{N}}$ in $C_o(X_k)$ is given, $k = 1, \ldots, m$. Compute the basis constant of the corresponding basis in $C_o(X_1 + \ldots + X_m)$ and in the space (2). Estimate the basis constant of a basis in $C_o(X_1 \times \ldots \times X_m)$.

(C) Let (φ_n) and (ψ_n) be Schauder bases in $C_o(X)$, $C_o(Y)$, respectively. Suppose that there exist sequences $\varrho_n\downarrow 0$ and $\sigma_n\downarrow 0$ such that

$$\| f - P_n f \| = O(\omega_f(\varrho_n)), \qquad \| g - Q_n g \| = O(\omega_g(\sigma_n))$$

for f in $C_o(X)$, $g \in C_o(Y)$. Let P_n, Q_n, S_n denote the partial sums for the bases (φ_n), (ψ_n) and $(\varphi_n \otimes \psi_m)$, respectively. Show that there exists a sequence $\tau_n\downarrow 0$ such that

$$\| h - S_n h \| = O(\omega_h(\tau_n)) \quad \text{for } h \text{ in } C_o(X \times Y).$$

(D) Let X be a closed subset of $\mathbb{R}^m$ and let $X_k = \pi_k(X) \subset \mathbb{R}$, where π_k is the k-th coordinate projection. Let $(\varphi_n^{(k)})_{n\in\mathbb{N}}$ be a basis in $C(X_k)$ constructed as in 2.5.1. Suppose that the functions $\operatorname{rest}_X \varphi_{n_1, \ldots, n_m}$, where $\varphi_{n_1, \ldots, n_m}$ is given by (4), form a basis in $C(X)$. Show that $X = X_1 \times \ldots \times X_m$.

1.4.10. <u>Notes</u>. L.A.Gurevič ['54] has constructed the basis in $C(X_1 \times \ldots \times X_m)$ described in Exercise (D); yet, he claimed that this yields a basis in $C(X)$ for an arbitrary compact subset of $\mathbb{R}^m$.

1.4.3 was proved by B.Gelbaum and J.Gil de Lamadrid ['61] and

Gil de Lamadrid ['65]. For a direct proof of 1.4.5 for compact spaces with 1.4.9.(C)   see Semadeni ['63b].

1.5. Unconditional bases in $C(X)$.

1.5.1. <u>Definition</u>. A series Σf_n in a Banach space F is <u>unconditionally convergent</u> if it is convergent for any permutation of terms, i.e., $\Sigma f_{\pi(n)}$ is convergent for every bijection $\pi : \mathbb{N} \longrightarrow \mathbb{N}$. A basis (e_n) in F is <u>unconditional</u> if for any $f = \Sigma a_n e_n$ the convergence of the series is unconditional. Σf_n is <u>absolutely convergent</u> iff $\Sigma \| f_n \| < \infty$.

It is obvious that the basis of unit vectors in $C_o(\mathbb{N})$ and the basis $e_\infty, e_1, e_2, \ldots$ in $C(\gamma\mathbb{N})$ are unconditional.

By a celebrated theorem of A.Dvoretzky and C.A.Rogers (see e.g. Lindenstrauss and Tzafiri ['77], Theorem 1.c.3), in each infinite dimensional Banach space there exists a series which is unconditionally convergent but not absolutely.

1.5.2. <u>Theorem</u>. Let F be a Banach space and $f_n \in F$ for n in $\mathbb{N}$. Then the following conditions are equivalent:

(i) Σf_n is unconditionally convergent,

(ii) Each subseries Σf_{n_k} (where $n_1 < n_2 < \ldots$) is convergent,

(iii) The series $\Sigma \eta_n f_n$ is convergent for every sequence (η_n) with $\eta_n = \pm 1$,

(iv) For every $\varepsilon > 0$ there exists an n_ε such that for any finite set $K \subset \mathbb{N}$ the condition $\min K > n_\varepsilon$ implies $\| \Sigma_{k \in K} f_k \| < \varepsilon$.

(v) For every bounded sequence (a_n) the series $\Sigma a_n f_n$ is convergent.

Proof. Condition (ii) means that each series $\Sigma \delta_n f_n$, where $\delta_n = 0$ or 1 , is convergent. Each series $\Sigma \eta_n f_n$ is the difference of two series of the form $\Sigma \delta_n f_n$; on the other hand, each series $\Sigma \delta_n f_n$ is the sum $\frac{1}{2} \Sigma f_n + \frac{1}{2} \Sigma \eta_n f_n$. Thus, the equivalence (ii) $\Leftrightarrow$ (iii) is obvious. The implications (v) $\Rightarrow$ (iii) and (iv) $\Rightarrow$ (ii) are also obvious.

(ii) $\Rightarrow$ (iv). If (iv) is not satisfied, then there exist: an $\varepsilon > 0$, a sequence $n_1 < n_2 < \ldots$ and a sequence of finite sets $K_m \subset \mathbb{N}$ such that $n_m < \min K_m < \max K_m < n_{m+1}$ and $\| \Sigma_{k \in K_m} f_k \| < \varepsilon$. Then $K = \cup K_n$ yields a subseries $\Sigma_{k \in K} f_k$ which is not convergent.

(i) $\Rightarrow$ (iv). Let K_n be as above. Let $\pi : \mathbb{N} \longrightarrow \mathbb{N}$ be a permutation such that the segment $\{n_m+1, \ldots, n_m + k_m\}$, where

$k_m = \text{card}(K_m)$, is mapped onto K_m. Then $\Sigma_{j=1}^{\infty} f_{\pi(j)}$ does not satisfy the Cauchy condition.

(iv) $\Rightarrow$ (i). Choose $\varepsilon > 0$ and then n_ε as in (iv). Let $\pi : \mathbb{N} \longrightarrow \mathbb{N}$ be a bijection. Denote $m = \sup\{\pi(k) : k \le n_\varepsilon\}$. Then any sum $\Sigma_{k=p+1}^{q} f_{\pi(k)}$ (where $m < p < q$) which appears in the Cauchy condition for $\Sigma f_{\pi(k)}$ is of the form $\| \Sigma_{j=1}^{q-p} f_{m_j} \|$ with $m_1 = \inf\{\pi(k) : p < k \le q\} > n_\varepsilon$. Therefore $\| \Sigma_{k=p+1}^{q} f_{\pi(k)} \| < \varepsilon$ for $q > p > m$.

(ii), (iii) and (iv) imply (v). Let B the closed unit ball of l^{∞}, i.e.,

$$B = \{ (a_n) : -1 \le a_n \le 1 \} = [-1,1]^{\omega}.$$

The extreme points of B are sequences of ± 1 ($\#\,4.4.4$). Thus, condition (iii) means that to each sequence (η_n) in ext B there corresponds an element $T(\eta_1, \eta_2, \dots) = \Sigma \eta_n f_n$ in F. We claim that

$$(1) \qquad r = \sup_{\eta_n = \pm 1} \| \sum_{n=1}^{\infty} \eta_n f_n \| < \infty.$$

Indeed, by (iv), there is an index n_ε such that $\| \Sigma_{k=n+1}^{m} \delta_k f_k \| < \varepsilon$ for $m > n \ge n_\varepsilon$, where (δ_k) is any sequence of zeros and ones. Hence by (ii), the series $\delta_{n+1} f_{n+1} + \delta_{n+2} f_{n+2} + \dots$ is convergent and

$$\| \sum_{k=n+1}^{\infty} \delta_k f_k \| \le \varepsilon \qquad \text{for} \quad n \ge n_\varepsilon.$$

Since $\sup_{\delta_k = 0,1} \| \Sigma_{k=1}^{n_\varepsilon} \delta_n f_n \|$ is finite, we get

$$(2) \qquad r' = \sup_{\delta_n = 0,1} \| \sum_{n=1}^{\infty} \delta_n f_n \| < \infty.$$

Thus, $r \le 2r' < \infty$. Condition (1) means that T transforms ext B into the ball $\{ f \in F : \| f \| \le r \}$. Consequently, T may be extended to a linear operator

$$T : l^{\infty} \longrightarrow F, \quad T(a_1, a_2, \dots) = \Sigma a_n f_n, \qquad \| T \| \le r.$$

This yields (v). $\square$

1.5.3. <u>Corollary</u>. If Σf_n is unconditionally convergent, then for any sequence (a_n) in l^{∞} we have

(3) $\quad \| \sum_{n=1}^{\infty} a_n f_n \| \leq \sup_n |a_n| \sup_{\eta_n = \pm 1} \| \sum_{n=1}^{\infty} \eta_n f_n \| .$ □

1.5.4. Proposition. If the series Σf_n is unconditionally convergent, then its sum does not depend on the permutation π.

Proof. This follows easily from (iv). □

1.5.5. Corollary. Let (e_n) be an unconditional basis in F. Then the set A of all sequences such that $\Sigma a_n e_n$ converges is a vector-lattice ideal in R^ω, i.e., the conditions $(a_n) \in A$ and $|b_n| \leq |a_n|$ for $n = 1, 2, \ldots$ imply $(b_n) \in A$. □

1.5.6. Proposition. Let F be a Banach lattice. If $\Sigma |f_n|$ converges, then Σf_n converges unconditionally.

Proof. We apply the condition (iv):

$$\| \sum_{k \in K} f_k \| = \| | \sum_{k \in K} f_k | \| \leq \| \sum_{k \in K} |f_k| \| \leq \| \sum_{k=\min K}^{\max K} |f_k| \| \leq \epsilon . \quad □$$

Note that in each space $L^1(\mu)$ the convergence of $\Sigma |f_n|$ is equivalent to the absolute convergence and hence the converse of 1.5.6 fails.

1.5.7. Proposition (W.Sierpiński). In $C(X)$ the unconditional convergence of Σf_n is equivalent to the convergence of $\Sigma |f_n|$.

Proof. According to the preceding proposition, we have to show only that the convergence of each $\Sigma f_{\pi(n)}$ implies the convergence of $\Sigma |f_n|$. Suppose that $\Sigma |f_n|$ is not uniformly convergent. Then there are: $\epsilon > 0$, sequences $p_1 < q_1 < p_2 < q_2 < \ldots$ and points x_n in X such that

$$\sum_{k=p_n}^{q_n} |f_k(x_n)| \geq \epsilon .$$

Denote $\gamma_k = \text{sgn } f_k(x_n)$. Then $\| \Sigma_{k=p_n}^{q_n} \gamma_k f_k \| \geq \epsilon$ for $n = 1, 2, \ldots$ □

1.5.8. Proposition (A.Pełczyński). The space $C(\omega^\omega + 1)$ of continuous functions on the compact ordinal $\omega^\omega + 1$ (see ♯ 8.6.1) cannot be embedded into a Banach space with an unconditional basis. Consequently, a space $C(X)$ has an unconditional basis if and only if X is homeomorphic to a compact ordinal α smaller than ω^ω.

This was proved by Pełczyński in his Thesis in 1958. A related theorem was proved by Lindenstrauss and Pełczyński ['68], p.297. See also J.R.Holub ['70], F.G.Arutjunjan ['72], A.Pełczyński and W.Szlenk ['65].

1.5.9. Exercises. (A) In each infinite-dimensional space $C(X)$ and $L^1(\mu)$ construct a series Σf_n which is unconditionally convergent and not absolutely convergent.

(B) Let Σf_n be an unconditionally convergent series in a Banach space F. Then the set A of all sums Σf_{n_k} $(n_1 < n_2 < \ldots)$ is norm-compact in F. (Hint: consider the sequence (δ_n) in 1.5.2 as an element of 2^ω and prove that A is a continuous image of 2^ω.)

(C) In the space $C(\gamma\mathbb{N})$ consider the sequences

$$g_n = \underbrace{(1,\ldots,1,0,0,\ldots)}_{n} \quad \text{and} \quad h_n = \underbrace{(0,\ldots,0,1,1,\ldots)}_{n-1}.$$

Thus, $g_n = e_1 + \ldots + e_n$ $(n \geq 1)$, $h_1 = e_\infty$ and $h_n = e_\infty - g_{n-1}$ $(n \geq 2)$, where e_n, e_∞ are as in 1.2.2. (a) Show that $(g_n)_{n \in \mathbb{N}}$ is a conditional (i.e., not unconditional) normalized basis in $C_o(\mathbb{N})$. (b) Find formulas expressing the coefficients a_n of $\Sigma a_n e_n$ in terms of the coefficients c_n of $\Sigma c_n g_n$ and vice versa. (c) Find the sequence (g_n^*) in $l^1(\mathbb{N})$ biorthogonal to (g_n) and show that (g_n^*) augmented with the first unit vector is a conditional basis in $l^1(\mathbb{N})$. (d) Show that $(h_n)_{n \in \mathbb{N}}$ is a monotone normalized conditional basis in $C(\gamma\mathbb{N})$ and find formulas for coefficients. (e) Show that the sequence (h_n^*) in $l^1(\gamma\mathbb{N})$ biorthogonal to (h_n) is a conditional basis in $l^1(\mathbb{N})$ but not in $l^1(\gamma\mathbb{N})$.

(D) Let (e_n) be a basis such that (e_n^*) is an unconditioanl basic sequence. Then the basis (e_n) is unconditional.

1.5.11. Notes. 1.5.7 was proved by W.Sierpiński ['10] (cf. W.Orlicz ['55], p.405).

The concept of unconditional convergence of a series in a Banach space and basic properties are due to W.Orlicz; for references and more information, see Day ['73] and Lindenstrauss and Tzafiri ['77], ['79].

Chapter 2

BASES OF BROKEN-LINE FUNCTIONS OF A SINGLE VARIABLE

In this chapter we shall deal with most important systems of piecewise affine functions in $C(I)$, where $I = [0,1]$, namely with the Haar, Walsh, Faber-Schauder, and Franklin systems.

2.1. Preliminaries on dyadic indices

2.1.1. <u>Definitions</u>. If $n \geq 1$, let D_n denote the set consisting of all numbers of the form $(2k - 1)/2^n$, $k = 1, \ldots, 2^{n-1}$. Thus

$$D_1 = \{\tfrac{1}{2}\}, \quad D_2 = \{\tfrac{1}{4}, \tfrac{3}{4}\}, \quad D_3 = \{\tfrac{1}{8}, \tfrac{3}{8}, \tfrac{5}{8}, \tfrac{7}{8}\} \quad \text{etc.}$$

We shall also need the sets $D_o = \{0,1\}$, $\hat{D}_n = D_o \cup \ldots \cup D_n$ and

$$D = \bigcup_{n=1}^{\infty} D_n, \qquad \tilde{D} = \bigcup_{n=0}^{\infty} D_n = D \cup D_o.$$

The elements of the set $\tilde{D}$ are called <u>dyadic numbers</u> and those in D - <u>proper dyadic numbers</u>. A dyadic number is of <u>order</u> n iff it belongs to D_n. When we speak of arranging the elements of $\tilde{D}$ into a single sequence

$$0, 1, \tfrac{1}{2}, \tfrac{1}{4}, \tfrac{1}{4}, \tfrac{3}{4}, \tfrac{1}{8}, \ldots$$

we shall mean the lexicographic order: the elements of D_{n-1} are earlier that those of D_n, and within each D_n we keep the natural order. It is clear that $\text{card}(D_o) = 2$, $\text{card}(D_n) = 2^{n-1}$, and $\text{card}(\hat{D}_n) = 2^n + 1$.

2.1.2. <u>Definitions</u>. To each τ in D_n $(n \geq 1)$ there corresponds a closed interval $I_\tau = [\tau^-, \tau^+]$ of length $1/2^{n-1}$ and center τ. Thus, $\tau = \tfrac{1}{2}(\tau^- + \tau^+)$, $\tau^- = \tau - 2^{-n}$, $\tau^+ = \tau + 2^{-n}$, $(\tfrac{1}{2})^- = 0$, $(\tfrac{1}{2})^+ = 1$, $(\tfrac{1}{4})^- = 0$, $(\tfrac{1}{4})^+ = \tfrac{1}{2}$, $(\tfrac{3}{4})^- = \tfrac{1}{2}$, $(\tfrac{3}{4})^+ = \tfrac{1}{2}$ etc. If $\tau \in D_n$, then τ^- and τ^+ belong to $\hat{D}_{n-1}$.

We also define two unary operations on D, the sharp $\#$ and the flat $\flat$. If $\tau \in D_n$, $n \geq 1$, let

$$\flat\tau = \tau - 2^{-n-1}, \quad \sharp\tau = \tau + 2^{-n-1}.$$

Thus, $\flat\tau = \frac{1}{2}(\tau^- + \tau)$, $\sharp\tau = \frac{1}{2}(\tau + \tau^+)$. Clearly, $\flat: D_n \longrightarrow D_{n+1}$ and $\sharp: D_n \longrightarrow D_{n+1}$ are injections $(n \geq 1)$ and for each κ in D_{n+1} there is a unique τ in D_n such that either $\kappa = \flat\tau$ or $\kappa = \sharp\tau$; this τ will be denoted by $\natural\kappa$. Thus, $\natural: D_{n+1} \longrightarrow D_n$, $\natural\flat\tau = \natural\sharp\tau = \tau$. For $\tau = (2k-1)2^{-n}$, $1 \leq k \leq 2^{n-1}$ we have $\tau = \sharp\natural\tau$ if k is even and $\tau = \flat\natural\tau$ if k is odd.

2.2. The Haar system and a basis in $C(2^\omega)$.

2.2.1. <u>Definitions</u>. By the <u>Haar system</u> we shall mean the following sequence $(\chi_m)_{m \in \mathbb{N}}$ of functions defined for x in I: $\chi_1(x) = 1$ and

$$\chi_{2^n+k}(x) = \begin{cases} 1 & \text{for } x \text{ in } [\frac{2k-2}{2^{n+1}}, \frac{2k-1}{2^{n+1}}], \\ -1 & \text{for } x \text{ in } (\frac{2k-1}{2^{n+1}}, \frac{2k}{2^{n+1}}], \\ 0 & \text{for the remaining } x \text{ in } I, \end{cases}$$

where $n = 0, 1, \ldots$ and $k = 1, \ldots, 2^n$.
The graphs of first 5 Haar functions are shown below:

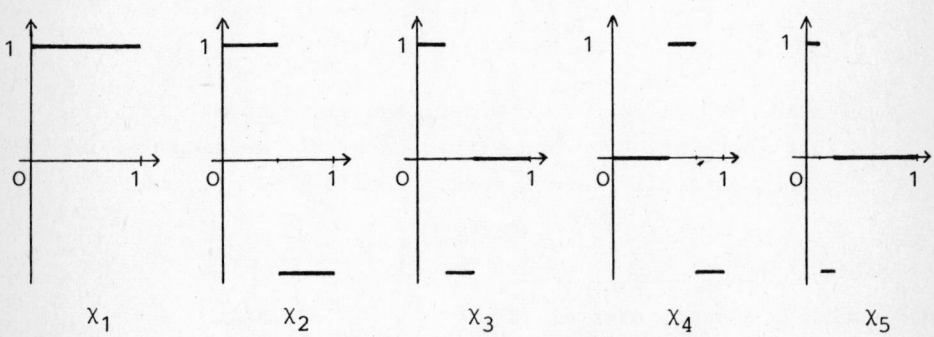

$$\chi_1 \qquad \chi_2 \qquad \chi_3 \qquad \chi_4 \qquad \chi_5$$

It is clear that $\int_0^1 \chi_n \chi_m dx = 0$ for $n \neq m$; thus (χ_m) is an orthogonal system. In many books (e.g. in Kaczmarz and Steinhaus ['35]) the Haar functions are normalized in $L^2(I)$, i.e., the above functions χ_m are multiplied by $\sqrt{2^n}$. We normalize them here in $L^\infty(I)$. Also the values of χ_m at jump points are sometimes defined differently. Yet, this is usually not important for us because in fact we consider χ_m as elements of $L^\infty(I)$ (it may be, however, significant for the Stieltjes integral, cf. 2.4.2).

Occasionally we shall consider the Haar functions indexed by dyadic numbers τ in $\tilde{D}$, $\tau \neq 0$, namely $\chi_1(x) = 1$ and, if $\tau \in D$,

$$\chi_\tau(x) = \begin{cases} 1 & \text{for } x \text{ in } [\tau^-, \tau], \\ -1 & \text{for } x \text{ in } (\tau, \tau^+], \\ 0 & \text{outside of } [\tau^-, \tau^+]. \end{cases}$$

Thus, for $n \geq 0$ and $k = 1, \ldots, 2^n$

$$\chi_{2^n+k} = \chi_\tau \quad \text{where} \quad \tau = \frac{2k-1}{2^{n+1}} \in D_{n+1}.$$

2.2.2. **Definitions**. It is natural to consider the Haar functions as defined on the Cantor set 2^ω. Specifically, let $\Theta : 2^\omega \longrightarrow I$ be the Lebesgue singular map $\Theta(\eta_1, \eta_2, \ldots) = \Sigma_{m=1}^\infty \eta_m 2^{-m}$, where $\eta_m = 0$ or 1. The figure below shows the corresponding map from the Cantor

set C onto I, extended piecewise affinely to the Lebesgue singular function from I onto I. For each m there is exactly one function $\tilde{\chi}_m$ in $C(2^\omega)$ such that $\tilde{\chi}_m(\eta) = \chi_m(\Theta(\eta))$ for all η in 2^ω except possibly of points such that $\Theta(\eta)$ is a jump point of χ_m.

Thus, if $\eta = (\eta_1, \eta_2, \ldots) \in 2^\omega$, then

$$\tilde{\chi}_1(\eta) = 1, \quad \tilde{\chi}_2(\eta) = 1 - 2\eta_1, \quad \tilde{\chi}_3(\eta) = (1 - \eta_1)(1 - 2\eta_2),$$

$$\tilde{\chi}_4(\eta) = \eta_1(1 - 2\eta_2), \quad \tilde{\chi}_5(\eta) = (1 - \eta_1)(1 - \eta_2)(1 - 2\eta_3).$$

Generally, if we denote $\varepsilon_0(\eta_j) = 1 - \eta_j$ and $\varepsilon_1(\eta_j) = \eta_j$, then

$$\tilde{\chi}_{2^n+k} = (1 - 2\eta_{n+1}) \prod_{j=1}^n \varepsilon_{a_{n-j}}(\eta_j),$$

where $k = 1, \ldots, 2^n$ and $k-1 = \Sigma_{j=0}^{n-1} a_j 2^j$ is the binary representation with coefficients $a_j = 0$ or 1. Thus, $\tilde{\chi}_{2^n+k}$ is the product $\varepsilon_a(1 - 2\eta_{n+1})$ where ε_a is the characteristic function of the set

$$\text{supp}(\tilde{\chi}_{2^n+k}) = \bigcap_{j=1}^n \{\eta \in 2^\omega : \eta_j = a_{j-1}\} \quad \text{and} \quad a = (a_1, \ldots, a_n).$$

2.2.3. <u>Lemma</u>. The linear span E_n of 2^n first Haar functions $\chi_1, \ldots, \chi_{2^n}$ in $L^\infty(I)$ coincides with the set E_n' of all functions which are constant on each of the intervals I_τ, $\tau \in D_n$, obtained by dividing I into 2^n equal intervals.

Proof. Obviously, $\chi_k \in E_n'$ for $k = 1, \ldots, 2^n$. Hence $E_n' \supset E_n$. Since $\dim E_n = 2^n$ and $\chi_1, \ldots, \chi_{2^n}$ are linearly independent, $E_n' = E_n$. $\square$

2.2.4. <u>Proposition</u>. The Haar system $(\chi_m)_{m \in \mathbb{N}}$ is a monotone basic sequence in $L^\infty(I)$.

Proof. Let $s_m = a_1\chi_1 + \ldots + a_m\chi_m$ and $s_{m+1} = s_m + a_{m+1}\chi_{m+1}$. If $\chi_{m+1} = \chi_\tau$ for some τ in D, then s_{m+1} coincides with s_m everywhere with except of I_τ; s_m is equal to a constant c on I_τ, whereas s_{m+1} is equal to $c+a_{m+1}$ on $[\tau^-, \tau]$ and to $c-a_{m+1}$ on $[\tau, \tau^+]$. Consequently, $\| s_{m+1} \| \geq \| s_m \|$. $\square$

2.2.5. <u>Proposition</u>. The system $(\tilde{\chi}_m)$ is a monotone basis in $C(2^\omega)$.

Proof. By the preceding proposition, $(\tilde{\chi}_m)$ is a monotone basic sequence in $C(2^\omega)$. By 2.2.3, E_n is a subalgebra of $L^\infty(I)$. Consequently, the linear span of $(\tilde{\chi}_m)_{m \in \mathbb{N}}$ is a subalgebra of $C(2^\omega)$ separating 2^ω; hence, by the Weierstrass-Stone theorem, $(\tilde{\chi}_m)_{m \in \mathbb{N}}$ is a basis in $C(2^\omega)$. $\square$

2.2.6. <u>Proposition</u>. The system $(\tilde{\chi}_m)$ is orthogonal in $L^2(2^\omega, \mu)$ and is a basis in $L^1(2^\omega, \mu)$, where μ is the Haar measure on 2^ω (product of $(\frac{1}{2}, \frac{1}{2})$-measures on axes of 2^ω).

Proof. By 1.1.6, the system biorthogonal is a basic sequence in the space $M(2^\omega)$ of Radon measures on 2^ω and hence also in its subspace $L^1(2^\omega)$ of measures absolutely continuous with respect to μ. The Haar system is orthogonal in $L^2(I)$. Moreover, the function $\Theta : 2^\omega \longrightarrow I$ is one-to-one except of a countable set and is measure-preserving. Therefore the spaces $L^2(I)$ and $L^2(2^\omega)$ are canonically isomorphic. Consequently, the system $(\tilde{\chi}_m)_{m \in \mathbb{N}}$ is orthogonal in $L^2(2^\omega)$ and dense in $L^1(2^\omega)$. $\square$

2.2.7. <u>Corollary</u>. The Haar system $(\chi_m)_{m \in \mathbb{N}}$ is a basis in $L^1(I)$. $\square$

2.2.8. <u>Definitions</u>. The <u>Rademacher functions</u> $r_1, r_2, \ldots$ on I are defined as $r_n(x) = \mathrm{sign}(\sin 2^n \pi x)$, $x \in I$. Obviously, $r_1 = \chi_2$, $r_2 = \chi_3 + \chi_4$, and generally

$$r_n = \sum_{k=1}^{2^{n-1}} \chi_{2^{n-1}+k}$$

everywhere except of the jump points; thus, the equalities may be understood in $L^\infty(I)$.

The __Walsh functions__ in $L^\infty(I)$ are defined as all possible products of Rademacher functions (note that the square of each r_n is equal to 1 a.e.). It is easy to see that these products are orthogonal.

There are two canonical orders of the set of all Walsh functions. Historically, the first was the original order of J.L.Walsh ['23], leading to the __Walsh system__. In this order the n-th function $(n = 0, 1, \ldots)$ changes the sign exactly n times in I. The second order, leading to the __Walsh-Paley system__ (see S.Kaczmarz ['29], R.Paley ['32]), is lexicographic with respect to indices of the Rademacher functions involved. The tabulation below shows the first 4 functions of the Walsh and Walsh-Paley systems and their signs on the 4 intervals in the partition $0 < 1/4 < 1/2 < 3/4 < 1$:

Walsh		Walsh-Paley	
$w_0' = 1$	+ + + +	$w_0 = 1$	+ + + +
$w_1' = r_1$	+ + - -	$w_1 = r_1$	+ + - -
$w_2' = r_1 r_2$	+ - - +	$w_2 = r_2$	+ - + -
$w_3' = r_2$	+ - + -	$w_3 = r_1 r_2$	+ - - +

Continuing these patterns, we get $w_4' = w_3' r_3 = r_2 r_3$, $w_5' = w_2' r_3 = r_1 r_2 r_3$, $w_6' = w_1' r_3 = r_1 r_3$, $w_7' = w_0' r_3 = r_3$, $w_8' = w_7' r_4$ etc. and $w_4 = r_3$, $w_5 = r_1 r_3$, $w_6 = r_2 r_3$, $w_7 = r_1 r_2 r_3$, $w_8 = r_4$ and generally $w_{2^n+k} = w_k r_{n+1}$ for $k = 1, \ldots, 2^n$. The reader will write for himself the analogous distribution of signs on 8 intervals of the partition of I to 8 equal parts. In this way one gets two 8×8 Walsh matrices of + and - in two different orders of rows. Clearly, there are 2^n products $\prod_{k=1}^n r_k^{\alpha_k}$ with $\alpha_k = 0$ or 1. Thus, in n-th step one gets 2^n orthogonal functions $w_0', \ldots, w_{2^n-1}'$ and 2^n orthogonal functions $w_0, \ldots, w_{2^n-1}$ in E_n.

The Walsh functions are linear combinations of Haar functions and vice versa. Yet, properties of these systems are quite different. In L_2-normalization, the Walsh functions are bounded and the Haar functions are not. For each f in $C(I)$ the Haar series of f is uniformly convergent whereas its Walsh series may be divergent even pointwise.

Let s_n^W and s_n^H denote the n-th partial sum of the Walsh-Paley and Haar series of f in $C(I)$, respectively. Then $s_{2^n}^W = s_{2^n}^H$ because first 2^n functions of (χ_m) and (w_m) span the same subspace E_n and a partial sum of an orthogonal series is obtained by orthogonal projections of f onto the subspace. Thus, $\|f - s_{2^n}^W\|_\infty \longrightarrow 0$ for each f in $C(I)$ and yet $\|f - s_m^W\|_\infty$ need not tend to 0.

2.2.9. The n-th Rademacher function r_n in $L^\infty(I)$ corresponds via the canonical map $\theta : 2^\omega \longrightarrow I$ (considered in 2.2.2) to the function $\tilde{r}_n$ in $C(2^\omega)$ defined as $\tilde{r}_n(\eta) = 1 - 2\eta_n$, i.e., $\tilde{r}_n(\eta) = 1$ if $\eta_n = 0$ and $\tilde{r}_n(\eta) = -1$ if $\eta_n = 1$. It is obvious that $\tilde{r}_n$ is a character of 2^ω, i.e., a homomorphism from 2^ω to the toroidal group $\{z \in \mathbb{C} : |z| = 1\}$; 2^ω is the product of ω copies of the cyclic group $Z_2 = \{0,1\}$.

The functions $\tilde{w}_1 = 1$, $\tilde{w}_2 = \tilde{r}_1$, $\tilde{w}_3 = \tilde{r}_2$, $\tilde{w}_4 = \tilde{r}_1\tilde{r}_2$, $\tilde{w}_5 = \tilde{r}_3$, ... corresponding to the Walsh-Paley functions are also characters on 2^ω and obviously are orthogonal with respect to the normalized Haar measure μ on 2^ω. Moreover, the product of two Walsh functions is a Walsh function. Consequently, by the Weierstrass-Stone theorem, the set of linear combinations of $(\tilde{w}_n)_{n=1,2,\dots}$ is dense in $C(2^\omega)$ and in $L^2(2^\omega)$.

2.2.10. In the space $C(2)$, where $2 = \{0,1\}$, a basis consists of the constant function 1 and the function $r(i) = 1-2i$ for $i \in \{0,1\}$. It is clear that the Walsh functions $\tilde{w}_1, \dots, \tilde{w}_{2^n}$ considered as functions on 2^n, form a basis in $C(2^n)$, which is the tensor product of n two-element bases $(1,r)$ in $C(2)$. The Walsh system $(\tilde{w}_n)_{n=1,2,\dots}$ in $C(2^\omega)$ may be regarded as the tensor product of ω copies of the basis $(1,r)$ in $C(2)$ (cf. 1.4.7). Yet, it is not a basis in $C(2^\omega)$.

2.2.11. <u>Exercises</u>. (A) Let T be the partition $0 = t_0 < t_1 < \dots < t_n = 1$ of I and let E_T be the subspace of $L^\infty(I)$ consisting of functions constant on each interval $[t_0, t_1], \dots, [t_{n-1}, t_n]$. If $f \in C(I)$, then $\mathrm{dist}(f, E_T) \le \omega_f(\delta/2)$.

(B) Represent the Walsh functions w_m' and w_m as linear combinations of Haar functions (note the distribution of signs $\pm$ in these sums). Conversely, represent Haar functions as linear combinations of Walsh functions (Hint: the Walsh $2^n \times 2^n$ matrices of ± 1, divided by suitable constants, are orthogonal matrices).

(C)   Let $m = 2^{k_1} + 2^{k_2} + \ldots + 2^{k_j}$ (where $j \geq 1$, $k_1 > k_2 > \ldots > k_j \geq 0$) be the binary representation of a number $m \geq 1$.   Show that

$$w_m = \prod_{i=1}^{j} r_{k_i+1} .$$

(D)   The L^2-normalized sequences (χ_n) and (w_n) tend weakly to 0 in $L^1(I)$.

2.2.12. <u>Notes</u>.   The Haar system is due to A.Haar ['10]. A detailed survey of results on Haar series is given by Golubov ['71].

The Haar system is an unconditional basis in $L^p(I)$ for $1 < p < \infty$ (J.Marcinkiewicz ['37]), but it is not an unconditional basis in $L^1(I)$; in fact, there is no unconditional basis in $L^1(I)$ (A.Pełczyński ['60]; see also Lindenstrauss and Tzafriri ['79], p.24). Concerning Walsh functions, see Balašov and Rubinšteǐn ['71] and Beauchamp ['75].

<u>2.3. Broken-line interpolation and Schauder hat functions with arbitrary nodes.</u>

2.3.1.   In the previous subsection we considered the Haar functions, which are piecewise constant. The remaining part of this chapter will be devoted to <u>piecewise affine</u> functions, i.e., <u>broken-</u> <u>-line</u> or <u>polygonal functions</u>, in particular to   the Faber-Schauder hat functions and the Franklin functions. Therefore we shall now recall some general properties of polygonal functions.

Let $a = t_0 < t_1 < \ldots < t_n = b$ be a partition of an interval $[a,b]$, $n \geq 1$,   and let $T = \{t_j\}_{j=0}^{n}$. B_T will denote the subspace of $C(I)$ consisting of polygonal functions with <u>nodes</u> (called also <u>knots</u>, <u>breakpoints</u> or <u>data points</u>) at $t_0, \ldots, t_n$.   For each sequence $u = (u_0, u_1, \ldots, u_n)$ of real numbers there is exactly one function p_u in B_T such that $p_u(t_j) = u_j$ for $j = 0, \ldots, n$.   This function is given by the formula

(1) $p_u(x) = \dfrac{u_j - u_{j-1}}{t_j - t_{j-1}} (x - t_{j-1}) + u_{j-1}$ for x in $[t_{j-1}, t_j]$.

If $f \in C[a,b]$,   Let $\Gamma_T f$ denote the polygonal function which inter- polates f at $n+1$ given points $t_0, \ldots, t_n$.   Thus

$$\Gamma_T f = p_u \quad \text{where} \quad u = (f(t_0), \ldots, f(t_n)).$$

It is clear that $\Gamma_T : C[a,b] \longrightarrow B_T$ is a projection and $\| \Gamma_T \| = 1$.

2.3.2. <u>Proposition</u>. Let $f \in C[a,b]$. Then

$$\| f - \Gamma_T f \| \leq \omega_f(\delta),$$

where $\omega_f(\delta)$ is the modulus of continuity of f and $\delta = \sup | t_{j+1} - t_j |$.

Proof. Let $x \in [t_{j-1}, t_j]$. Then $\Gamma_T f(x)$ is a number between $f(t_{j-1})$ and $f(t_j)$. Consequently

$$| f(x) - \Gamma_T f(x) | \leq \sup\{ | f(x) - f(x') | : x, x' \in [t_{j-1}, t_j] \} \leq \omega_f(\delta). \quad \Box$$

2.3.3. <u>Definitions</u>. The <u>hat functions</u> for the partition T are the functions $h_{t_j}^T$ in B_T determined by the conditions $h_{t_j}^T(t_k) = \delta_{jk}$, $j,k = 0,\ldots,n$ (see the figure below). Thus, if $j \geq 2$,

$$h_{t_j}^T(x) = \begin{cases} \dfrac{x - t_{j-1}}{t_j - t_{j-1}} & \text{for } x \in [t_{j-1}, t_j], \\[2mm] \dfrac{t_{j+1} - x}{t_{j+1} - t_j} & \text{for } x \in [t_j, t_{j+1}], \\[2mm] 0 & \text{otherwise.} \end{cases}$$

For $j = 0$ and $j = n$ we have "restricted hats", namely

$$h_a^T(x) = \begin{cases} \dfrac{t_1 - x}{t_1 - a} & \text{for } x \in [a, t_1], \\[2mm] 0 & \text{otherwise,} \end{cases}$$

and analogously for h_b^T.

The functions $h_{t_j}^T$, $j = 0,\ldots,n$ form a <u>partition of unity</u>, that is (cf. #20.4.1),

$$\sum_{j=0}^{n} h_{t_j}^T = 1 \quad \text{and} \quad h_{t_j}^T \geq 0.$$

Moreover, for each sequence $u = (u_0,\ldots,u_n)$ in $\mathbb{R}^{n+1}$ the function (1) can be represented as

$$p_u = \sum_{j=0}^{n} u_j h_{t_j}^T.$$

In particular,

$$\Gamma_T f = \sum_{j=0}^{n} f(t_j) h_{t_j}^T \qquad \text{for} \quad f \quad \text{in} \quad C[a,b].$$

2.3.4. Definitions. Let $T = (t_n)_{n=0,1,\ldots}$ be a sequence dense in $[a,b]$ with $t_o = a$, $t_1 = b$. Denote $T_n = \{t_o,\ldots,t_n\}$ and $\varphi_o^T = h_{t_o}^{T_1}$, $\varphi_1^T = h_{t_1}^{T_1}$ and $\varphi_n^T = h_{t_n}^{T_n}$ for $n \geq 2$. Thus, $\varphi_n^T \in B_{T_n}$ for $n \geq 1$, $\varphi_n^T(t_n) = 1$ for $n \geq 0$, and $\varphi_n^T(t_k) = 0$ for $0 \leq k < n$. The functions $\varphi_o^T, \varphi_1^T, \ldots$ are the _Schauder hat functions_ for the sequence (t_n).

2.3.5. Proposition. The sequence $(\varphi_n^T)_{n=0}^{\infty}$ is a basis in $C[a,b]$.

Proof. Let $a_o = f(t_o)$, $a_1 = f(t_1)$ and

(2) $\qquad a_{n+1} = f(t_{n+1}) - s_n(t_{n+1})$, where $s_n = \sum_{k=0}^{n} a_k \varphi_k^T$.

The figure below shows graphs of f and s_3 (continuous broken line)

and of s_4 (dotted line). Obviously, $s_n \in B_{T_n}$ for $n \geq 1$ and $s_n = \Gamma_{T_n} f$. Consequently, by 2.3.2, $\| f - s_n \| \longrightarrow 0$ as $n \longrightarrow \infty$. It is clear that the coefficients a_n in the series $f = \Sigma_{n=0}^{\infty} a_n \varphi_n^T$ are unique (cf. 1.3.2). ▯

2.3.6. Corollary. Omitting φ_o^T and/or φ_1^T in the sequence (φ_n^T) we get a basis in the space of functions in $C[a,b]$ which vanish at the point a and/or at the point b. ▯

In 4.5.5 we shall show that in $C[a,b]$ there is an interpolating basis consisting of polynomials.

2.3.7. Let $C^{(1)}[a,b]$ denote the space of all functions f having continuous first derivative in $[a,b]$, with the norm

$$\| f \|^{(1)} = \max(|f(a)|, \sup_{a < x < b} |f'(x)|).$$

If $g \in C[a,b]$, then $\Phi g.x = \int_a^x g(t)dt$ defines a linear isometry Φ from $C[a,b]$ onto $\{f \in C^{(1)}[a,b] : f(a) = 0\}$. Consequently, the sequence $1, \Phi(\varphi_o^T), \Phi(\varphi_1^T), \ldots$ is a monotone basis in $C^{(1)}[a,b]$.

2.3.8. <u>Exercises</u>. (A) If f increasing [decreasing], then so is $\Gamma_T f$.

(B) If f is convex [concave], then so is $\Gamma_T f$.

(C) What are the relations between φ_n^T and $h_{t_j}^{T_n}$ (j < n) ?

(D) If P : C(X) $\longrightarrow$ G is any projection, then

$$\| f - Pf \| \leq (1 + \| P \|) \mathrm{dist}(f,g).$$

(E) If $g \in C^{(1)}[a,b]$, then $\| g' - (\Gamma_T g)' \|_{L^\infty} \leq 2\,\mathrm{dist}(g',E_T)$, where E_T is the set of functions constant on each interval $[t_{j-1},t_j]$.

(F) If $f \in C^{(2)}[a,b]$, then $\| f - \Gamma_T f \|_\infty \leq \frac{1}{8}\delta^2 \| f'' \|$, where $\delta = \max|t_j - t_{j-1}|$.

(G) Let $g \in C(I)$ and $x_o \in [0,\frac{1}{2}]$. Let p be an affine function interpolating g at points x_o and $2x_o$. Then $|g(x)-p(x)| \leq 2\omega_g(x_o)$ for x in $[0,x_o]$.

(H) Show that the sequence $\| \varphi_1^T + \ldots + \varphi_n^T \|$ is unbounded.

(I) Show that for each T there exists a sequence (ε_n) such that $\varepsilon_n = \pm 1$ and the functions

$$s_n = \sum_{k=0}^{n} \varepsilon_k \varphi_k^T \quad \text{and} \quad s_n' = \sum_{k=1}^{n} \varepsilon_k \varphi_k^T$$

satisfy $\| s_n \| = 1, \quad \| s_n' \| = 1, \quad n = 1,2,\ldots$

(J) Let $\beta > \alpha > 0$ and $\gamma = \frac{1}{2}(\alpha+\beta)$, $\eta = \alpha - \gamma$. Denote $\psi_o = \gamma$, $\psi_n = \gamma + \eta s_n'$ where s_n' is as in Exercise (G). Show that (ψ_n) is a basis in C[a,b] and $\alpha \leq \psi_n(x) \leq \beta$ for x in [a,b] (hint: Exercise 1.1.13(E)).

(K) Show that for each T the sequence φ_n^T does not tend weakly to O but there exists a subsequence $\varphi_{n_k}^T$ tending weakly to O in C[a,b] (hint: f_n tends weakly to O in C(X) if $\| f_n \|$ are bounded and $f_n(x) \longrightarrow$ O for x in X, see #19.3.1).

(L) Let (μ_n) be the system biorthogonal to (φ_n^T). Then (μ_n) does not tend weakly to O (though the biorthogonal system to any normalized basis tends *weakly to O).

2.3.9. <u>Notes</u>. The basis (φ_n^T) was constructed by J.P.Schauder ['27] (see also 2.4.10).

Special cases of (H), (I), (K), (L) are in C.Foiaş and I.Singer ['65]; (I) and (J) - in Meletidi ['71]. (G) is from Warren ['71].

2.4. The Faber-Schauder system of dyadic hat functions in C(I)

2.4.1. We shall now consider the interval $I = [0,1]$ with the specified countable sequence T of all elements of the set D ordered as in 2.1.1. The corresponding basis $(\varphi_n)_{n=0,1,\ldots}$ in C(I) consisting of hat functions φ_n^T is of particular importance. It will be referred to as the <u>Faber-Schauder system</u> of dyadic hat functions in C(I).

In these lecture notes, we normalize (φ_n) in C(I) and assume that φ_0 is also a hat function. Thus,

(1) $\varphi_0(x) = 1 - x, \quad \varphi_1(x) = x,$

(2) $\varphi_{2^n+k}(x) = \max[0, 1 - 2^{n+1}|x - \frac{2k-1}{2^{n+1}}|], \quad n = 0,1,\ldots \quad k = 1,\ldots,2^n,$

(3) $\|\varphi_m\| = 1 \quad$ for $\quad m = 0,1,2,\ldots$

The first 5 functions are shown below:

We shall also use dyadic indices. If $\tau = (2k-1)2^{-n-1}$ is in D_{n+1}, then φ_τ means the same as φ_{2^n+k} and $a_\tau = a_{2^n+k}$. For the partial sums we shall use two kinds of symbols

$$s_m = \sum_{k=0}^{m} a_k \varphi_k \quad \text{and} \quad s^{(n)} = \sum_{k=0}^{2^n} a_k \varphi_k = s_{2^n} = \sum_{j=0}^{n} \sum_{\tau \in D_n} a_\tau \varphi_\tau$$

$(m = 0,1,\ldots, \quad n = 0,1,\ldots)$. Thus, $s^{(0)} = s_1 = a_0 \varphi_0 + a_1 \varphi_1,$ $s^{(1)} = s_2 = s^{(0)} + a_2 \varphi_2, \quad s^{(2)} = s^{(1)} + a_3 \varphi_3 + a_4 \varphi_4 \quad$ etc.

2.4.2. The coefficients (a_m) of the expansion $f = \sum_{m=0}^{\infty} a_m \varphi_m$ can be computed in the way described in 2.3.5. In case of dyadic nodes, the formulas are particularly simple:

(4) $a_0 = f(0), \quad a_1 = f(1),$

(5) $a_\tau = f(\tau) - \frac{1}{2}[f(\tau^-) + f(\tau^+)] \quad$ for τ in $D_n, \quad n \geq 1.$

Indeed, by formula (2) in 2.3.5, $a_\tau = f(\tau) - s^{(n-1)}(\tau)$ for τ in D_n. The partial sum $s^{(n-1)}$ is affine on the interval $I_\tau = [\tau^-, \tau^+]$ and has values $f(\tau^-)$ and $f(\tau^+)$ at the end points of I_τ. Since τ is the center of I_τ, $s^{(n-1)}(\tau)$ is equal to the arithmetic mean of $f(\tau^-)$ and $f(\tau^+)$.

Formula (5) may be written as

$$(6) \qquad a_{2^n+k} = f\left(\frac{2k-1}{2^{n+1}}\right) - \frac{1}{2}\left[f\left(\frac{k-1}{2^n}\right) + f\left(\frac{k}{2^n}\right)\right].$$

In other words, $a_\tau = \int_I f d\mu_\tau$, where $\mu_\tau = \delta_\tau - \frac{1}{2}(\delta_{\tau^-} + \delta_{\tau^+})$ and δ_x denotes the Dirac unit mass at x. Replacing $\int_I f d\mu_\tau$ by the Stieltjes integral with respect to a function of bounded variation we get, for each τ in D,

$$a_\tau = -\frac{1}{2}\int_0^1 f d\chi_\tau^o, \quad \text{where} \quad \chi_\tau^o(x) = \begin{cases} \chi_\tau(t) & \text{for} \quad 0 < t < 1 \\ \\ 0 & \text{for} \quad t = 0,1 \end{cases}$$

Integration by parts gives

$$(7) \qquad a_\tau = \frac{1}{2}\int_0^1 \chi_\tau df \qquad \text{for} \quad \tau \text{ in } D$$

(since f is continuous, in this integral we do not have to bother with values of χ_τ at 0 and 1).

2.4.3. <u>Proposition</u>. If $f = \Sigma a_m \varphi_m$ in $C(I)$ is concave (or convex), then the series $\Sigma a_m \varphi_m$ is unconditionally convergent in $C(I)$.

Proof. From (5) it follows that $a_m \geq 0$ for $m \geq 2$. We now apply Sierpiński's criterion 1.5.7. $\square$

2.4.4. The coefficient a_τ in formula (5) may be interpreted as $a_\tau = -\frac{1}{2}\Delta^2 f$, where $\Delta^2 f$ is the <u>second difference</u> at points τ^-, τ, τ^+, i.e.,

$$\Delta^2 f = [f(\tau^+) - f(\tau)] - [f(\tau) - f(\tau^-)] = f(\tau^+) + f(\tau^-) - 2f(\tau).$$

Consequently, in order to estimate a_τ it is convenient to use the <u>second modulus of continuity</u>:

$$\omega_f^{(2)}(\delta) = \sup\{|f(x_1) + f(x_2) - 2f\left(\frac{x_1+x_2}{2}\right)| : |x_1 - x_2| \leq \delta\}.$$

In particular, we have the obvious inequality

$$(8) \qquad |a_\tau| \leq \frac{1}{2} \omega_f^{(2)} (\frac{1}{2^n}) \qquad \text{for} \qquad \tau \in D_n, \quad n \geq 1.$$

2.4.5. Proposition. Let $f \in C(I)$ and

$$(9) \qquad \sum_{n=1}^{\infty} \omega_f^{(2)} (\frac{1}{2^n}) < \infty.$$

Then the series $\Sigma a_n \varphi_n$ converges unconditionally in $C(I)$.

Proof. Applying (8) we get

$$\sum_{m=2^j+1}^{\infty} |b_m \varphi_m(x)| = \sum_{n=j}^{\infty} \sum_{k=1}^{2^n} |b_{2^n+k} \varphi_{2^n+k}(x)| \leq \frac{1}{2} \sum_{n=j}^{\infty} \omega_f^{(2)} (\frac{1}{2^n}).$$

Thus, the series $\Sigma |a_m \varphi_m|$ is uniformly convergent and we apply 1.5.7. □

2.4.6. Proposition. If f has bounded second derivative, then $\Sigma a_m \varphi_m$ converges absolutely in $C(I)$.

Proof. Denote $c = \sup_{x \in I} |f''(x)|$. Then $\omega_f^{(2)} (\delta) \leq 2c\delta^2$. Hence

$$\sum_{m=2}^{\infty} \| a_m \varphi_m \| = \sum_{n=1}^{\infty} \sum_{\tau \in D_n} |a_\tau| \leq \sum_{n=1}^{\infty} 2^n \max_{\tau \in D_n} |a_\tau| \leq \sum_{n=1}^{\infty} 2^{n-1} \omega_f^{(2)} (2^{-n})$$

$$\leq \sum_{n=1}^{\infty} 2^{n-1} \cdot 2c \cdot 2^{-2n} = \frac{1}{2} c. \quad □$$

2.4.7. Z.Ciesielski (['59] and other papers) assumed that the first function in (φ_n) is constant (the change is not essential because the linear span of first two functions remains the same). In this way one gets a particularly neat representation of the Faber-Schauder functions as indefinite integrals of L^2-normalized Haar functions $\bar{\chi}_m$:

$$(10) \qquad \bar{\varphi}_0(x) = 1, \quad \bar{\varphi}_1(x) = x, \quad \bar{\varphi}_{2^n+k}(x) = \int_o^x \bar{\chi}_{2^n+k}(t)dt$$

for $n = 0,1,\ldots,$ $k = 1,\ldots,2^n$, $\| \bar{\varphi}_{2^n+k} \| = 1/2\sqrt{2^n}$. The function $\bar{\varphi}_0$ plays the role of the integration constant. In this setting, formulas (4) and (5) become particularly simple:

$$\bar{a}_o = f(0), \quad \bar{a}_1 = f(1) - f(0) = \int_o^1 \bar{\chi}_1 df, \quad \bar{a}_n = \int_o^1 \bar{\chi}_n df.$$

Thus,

$$f(x) = f(0) + \sum_{n=1}^{\infty} [\int_o^1 \bar{\chi}_n(t)df(t)]\bar{\varphi}_n(x).$$

In these lecture notes we prefer to keep the general pattern of hat functions (which will be replaced by pyramidal functions in case of several variables, see next chapter). Note that $\bar{\varphi}_0(x) = 1$ may be regarded as a prism-pyramidal function (in the sense of 4.2.2) obtained by passage from dimension 0 to dimension 1.

2.4.8. <u>Exercises</u>. (A) Let (τ_n) be a sequence of dyadic numbers such that the supports of φ_{τ_n} are nonoverlapping, e.g., $\tau_n = 1 - 3 \cdot 2^{-2}$, $n \geq 1$. Then the subsequence (φ_{τ_n}) of the Faber-Schauder system is equivalent to the unit-vector basis (e_n) of $C_0(\mathbb{N})$, and the correspondence $e_n \longmapsto \varphi_{\tau_n}$ can be extended to a linear homeomorphism from $C_0(\mathbb{N})$ onto the closure of the span of (φ_{τ_n}).

(B) Construct a sequence v_n of dyadic numbers such that $v_1 = 1/2$, $\varphi_{v_n}(v_{n+1}) > 1 - 2^{-n}$ and neither the sequence (φ_{v_n}) nor any of its subsequences $(\varphi_{v_{n_k}})$ is equivalent to the unit-vector basis of $C_0(\mathbb{N})$.

(C) Let $\tau \in D_n$ and $f \in C(I)$. Show that $f(\tau)$ can be written as a linear combination of $n+2$ coefficients of the expansion $f = \Sigma_{m=0}^{\infty} a_m \varphi_m$.

(D) Let the series $\Sigma a_n \varphi_n$ be uniformly convergent and let its sum f be in $C^{(1)}(I)$. Then the series of derivatives $\Sigma a_n \varphi_n'$ is uniformly convergent to f' on $I \smallsetminus D$.

(E) Let $\bar{\varphi}_n$ be as in (10). Show that for each t the number $\bar{\varphi}_n(t)$ is the n-th Fourier-Haar coefficient of the characteristic function $I_{[0,t]}$ of the interval $[0,t]$.

(F) Show that

$$(11) \qquad \sum_{m=1}^{\infty} \bar{\varphi}_m(t)\bar{\varphi}_m(u) = \int_0^1 I_{[0,t]}(x) I_{[0,u]}(x) dx = \min(t,u)$$

(hint: apply (E) and Parseval's identity).

2.4.9. We shall now show an application of the Faber-Schauder system to a construction (due to Z.Ciesielski ['61]) of the Wiener stochastic process representing <u>Brownian motion</u>, which is characterized as a Gaussian process with the convariance equal to

$$E(X(t)X(u)) = \min(t,u) \quad \text{for} \quad t,u \in I.$$

Let (Ω, P) be any probability space with any sequence (ξ_n) of independent Gaussian random variables such that $E(\xi_n) = 0$, $E(\xi_n^2) = 1$ for $n = 1, 2, \ldots$ Then for any positive number a and fixed n we have

$$P\{|\xi_n(\omega)| > a\} = \frac{1}{\sqrt{2\pi}} \int_{|x|>a} e^{-x^2/2} dx \leq \frac{2}{\sqrt{2\pi}} \int_a^\infty \frac{x}{a} e^{-x^2/2} dx = \frac{2}{a\sqrt{2\pi}} e^{-a^2/2}.$$

Let $a = \sqrt{\gamma \log n}$ with $\gamma > 2$. Then

$$\sum_{n=1}^\infty P\{|\xi_n| > \sqrt{\gamma \log n}\} \leq \sum_{n=1}^\infty \frac{\text{const}}{\sqrt{\log n}\ n^{\gamma/2}} < \infty.$$

Consequently, by the Borel-Cantelli lemma, the condition

$$(12) \quad \xi_n = O(\sqrt{\log n}) \quad \text{as} \quad n \longrightarrow \infty$$

is satisfied with probability 1. Thus, for almost all ω in Ω, the Schauder series $\sum_{m=0}^\infty |\xi_m(\omega)\bar\varphi_m(t)|$, with coefficients satisfying (12) and with $\bar\varphi_m$ given by (10), is uniformly and unconditionally convergent. Indeed,

$$\sum_{m=2}^\infty |\xi_m \bar\varphi_m(t)| = \sum_{n=0}^\infty \sum_{m=2^n+1}^{2^{n+1}} |\xi_m| 2^{-1-n/2} \varphi_m(t)$$

$$\leq \sum_{n=0}^\infty 2^{-n/2} c(\omega) \sum_{m=2^n+1}^{2^{n+1}} \sqrt{\log_2 m}\ \varphi_m(t)$$

$$\leq c(\omega) \sum_{n=0}^\infty 2^{-n/2} \sqrt{\log_2 2^{n+1}} = c(\omega) \sum_{n=0}^\infty \frac{\sqrt{n+1}}{\sqrt{2^n}} < \infty.$$

Hence

$$(13) \quad X(\omega,t) = \sum_{m=0}^\infty \xi_m(\omega)\bar\varphi_m(t), \quad t \in I$$

is (for almost all ω) a continuous function of t. Moreover, since (ξ_n) are independent, for each t the function $X(\cdot,t)$ is a Gaussian random variable with $E(X) = 0$ and, by Exercise (F),

$$E(X(t)X(u)) = \int_\Omega \sum_{m=0}^\infty \sum_{n=0}^\infty \xi_m(\omega)\xi_n(\omega)\bar\varphi_m(t)\bar\varphi_n(u)\,dP(\omega)$$

$$= \sum_{m=0}^\infty \bar\varphi_m(t)\bar\varphi_m(u) = \min(t,u).$$

Thus, (13) is a Wiener process. It is a family of random variables indexed by t; it may also be regarded as a measurable function from

Ω into C(I). This measurable function carries the measure P on Ω to a measure on C(I), called the <u>Wiener measure</u> on C(I).

2.4.10. <u>Notes</u> (see also 2.6.8). The dyadic hat-function system was first considered by G.Faber ['10], p.112, who also wrote the formulas (4)-(5) for coefficients. This paper was forgotten and only the (more general) construction of Schauder ['27] was quoted. Z.Cie-sielski in a series of papers ['59], ['60a], ['60b], [61] (followed by ['63], ['66], ['69] on the Franklin system) has shown the advantages of the dyadic nodes, which became standard. See also Golubov ['71].

V.A.Matveev ['67] proved that

$$(14) \qquad \| f - s_m \| \le 4\omega_f^{(2)}(\tfrac{1}{m}) \qquad \text{for} \quad f \quad \text{in} \quad C(I), \quad n \ge 1;$$

in particular, if f" is bounded, then $\omega_f^{(2)} = O(\delta^2)$ and hence $\| f - s_m \| = O(m^{-2})$.

A significant theorem of Ciesielski ['60a] states that a function f in C(I) satisfies the Hölder conditions with some exponent α $(0 < \alpha < 1)$ if and only if

$$a_n = O(\frac{1}{n^\alpha}).$$

In 3.6.2 a more general theorem will be proved.

P.L.Ul'janov ['70] has shown that each function f in $L^p(I)$ $0 < p < \infty$, can be represented by a Faber-Schauder series $\Sigma a_n \varphi_n$ convergent in the L^p-norm, but this representation is not unique.

By the Krein-Milman-Rutman theorem, for each $\varepsilon > 0$ there is a basis in C(I) consisting of algebraic or trigonometric polynomials, with basis constant less than $1 + \varepsilon$. (For more on this topic, see 4.5.5 and 4.5.8).

2.5. Basis in C(Y) where $Y \subset [a,b]$.

2.5.1. <u>Proposition</u>. Let Y be a closed subset of an interval [a,b]. Then there exist: a sequence (y_n) dense in Y and a monotone interpolating basis in C(Y) with nodes (y_n).

Proof. We may assume that Y is infinite and $a = \inf Y$, $b = \sup Y$. The set $[a,b] \setminus Y$ is the union of a finite or infinite sequence of disjoint open intervals (a_n, b_n) (if Y has isolated points, it may happen that $a_n = b_m$ for some $n \ne m$). There exist sequences $(y_n)_{n=0,1,\ldots}$ and $(t_n)_{n=0,1,\ldots}$ satisfying the following conditions:

(i) (y_n) is dense in Y, (t_n) is dense in $[a,b]$;

(ii) $y_n \neq y_m$ and $t_n \neq t_m$ for $n \neq m$;

(iii) $y_o = t_o = a$, $y_1 = t_1 = b$;

(iv)   for each n there are p_n, q_n such that $a_n = y_{p_n}$, $b_n = y_{q_n}$;

(v)   there is a sequence $k_o < k_1 < k_2 < \ldots$ such that $y_n = t_{k_n}$ for $n = 0, 1, \ldots$;

(vi)   if $t_m \in (a_j, b_j)$, then the number t_m appears in the sequence (t_n) later than the numbers a_j and b_j, i.e.,

if $t_{k_{p_n}} = y_{p_n} < t_m < y_{q_n} = t_{k_{q_n}}$, then $m > k_{p_n}$ and $m > k_{q_n}$.

Let $(\varphi_n^T)_{n=0,1,\ldots}$ be the Schauder basis in $C[a,b]$ with nodes (t_n). If $t_m \notin Y$, then $t_m \in (a_j, b_j)$ for some j and, by (vi), we infer that $\varphi_m^T(a_j) = 0$, $\varphi_m(b_j) = 0$ and $\text{rest}_Y \varphi_m^T = 0$. It is easy to verify (cf. Proposition 4.2.9) that the functions $\psi_n = \text{rest}_Y \varphi_{k_n}^T$ form a basis in $C(Y)$ with nodes (y_n). □

2.5.2. Corollary. Let Y be a 0-dimensional metrizable compact space (in particular, Y may be any countable compact space). Then $C(Y)$ has an interpolating monotone basis. □

2.5.3. Exercises. (A) What happens with the construction in 2.5.1 if (y_n) is an arbitrary sequence dense in Y and (iv) does not hold ?

(B) Let $Y = \{1/n : n = 1, 2, \ldots\} \cup \{0\}$. Compare the above basis in $C(Y)$ with the standard basis $(e_\infty, e_1, e_2, \ldots)$ in $C(\gamma \mathbb{N})$.

(C) Let X be a countable compact space. Show that there exists a sequence $(A_n)_{n \in \mathbb{N}}$ of open-closed subsets of X such that their characteristic functions χ_{A_n} form a basis in $C(X)$. Hints: Mazurkiewicz-Sierpiński theorem ($\sharp 8.6.10$), Cantor normal form of an ordinal ($\sharp 8.6.4$), countable limit ordinals are cofinal with ω, basis in $C(\gamma Y)$ (see 1.2.2), products and c_o-products (1.4.1 and 1.4.2.)

2.5.4. Notes. A construction of a basis in $C(Y)$ using Schauder hat functions can be found in Gurevič ['54].

2.6. The Franklin functions.

2.6.1. <u>Definitions</u>. Let $\mathcal{T} = (t_n)_{n=0,1,\ldots}$ be a sequence dense in $[a,b]$ with $t_o = a$, $t_1 = b$, and let $\varphi_n^{\mathcal{T}}$ be the corresponding Schauder hat functions defined in 2.3.4. We shall consider the sequence $1, \varphi_1^{\mathcal{T}}, \varphi_2^{\mathcal{T}}, \ldots$ with $\varphi_o^{\mathcal{T}}$ replaced by the constant function 1 (see 2.4.7). Since these functions are linearly independent, the well-known process of Schmidt orthogonalization applied to $1, \varphi_1^{\mathcal{T}}, \varphi_2^{\mathcal{T}}, \ldots$ yields an orthogonal system

(1) $f_n^{\mathcal{T}} = \lambda_{On} + \sum_{j=1}^{n} \lambda_{jn} \varphi_j^{\mathcal{T}}$, where $\lambda_{jj} > 0$.

The functions $f_n^{\mathcal{T}}$ $(n = 0,1,\ldots)$ are called the <u>Franklin functions</u> corresponding to the sequence $\mathcal{T} = (t_n)_{n=0,1,\ldots}$. Obviously, $f_n^{\mathcal{T}} \in B_{T_n}$, where $T_n = \{t_o, \ldots, t_n\}$, i.e., each $f_n^{\mathcal{T}}$ is a broken-line function. The coefficients λ_{jn} can be expressed in terms of the scalar products $\int_o^1 \varphi_k^{\mathcal{T}}(x) \varphi_m^{\mathcal{T}}(x) dx$; yet, this representation is not convenient in computations.

The Franklin functions corresponding to the sequence of dyadic numbers (i.e., the functions obtained by the Schmidt orthogonalization of the functions $\bar{\varphi}_n$ considered in 2.4.7) will be denoted by f_n^F $(n = 0,1,\ldots)$.

2.6.2. <u>Theorem</u> (Ciesielski's inequality). Let T be a partition

(2) $a = x_o < x_1 < \ldots < x_n = b$, $n \geq 1$.

Let $P_T : C[a,b] \longrightarrow B_T$ be the orthogonal projection in $L^2[a,b]$ and let $\tilde{p} = P_T(f)$ for some f in $C[a,b]$. In other words, $\tilde{p}$ is the function in B_T which minimalize the L_2-distance of f from B_T:

(3) $\| f - \tilde{p} \|_2 = \inf\{\| f - p \|_2 : p \in B_T\}$.

Then

(4) $\| \tilde{p} \|_\infty \leq 3 \| f \|_\infty$,   i.e., $\| P_T \| \leq 3$.

Proof. Let p in B_T be of the form $p_u = \sum_{j=0}^{n} u_j h_{x_j}^T$ (see 2.3.1). Then

$$\alpha(u) = \| f - p_u \|_2^2 = \int_a^b | f(x) - p_u(x) |^2 dx$$

is a polynomial of degree 2 of $n+1$ variables $u_o, \ldots, u_n$. Obviously, $\|p_u\| = \max\limits_{0 \le j \le n} |u_j|$. Solving the system of $n+1$ linear equations $\partial \alpha / \partial u_j = 0$ we get the function $\tilde{p}$ in (3). To compute $\partial \alpha / \partial u_j$ we split $\int_a^b$ to n integrals over the intervals $I_j = [x_{j-1}, x_j]$. Note that the integral $\int_{I_j} [f(x) - p_u(x)]^2 dx$ depends only on variables u_{j-1} and u_j, so the remaining derivatives vanish. In this way we get a system of $n+1$ equations

$$2u_o + u_1 = \frac{6}{\delta_1^2} \int_{I_1} f(x)(x_1 - x)dx,$$

$$c_j u_{j-1} + 2u_j + c_j' u_{j+1} = b_j \int_{I_j} f(x)(x - x_{j-1})dx + b_j' \int_{I_{j+1}} f(x)(x_{j+1} - x)dx$$

$$\text{for } j = 1, \ldots, n-1,$$

$$u_{n-1} + 2u_n = \frac{6}{\delta_n^2} \int_{I_n} f(x)(x - x_{n-1})dx,$$

where $\delta_j = x_j - x_{j-1}$, $c_j = \delta_j(\delta_j + \delta_{j+1})^{-1}$, $c_j' = 1 - c_j$, $b_j = 6\delta_j^{-1}(\delta_j + \delta_{j+1})^{-1}$, $b_j' = 6\delta_{j+1}^{-1}(\delta_j + \delta_{j+1})^{-1}$. Each of the integrals $\int_{I_j}$, $\int_{I_{j+1}}$ on the right-hand side is less than or equal to $\frac{1}{2}\|f\| \delta_j^2$, $\frac{1}{2}\|f\| \delta_{j+1}^2$, respectively; here $\|f\| = \|f\|_\infty$. Therefore each left-hand side of the above $n+1$ equations does not exceed $3\|f\|$. Consequently, $|2u_j| \le 3\|f\| + \|\tilde{p}\|$ and hence $2\|\tilde{p}\| \le 3\|f\| + \|\tilde{p}\|$ and $\|\tilde{p}\| \le 3\|f\|$. $\square$

2.6.3. <u>Corollary</u>. Let $S_n^{\mathcal{T}} : C[a,b] \longrightarrow B_{\mathcal{T}_n}$ be the operator of the n-th partial sum of the Fourier-Franklin series $(n \ge 0)$. Thus, $S_n^{\mathcal{T}}(g) = \Sigma_{k=0}^n a_k f_k^{\mathcal{T}}$ where $a_k = \int_a^b g f_k^{\mathcal{T}} dx$. Then

$$\|S_n^{\mathcal{T}}\| \le 3 \quad \text{for } n = 0, 1, \ldots \quad \square$$

2.6.4. <u>Corollary</u>. Let $f \in C[a,b]$ and let p be a function in $B_{\mathcal{T}_n}$ such that $\|f - p\|_\infty$ is minimal. Then

$$\|f - S_n^{\mathcal{T}} f\|_\infty \le 4\|f - p\|_\infty \le 4\omega_f(\delta), \quad \text{where } \delta = \max|t_j - t_{j-1}|.$$

Proof. $\|f - S_n^{\mathcal{T}} f\| \le \|f - p\| + \|p - S_n^{\mathcal{T}} p\| + \|S_n^{\mathcal{T}}(p-f)\| \le 4\|f - p\|$. $\square$

2.6.5. <u>Corollary</u>. The Franklin system $(f_n^{\mathcal{T}})$ is a basis in $C[a,b]$. $\square$

2.6.6. Let $[a,b] = I$. Since the functions f_n^T are orthogonal, we have a formula for coefficients:

$$\mu_n(f) = \int_0^1 f(x) f_n^T(x) dx.$$

On the other hand, $C(I)$ is dense in $L_1(I,\lambda)$, where λ is the Lebesgue measure. Therefore the closed linear span of the measures μ_n in the space $M(I)$ of Radon measures on I consists of all measures $d\nu = hd\lambda$ with $h \in L_1(I,\lambda)$.

2.6.7. Exercises. (A) Modify the construction of the Franklin functions f_n^F by discarding the function φ_1. Show that: (a) the modified functions f_n^P satisfy the condition $f_n^P(0) = f_n^P(1)$ and can be extended to periodic functions on R, (b) the functions f_n^P form a basis in the space of all f in $C(I)$ satisfying $f(0) = f(1)$.

2.6.8. Notes. The functions f_n^T were introduced by P.H.Franklin ['28] (see also Kaczmarz and Steinhaus ['35], Section 4.4). The above simple proof that the Franklin system is a basis in $C(I)$ is due to Ciesielski ['63]. See also 3.9.2, 3.9.5, de Boor ['78], p.43 and Oskolkov ['79].

In 1975 Z.Ciesielski (unpublished) presented a proof that for any partition $T : 0 = x_0 < \ldots < x_n = 1$ of $[0,1]$ the C-norm of the orthogonal projection P_T onto B_T is strictly less than 3, that is, in the inequality (4) one has $\| P_T \| < 3$. Yet, P.Oswald ['77] and K.I.Oskolkov ['79] proved independently that the constant 3 is best possible. Specifically, $\sup \| P_T \| \geq 3 - 4 \cdot 3^{-1} 2^{-n}$ where sup is taken over all partitions of I with n distinct knots between 0 and 1, and the supremum is approached when the knots tend to 0 so that $t_{k+1} - t_k$ increases geometrically. One the other hand, Ciesielski ['75] has shown that for the unifom partitions, i.e., with constant difference $t_{k+1} - t_k$, one has $2 - \varepsilon_n < \| P_T \| < 2$, where $\varepsilon_n \longrightarrow 0$ as the number of knots tends to ∞.

Matveev's estimation (13) in 2.4.10 together with 2.6.2 imply that $\| f - \Sigma_{k=1}^m a_k f_k^F \| \leq 7 \omega_f^{(2)}(m^{-1})$; a similar inequality was proved by Ul'janov ['70].

It should be noted that 2.6.5 is a very special property of polygonal functions. W.Szlenk ['66] has constructed, for any given sequence $\varepsilon_n \longrightarrow 0$, a basis (g_n) in $C(I)$ such that $\| g_n - \varphi_n \| < \varepsilon_n$ where φ_n is the Faber-Schauder system, and yet the Schmidt ortho-gonalization of (g_n) is not a basis in $C(I)$. For more general theorems in this line, see A.M.Olevskiĭ ['70] and Z.A.Čanturija ['71].

A.M.Olevskiĭ ['75], p.22, showed that for any modulus of continuity

ω on I (i.e., continuous increasing subadditive function ω such that $\omega(0)=0$) there exists a uniformly bounded orthogonal system (g_m) such that each f in H^ω has a uniformly convergent Fourier series with respect to (g_m).

There are various theorems akin to 2.6.2 which estimate the C[a,b] - norm of an orthogonal projection. Let us mention the result of S.M.Lozinskiĭ (see I.K.Daugavet ['77], p.179): Let T_n be the subspace of all trigonometric polynomials of order n in C[0,2π]. Let P be any projection from C[0,2π] onto T_n and let S_n be the corresponding projection, i.e., the operator of n-th partial sum of the Fourier series. Then $\|P\| \geq \|S_n\|$.

2.7. Hat-function bases in $C_0(\mathbb{R})$.

2.7.1. We shall deal with the space $C_0(\mathbb{R})$ of functions continuous on $\mathbb{R}$ and tending to O as $|x| \longrightarrow \infty$. Let

$$\varphi_{k+\tau}(x) = \varphi(2^n(x-k-\tau))$$

where $k \in \mathbb{Z}$, $\tau \in D_n$, $\tau \neq 1$ and $\varphi(x) = \max(0, 1-|x|)$. Thus, $\varphi_{k+\tau}(x) = \varphi_\tau(x-k)$, where φ_τ is the Faber-Schauder dyadic function with peak at τ (extended to $\mathbb{R}$ by setting $\varphi_\tau(x) = O$ for x in $\mathbb{R} \setminus I$).

Denote $V = Z + \tilde{D} = \{k+\tau : k \in \mathbb{Z}, \tau \in \tilde{D}\}$. It is easy to see that the system $(\varphi_v)_{v \in V}$, suitably ordered, is a basis in $C_0(\mathbb{R})$. Let us consider the order determined by the partition $V = V_0 \cup V_1 \cup \ldots$ where $V_0 = \{0,1\}$, $V_1 = \{-1, \frac{1}{2}, 2\}$, and V_n consists of the following numbers: $-n, -n+1+D_1, -n+2+D_2, \ldots, -n+k+D_k, \ldots, -1+D_{n-1}, D_n, 1+D_{n-1}, 2+D_{n-2}, \ldots, n-1+D_1, n+1$ (the symbol $a+D_k$ stands here for the increasing sequence of all numbers $a+\tau$ with $\tau \in D_k$).

We now define an order on V assuming that the elements of V_k are earlier than the elements of V_n if $0 \leq k < n$, and within each V_n we keep the usual order. In this way we get: a sequence $(v_n)_{n \in \mathbb{N}}$ of all elements of V and the corresponding sequence φ_{v_n} of functions. In the figure below we have first 11 functions (the index of a function is marked at its peak).

Thus, if $v,w \in V_n$, then the supports of φ_v and φ_w are non-overlapping; consequently

$$\| \sum_{v \in V_n} a_v \varphi_v \| = \sup_{v \in V_n} |a_v|.$$

The coefficients a_v of a function f in $C_o(\mathbb{R})$ are equal to

$$a_k = f(k) \quad \text{for } k \text{ in } Z,$$

$$a_{k+\tau} = f(k+\tau) - \frac{1}{2}[f(k+\tau^-) + f(k+\tau^+)] \quad \text{for } k \text{ in } Z, \ \tau \text{ in } D.$$

2.7.2. Let f_{v_n} be the Schmidt orthogonalization of $\varphi_{v_1}, \varphi_{v_2}, \ldots$ in $L^2(\mathbb{R})$. Thus, each f_v is a polygonal function of compact support and

$$\int_{-\infty}^{\infty} f_v(x) f_w(x) \, dx = 0 \quad \text{for } v \neq w, \qquad \int_{-\infty}^{\infty} [f_v(x)]^2 dx = 1.$$

2.7.3. <u>Proposition</u>. For a given sequence $T : x_o < x_1 < \ldots < x_n$ let $B_T^{\mathbb{R}}$ be the space of all functions p in $C_o(\mathbb{R})$ such that $p(x) = 0$ for $x < x_o$ and for $x > x_n$ and p is affine on each interval $I_j = [x_{j-1}, x_j]$, $j = 1, \ldots, n$. Let

$$P_T : C_o(\mathbb{R}) \longrightarrow B_T^{\mathbb{R}}$$

be the orthogonal projection in the space $L^2(\mathbb{R})$. Then

(1) $$\| P_T(f) \|_\infty \leq 3 \| f \|_\infty .$$

Proof. We begin as in 2.6.2 with $[a,b] = [x_o, x_n]$. Consider the minimum of the function

$$\alpha(u) = \| f - p_u \|_2^2 = \int_{-\infty}^{\infty} |f(x) - p_u(x)|^2 dx,$$

where $u = (u_o, \ldots, u_n) \in \mathbb{R}^{n+1}$ with $u_o = 0$, $u_n = 0$. We now have $n-1$ equations $\partial \alpha / \partial u_j = 0$, $j = 1, \ldots, n-1$. Therefore the first and the last of the linear equations in 2.6.2 are discarded and we get

$$c_j u_{j-1} + 2u_j + c_j' u_{j+1} = b_j \int_{I_j} f(x)(x - x_{j-1}) dx + b_j' \int_{I_{j+1}} f(x)(x_{j+1} - x) dx$$

for $j = 1, \ldots, n-1$ with $u_o = u_n = 0$. The same argument as before yields $|u_j| \leq 3 \| f \|$ for $j = 1, \ldots, n-1$. $\square$

2.7.4. <u>Corollary</u>. The system $(f_v)_{v \in V}$ is a basis in $C_o(\mathbb{R})$. □

2.7.5. <u>Exercises</u>. (A) Let $x_o < x_1 < x_2$. Find the infimum of $\int_{x_o}^{x_2} |p(x)|^2 dx$ over all affine functions p on the interval $[x_o, x_2]$ such that $p(x_2) = 1$

(B) Find the infimum of $\int_{x_o}^{x_2} |p(x)|^2 dx$ over all polygonal functions on $[x_o, x_2]$ with one break point x_1 $(x_o < x_1 < x_2)$ and such that $p(x_2) = 1$.

(C) Let $f \in C[a,b]$ and $f(x) = 0$ for x in $I_1 \cup I_2$, where $a = x_o$, $b = x_n$, $I_j = [x_{j-1}, x_j]$. Let $\tilde{u}_o, \ldots, \tilde{u}_n$ be the solution of the system of equations considered in 2.6.2. (a) Show that $\tilde{u}_o$ need not be equal to O (hint: consider partition such that $(x_1 - x_o)/(x_2 - x_1)$ is small and an f in $C[a,b]$ of norm 1, vanishing on I_1 and such that $\int_{x_1}^{x_2} f dx$ is close to $x_2 - x_1$; apply Exercises (A) and (B)).
(b) Show that

$$|u_o| \leq (x_2 - x_1)/(x_2 - x_o) \| f \| .$$

(c) Suppose that one omits the breakpoint x_1 and gets the solution $\hat{u}_o, \hat{u}_2, \ldots, \hat{u}_n$ of the corresponding system of equations for the ortho-gonal projection from $C[x_o, x_n]$ onto the subspace of polygonal func-tions with breakpoints $x_o, x_2, \ldots, x_n$. Show that $\hat{u}_o \neq \tilde{u}_o$ in general (hint: Exercise (B)).

BASES OF PYRAMIDAL AND MULTIAFFINE FUNCTIONS OF SEVERAL VARIABLES

This chapter is devoted to the description of various bases in the space $C(I^d)$, where I^d is a d-dimensional cube, $d \geq 2$. In Sections 3.1-3.6 we deal with pyramidal functions, i.e., functions $f : I^d \longrightarrow R$ such that the graph of f in R^{d+1} is the lateral surface of a pyramide (on the support of f). In Section 3.7, the basis functions are piecewise multiaffine, i.e., products

$$h(x_1,\ldots,x_d) = (a_1x_1+b_1)(a_2x_2+b_2)\ldots(a_dx_d+b_d)$$

and finite sums of such products. In 3.8 such functions are used to construct bases of differentiable functions.

The reader may find it a good idea to interwine reading of 3.1 with 3.2 and 3.3.

3.1. The concept of a pyramidal basis.

3.1.1. <u>Definitions</u>. In this section the letter X stands for a convex polytope of dimension d, $X \subset R^d$, or more generally, for any d-dimensional polytope, not necessarily convex. In the latter case we assume that there is a simplicial complex C (see e.g. Grünbaum ['67], p.59) such that each member of C is contained in a d-simplex of C and X is the union of all those simplexes.

By a <u>triangulation</u> of X we shall mean a set T of d-simplexes such that

    (a) $\cup\{\Delta : \Delta \in T\} = X$,

    (b)   if $\Delta,\Delta' \in T$ and $\Delta \cap \Delta' \neq \emptyset$, then $\Delta \cap \Delta'$ is a common face of Δ and Δ'.

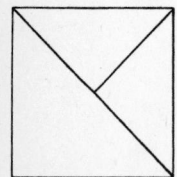

Thus, the set of 3 triangles shown in the picture on the left is not a triangulation of the square. From (b) it follows that the interiors of Δ and Δ' are disjoint; also, if v is a vertex of Δ and $v \in \Delta'$, then v must be a vertex of Δ'.

If X is not convex, we shall restrict ourselves to only such triangulations of X which are refinements of that given by C.

If T is a triangulation of X, the symbol $\text{vert}(T)$ will denote the set of vertices of T, i.e. the set of all vertices of all simplexes in T.

A function $\varphi : X \longrightarrow \mathbf{R}$ will be called <u>pyramidal</u> with respect to a triangulation T of X (shortly: T-pyramidal) if there exists a vertex v in $\text{vert}(T)$, called the <u>peak vertex</u> of φ, such that:

(i) $\varphi(v) = 1$,

(ii) $\varphi(w) = 0$ if $w \in \text{vert}(T)$ and $w \neq v$,

(iii) φ is affine on each simplex Δ in T, i.e.

$$\varphi(\Sigma c_i w_i) = \Sigma c_i \varphi(w_i) \quad \text{for} \quad w_i \in \text{vert}(\Delta), \quad c_i \geq 0, \quad \Sigma c_i = 1.$$

From (iii) it follows that the support of φ, i.e. the closure of $\{x \in X : \varphi(x) > 0\}$, can be represented as the union of all simplexes of T containing v. Note that this support need not be convex (see the shaded area on the picture).

Obviously, for each triangulation T of X and each v in $\text{vert}(T)$ there is exactly one T-pyramidal function with vertex v.

A triangulation T' will be called a <u>subtriangulation</u> of T (in symbol: $T' \leq T$) is each simplex Δ' in T' is contained in a simplex Δ in T; Consequently, each simplex in T is the union of some simplexes in T' and $\text{vert}(T) \quad \text{vert}(T')$. $T' < T$ will mean that T' is a subtriangulation of T and $T \neq T'$.

Let $T_1 > T_2 > \ldots$ be a sequence of triangulations of X. Denote $W_n = \text{vert}(T_n)$, $V_1 = W_1$, $V_{n+1} = W_{n+1} \quad W_n \quad (n \geq 1)$, $V = U_{n=1}^{\infty} V_n$.

A basis (φ_m) in $C(X)$ will be called <u>pyramidal</u> with respect to the sequence (T_n) (or shortly, pyramidal) iff the following conditions are satisfied:

(α)   for each m in $\mathbf{N}$ there exists a ν in $\mathbf{N}$ such that φ_m is pyramidal with respect to T_ν; the smallest ν with this property will be denoted by $\nu(m)$,

(β)   if $m < m'$, then $\nu(m) \leq \nu(m')$,

(γ)   each v in V is the peak vertex of exactly one function φ_m.

A pyramidal basis may or may not satisfy another condition

(δ)   if $v \in W_n$ and v is the peak vertex of φ_m, then $\nu(m) \leq n$.

If $v, w \in V$ and $\varphi_n(v) = 1$, $\varphi_m(w) = 1$, then $v < w$ will mean that $n < m$.

3.1.2. <u>Lemma</u>. For each sequence $T_1 > T_2 > T_3 > \ldots$ of triangulations of X there is a sequence (φ_m) in $C(X)$ satisfying conditions $(\alpha) - (\delta)$. $\square$

3.1.3. It will be proved in 3.1.6 that if a sequence $\varphi_1, \varphi_2, \ldots$ in $C(X)$ satisfies $(\alpha) - (\delta)$ and $\lim d_n = 0$, where $d_n = \sup\{\text{diam}\, \Delta : \Delta \in T_n\}$, then (φ_m) is, indeed, a basis in $C(X)$, i.e., each f in $C(X)$ can be uniquely written as a uniformly convergent series $\Sigma a_m \varphi_m$; moreover, (φ_m) is interpolating with nodes $v_1, v_2, \ldots$

Together with notation $\varphi_1, \varphi_2, \ldots$ we shall often use a modified one, with vertices as indices; φ_v will denote the pyramidal function φ_m with peak vertex v and a_v will denote the corresponding coefficient a_m. From (α) and (γ) it follows that $v < w$ implies $\varphi_w(v) = 0$.

Besides the symbol s_m for the partial sum $\Sigma_{k \leq m} a_k \varphi_k$, we shall use the symbol $s^{(n)}$ to denote the partial sum over all vertices of n first triangulations provided that condition (δ) is satisfied. Thus,

$$(1) \qquad s^{(n)} = \sum_{\nu(k) \leq n} a_k \varphi_k = \sum_{w \in W_n} a_w \varphi_w, \quad n \geq 1.$$

If $v \in W_n$, let $s_v = s^{(n-1)} + \Sigma a_w \varphi_w$ where the sum is extended over w in V_n satisfying $w \leq v$. The sum $s^{(n)}$ has two crucial properties:

1) $s^{(n)}$ is affine on each simplex of T_n,

2) $s^{(n)}(v) = f(v)$ for v in W_n.

These properties determine $s^{(n)}$ uniquely. Let also $s^{(o)} = 0$.

3.1.4. Let $v = v_k$, $w = v_m$, $k < m$, $\varphi_k(w) > 0$, $c = \varphi_k(w)^{-1}$. Then, by (β), $\nu(k) \leq \nu(m)$ and $\text{supp}\,\varphi_m$ is the union of Δ' in $T_{\nu(m)}$ such that $w \in \Delta'$. Obviously, $\varphi_k \geq c \varphi_m$ on each Δ'. Hence

$$(2) \qquad \text{if } w > v \text{ and } \varphi_v(w) > 0, \text{ then } \{x : \varphi_w(x) > 0\} \subset \{x : \varphi_v(x) > 0\}.$$

3.1.5. We shall derive formulas for coefficients in the sum (1). First let $v \in W_1$. Then

$$f(v) = s^{(1)}(v) = \sum_{w \in W_1} a_w \varphi_w(v) = a_v,$$

since $\varphi_w(v) = 0$ for $w \neq v$ and $\varphi_v(v) = 1$. Thus, the coefficients corresponding to the vertices of T_1 are just the values of the function f at these vertices, i.e.,

(3) $a_v = f(v)$ for v in W_1.

Now, let $v \in V_n = W_n \setminus W_{n-1}$, $n \geq 2$. Then

$$f(v) = s^{(n)}(v) = s^{(n-1)}(v) + \underset{w \in V_n}{\Sigma} a_w \varphi_w(v) = s^{(n-1)}(v) + a_v.$$

Hence,

(4) $a_v = f(v) - s^{(n-1)}(v)$ for v in V_n.

Thus, a_v is the difference between $f(v)$ and a certain convex combination $s^{(n-1)}(v)$ of values of f at vertices of a simplex Δ in T_{n-1}.

3.1.6. <u>Proposition</u>. Let $T_1 > T_2 > \ldots$ be a sequence of triangulations of X and let $(\varphi_m)_{m \in \mathbb{N}}$ be a sequence in $C(X)$ satisfying conditions $(\alpha) - (\delta)$. Then the following conditions are equivalent:

(i) (φ_m) is a basis in $C(X)$,

(ii) $d_n \longrightarrow 0$, where $d_n = \sup\{\text{diam } \Delta : \Delta \in T_n\}$,

(iii) for each x in X, $\overset{\infty}{\underset{n=1}{\cap}} T_n(x) = \{x\}$, where $T_n(x) = $ $\cup \{\Delta : \Delta \in T_n, x \in \Delta\}$.

Proof. (ii) $\Rightarrow$ (i) Obviously, (φ_m) satisfies condition (ii) of 1.3.2. Therefore it is enough to show that $\Sigma a_v \varphi_v$ converges to f uniformly, where a_v are given by (3) and (4). Let $x \in X$ and $n \in \mathbb{N}$. Then x belongs to some simplex Δ of T_n. Since $s^{(n)}$ is affine on Δ and coincides with f at each vertex of Δ, both numbers $f(x)$ and $s^{(n)}(x)$ are between the infimum and the supremum of f on Δ. Consequently,

(5) $|f(x) - s^{(n)}(x)| \leq \omega_f(\text{diam } \Delta) \leq \omega_f(d_n)$.

Similarly

(6) $|f(x) - s_v(x)| \leq \omega_f(d_n)$ for v in $V^{(n+1)}$, x in X.

Thus, $\| f - s_m \| \longrightarrow 0$ as $m \longrightarrow \infty$. Moreover,

(7) $|a_v| \leq \omega_f(d_{n-1})$ for v in $V^{(n)}$.

(i) $\Rightarrow$ (ii). Suppose that (ii) is not satisfied, i.e., there exists an $\varepsilon > 0$ such that $d_n \geq d_{n+1} \geq \varepsilon$ for $n = 1,2,\ldots$ There exist: a sequence $\Delta_n \supset \Delta_{n+1}$ of simplexes Δ_n in T_n and points x_n, y_n in Δ_n such that $\rho(x_n, y_n) \geq \varepsilon$, where ρ is the Euclidean distance. Consequently, $x_{n_k} \longrightarrow x_0$ and $y_{n_k} \longrightarrow y_0$ for some subsequence (n_k), and $\rho(x_0, y_0) \geq \varepsilon$. Clearly, the segment J joining x_0 and y_0 belongs to $\cap_{n=1}^{\infty} \Delta_n$. Therefore all functions φ_m and hence all functions $\Sigma a_m \varphi_m$ are affine on J. Thus, the sums $\Sigma a_m \varphi_m$ $(a_m \in \mathbf{R})$ cannot be dense in $C(X)$.

The implication (ii) $\Rightarrow$ (iii) is obvious. Finally, (iii) implies (ii), since from $\lim d_n > 0$ it follows that $J \subset \cap_{n=1}^{\infty} T_n(x)$. $\square$

3.1.7. If T' is a subtriangulation of T, then a function pyramidal with respect to T may happen to be pyramidal also with respect to T' (even if each Δ' in T' is a proper subset of some Δ in T, see the picture, where dotted lines belong to T'). T' will be called a <u>strict subtriangulation</u> of T (in symbol: $T \gg T'$) if, besides conditions (a) and (b), for each Δ in T there exists a vertex v of T' in the interior of Δ.

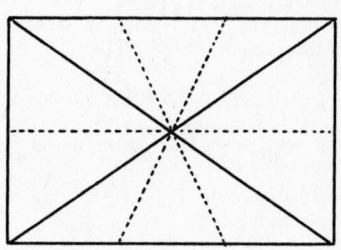

If $T \gg T'$ and φ is T-pyramidal, (ii) fails to hold and φ is not T'-pyramidal. Consequently, if (φ_m) is a pyramidal basis with respect to a sequence (T_n) of triangulations and $T_1 \gg T_2 \gg \ldots$, then each v in condition (α) is uniquely determined by m; moreover, the sequence (φ_m) is then determined by conditions $(\alpha) - (\delta)$ uniquely up to permutations of sets $\Phi_n = \{\varphi_m : v(m) = n\}$.

3.1.8. <u>Exercises</u>. (A) Show that $\omega_{s_v}(\sigma) \leq \omega_f(\sigma)$ for all f, v, σ.

(B) Show that the inequality (5) cannot be replaced by $|a_v| \leq c \omega_f(d_m)$ with some $c < 1$.

(C) If f satisfies the Lipschitz condition with a constant L, then $|a_v| \leq \frac{1}{2} L d_n$.

(D) Find example of a pyramidal basis not satisfying condition (δ) in 3.1.1.

3.2. The regular pyramidal basis in $C(\mathbf{I}^2)$

3.2.1. We now construct triangulations $T_1 > T_2 > \ldots$ of the square $\mathbf{I}^2$ and the corresponding pyramidal functions in a sequence of steps $\Omega(n,k)$, where $n = 1,2,\ldots$, $k = 1,2$, with recur instructions. First 6 steps are presented in a form of a tabulation. In each picture below, a heavy dot • denotes a new vertex while a normal dot · denotes a previously constructed vertex. In the last column, the shaded area is the support of a typical pyramidal function in a given step; strips are drawn parallel to the lines where φ_v is constant.

Step	Triangulations	set and number of new vertices	Typical function
$\Omega(1,1)$	Take the vertices of the square $\mathbf{I}^2$ Draw diagonals	$D_o \times D_o$ 4	
$\Omega(1,2)$	Add the center of $\mathbf{I}^2$ (intersection of diagonals)	$D_1 \times D_1$ 1	

After these two initial steps, from step $\Omega(2,1)$ on, the pattern is always the same. In each step $\Omega(n,k)$, $n \geq 2$, $k = 1,2$, one has to proceed as follows: (a) add centers of pairs of adjacent vertices of step $\Omega(n-1,k)$, (b) join these centers with the nearest vertices.

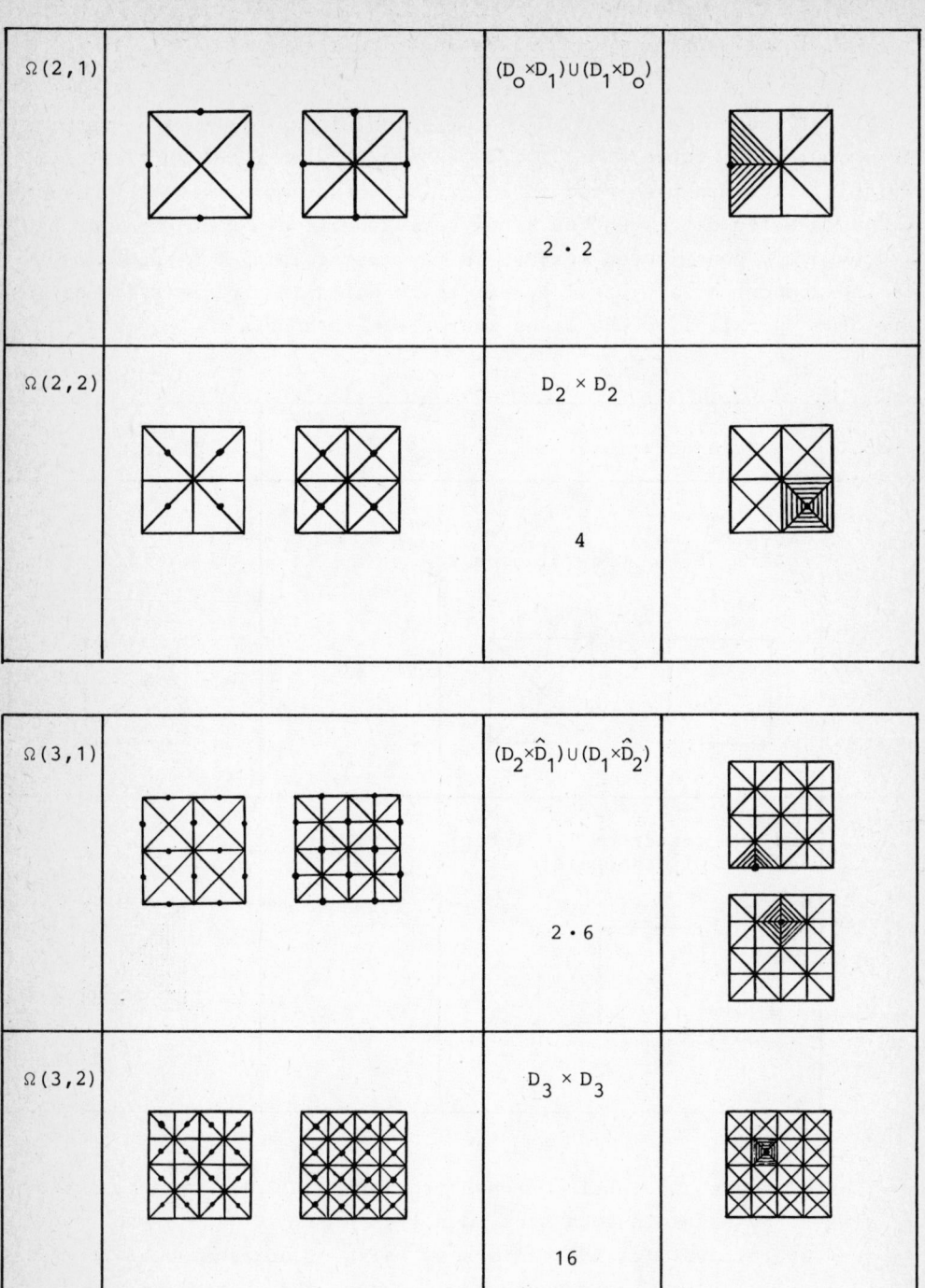

The triangulation obtained in step $\Omega(n,2)$ will be denoted by T_n. It has $2^{2n-1} + 2^n + 1$ vertices. Construction in step $\Omega(n,1)$ plays an auxiliary role (this will be seen more clearly for dimensions $d > 2$). Note that the edges and vertices of $\Omega(1,1)$ do not form a triangulation.

3.2.2. All pyramidal functions constructed for this sequence of triangulations can be obtained - by shrinking the bases of pyramides and shifting them - from two functions Θ and $\hat{\Theta}$ defined as

$$\hat{\Theta}(x) = \max[0, 1 - (|x_1| + |x_2|)] \quad \text{for} \quad x = (x_1,x_2) \quad \text{in} \quad \mathbf{R}^2,$$

$$\Theta(x) = \max[0, 1 - 2\max(|x_1|,|x_2|)] = \max[0, 1 - (|x_1+x_2| + |x_1-x_2|)].$$

In step $\Omega(n,1)$, $n \geq 2$, there are $2^{2n+3} + 2^{n-1}$ functions

$$\varphi_v(x) = \hat{\Theta}(2^{n-1}(x-v)), \quad v \in (D_{n-1} \times \hat{D}_{n-2}) \cup (\hat{D}_{n-2} \times D_{n-1}),$$

whereas in step $\Omega(n,2)$ there are 2^{2n-2} functions $\varphi_v(x) = \Theta(2^{n-1}(x-v))$, $v \in D_n \times D_n$.

3.2.3. The formulas for the coefficients a_v derived in 3.1.4 can now be written in a more explicit form. We recall that $W_n = \mathrm{vert}(T_n)$, $V_1 = W_1$ and $V_{n+1} = W_{n+1} \setminus W_n$. If $v \in V_1$, then $a_v = f(v)$. Moreover, $a_v = f(v) - s^{(n-1)}(v)$ for v in V_n, $n \geq 2$. Let $v = (\frac{1}{2},0) \in V_2$, say. The function $s^{(1)}$ is affine on the segment joining $(0,0)$ and $(1,0)$; hence

$$s^{(1)}(\tfrac{1}{2},0) = \tfrac{1}{2}[s^{(1)}(0,0) + s^{(1)}(1,0)] = \tfrac{1}{2}[f(0,0) + f(1,0)].$$

Similarly, for each v in V_n,

$$(1) \qquad a_v = f(v) - \tfrac{1}{2}[f(v') + f(v'')], \quad v = \tfrac{1}{2}(v' + v''),$$

where v',v'' are the vertices in step $\Omega(n,k)$ such that v is the center of v',v''. In other words, $-2a_v$ is the second difference $\Delta^2 f$ computed for points v',v,v''.

There are two possible ways of looking for explicit formulas for v' and v''. The first is to extend the operations $\tau \longrightarrow \tau^+, \tau \longrightarrow \tau^-$ on D (see 2.1.2) to the case of $D \times D$. If $v = (\tau_1,\tau_2) \in D_n \times D_m$ and $n \neq m$, define

$$v^- = (\tau_1^-, \tau_2), \quad v^+ = (\tau_1^+, \tau_2) \quad \text{if} \quad n > m,$$

and

$$v^- = (\tau_1, \tau_2^-), \quad v^+ = (\tau_1, \tau_2^+) \quad \text{if} \quad n < m.$$

Then

$$v = \frac{1}{2}(v^+ + v^-) \quad \text{and} \quad a_v = f(v) - \frac{1}{2}[f(v^+) + f(v^-)].$$

If $v = (\tau_1, \tau_2) \in D_n \times D_n$, $\tau_1 = (2k-1)/2^n$, $\tau_2 = (2j-1)/2^n$, then

$$a_v = f(v) - \frac{1}{2}[f(\tau_1^+, \tau_2^+) + f(\tau_1^-, \tau_2^-)] \quad \text{if} \quad k+j \quad \text{is even},$$

$$a_v = f(v) - \frac{1}{2}[f(\tau_1^+, \tau_2^-) + f(\tau_1^-, \tau_2^+)] \quad \text{if} \quad k+j \quad \text{is odd}.$$

Another approach (which is more useful for $d > 2$) is to use operations $\flat$ and $\sharp$ (see 2.1.2). Let $v = (\tau_1, \tau_2) \in D_n \times D_n$; v is a vertex added in step $\Omega(n,2)$, i.e. the center of a square in T_n. Denote

$$v^{\flat\flat} = (\tau_1^\flat, \tau_2^\flat), \quad v^{\flat\sharp} = (\tau_1^\flat, \tau_2^\sharp), \quad v^{\sharp\flat} = (\tau_1^\sharp, \tau_2^\flat), \quad v^{\sharp\sharp} = (\tau_1^\sharp, \tau_2^\sharp).$$

Obviously, these four points belong to $D_{n+1} \times D_{n+1}$ and are just points arising in step $\Omega(n+1,2)$. Thus, we get 4 operations

$$\gamma_{i,j} = \bigcup_{n=1}^{\infty} (D_n \times D_n) \longrightarrow \bigcup_{n=2}^{\infty} (D_n \times D_n), \quad i,j = \pm 1,$$

where $\gamma_{-1,-1}(v) = v^{\flat\flat}$, $\gamma_{-1,1}(v) = v^{\flat\sharp}$ etc.

In the figure shown on the right, the heavy dots denote the center v of one of the sequence of a triangulation T_n and the four points $v^{\flat\flat}$, $v^{\flat\sharp}$, $v^{\sharp\flat}$, $v^{\sharp\sharp}$. The remaining 16 dots show the results of iterating those 4 operations, i.e. applying them to the first 4 points.

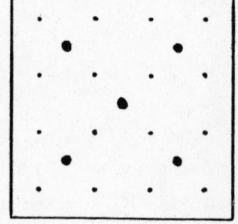

It is clear that all points of $\bigcup_{n=2}^{\infty}(D_n \times D_n)$, i.e. the peak vertices of functions of the type $\varphi_v = \theta(2^{n-1}(x-v))$, arising in steps $\Omega(n,2)$, $n = 2,3,\ldots$, arise from the single point

$(\frac{1}{2},\frac{1}{2})$ by iterating the operations $\gamma_{i,j}$ $(i,j = \pm 1)$. For each w in $D_n \times D_n$, $n \geq 2$, there is exactly one of the above operations $\gamma_{i,j}$ and exactly one v in $D_{n-1} \times D_{n-1}$ such that $w = \gamma_{i,j}(v)$. Denote $v = w^\natural$. Thus $\natural$ is a function from $U_{n=2}^{\infty}(D_n \times D_n)$ to $U_{n=1}^{\infty}(D_n \times D_n)$ and $\natural\gamma_{i,j}(v) = v$ for $i,j = \pm 1$. If $v \in D_n \times D_n$, $n \geq 2$, then $v^\natural$ is one of the points v',v'' in the formula (1) for the coefficient a_v:

(2) $\qquad a_v = f(v) - \frac{1}{2}[f(v^\natural) + f(v^*)],$

$\qquad$ where $v^* = v + (v - \natural)$ $\qquad$ and $\qquad v = \frac{1}{2}(v^\natural + v^*)$.

The coordinates of $v^\natural$ are $\tau_1^\natural, \tau_2^\natural$ and can be computed in the way described in 2.1.2.

$\qquad$ We assume that the operations $\gamma_{i,j}$ $(i,j = \pm 1)$ are not defined on $D_n \times D_m$ if $n \neq m$. In order to extend formula (2), however, we define four new operations $\gamma_{0,-1}, \gamma_{0,1}, \gamma_{-1,0}, \gamma_{1,0}$ where $\gamma_{0,-1}(v) = v^O = (\tau_1, \tau_2)$ and $\gamma_{0,1}(v) = v^O = (\tau_1, \tau_2)$ for $v = (\tau_1, \tau_2)$ in $D_n \times D_m$, $1 \leq n \leq m$, and analogously $\gamma_{-1,0}(v) = v^O = (\tau_1, \tau_2)$ and $\gamma_{1,0}(v) = v^O = (\tau_1, \tau_2)$ for v in $D_n \times D_m$, $n \geq m \geq 1$ (see fig.A below).

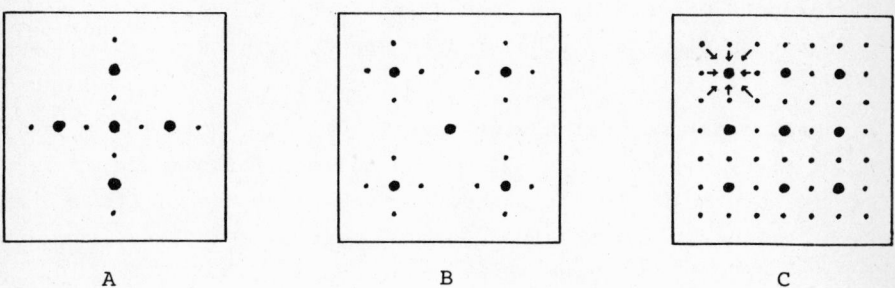

$\qquad\qquad$ A $\qquad\qquad\qquad\qquad\qquad$ B $\qquad\qquad\qquad\qquad\qquad$ C

Note that e.g. $\gamma_{1,j}(\gamma_{0,1}(v))$ $(j = 0, \pm 1)$ is not defined.

$\qquad$ If $v = (\tau_1, \tau_2)$ and both τ_1, τ_2 belong to the same D_n, all operations $\gamma_{i,j}(i,j = 0, \pm 1)$ can be applied to v (see fig.B). Yet, if $\tau_1 \in D_n$, $\tau_2 \in D_m$ and $n < m$, only $\gamma_{0,j}$ can be applied, and the first coordinate remains fixed. Therefore all further iterations of these operations cannot take the points out of the vertical line $\tau_1 = const$. Analogously, if $n > m$, all iterations lead to points on the line $\tau_2 = \text{const.}$

Fig.C gathers points presented separately in 3 previous pictures. The center v satisfies the condition $v = w^\natural$ for each of 8 heavy dots around. The arrows show the function $u \longrightarrow u^\natural$ for vertices of the next triangulation.

The functions $\gamma_{j,0}$ $(j = \pm 1)$ can also be applied to vertices on the horizontal sides of the original unit square, i.e. points in $D \times D_0$, whereas $\gamma_{0,1}$ can be defined on $D_0 \times D$. The functions

$$\natural : D_n \times D_0 \longrightarrow D_{n-1} \times D_0 \quad \text{and} \quad \natural : D_0 \times D_n \longrightarrow D_0 \times D_{n-1}, \quad n \geq 2,$$

can be defined analogously, and (2) still holds. There is no natural way, however, to define functions $\gamma_{i,j}$ on $D_0 \times D_0$; consequently, $\natural$ is not defined at $(\frac{1}{2}, \frac{1}{2})$, $(\frac{1}{2}, 0)$ and $(0, \frac{1}{2})$.

3.2.4. <u>Notes</u>. The above basis was described geometrically by H.W.Ellis and D.G.Kuehner ['60] and by J.Ryll ['73].

3.3. Description of a regular pyramidal basis in $C(\mathbf{I}^d)$, $d \geq 3$.

3.3.1. The preceding construction for $C(\mathbf{I}^2)$ helps us greatly to understand the analogous basis in $C(\mathbf{I}^3)$ and, more generally, in $C(\mathbf{I}^d)$. Yet, the case $d = 2$ may deceive a reader unaware of significant differences between dimensions 2 and 3. E.g., the center of $\mathbf{I}^d$, $d \geq 3$, is not an element of the simplex spanned by the point O and d unit vectors.

We shall construct triangulations $T_1 > T_2 > \ldots$ of $\mathbf{I}^3$ in steps $\Omega(n,k)$, $n = 1, 2, \ldots$, $k = 1, 2, 3$. In third column, for simplicity we shall write not the whole set of new vertices, but one of its typical subsets, the other being obtained just by permuting indices, e.g. if we write $D_1 \times D_1 \times D_0$, it should be understood

$$(D_1 \times D_1 \times D_0) \cup (D_1 \times D_0 \times D_1) \cup (D_0 \times D_1 \times D_1),$$

and then we write the number of new vertices in the last column as $3 \cdot 2$. The triangulation obtained in step $\Omega(n,3)$ is denoted by T_n.

Step	Description of new vertices	Set of new vertices	number of new vertices
$\Omega(1,1)$	The vertices of the cube $\mathbf{I}^3$	$D_o \times D_o \times D_o$	8
$\Omega(1,2)$	The centers of faces	$D_1 \times D_1 \times D_o$	$3 \cdot 2$
$\Omega(1,3)$	The center of the cube	$D_1 \times D_1 \times D_1$	1
$\Omega(2,1)$	The centers of edges (remaining vertices of cubes of the triangulation T_2)	$D_1 \times D_o \times D_o$	$3 \cdot 4$
$\Omega(2,2)$	The centers of faces of cubes of T_2	$D_2 \times D_2 \times \hat{D}_1$	$3 \cdot 12$
$\Omega(2,3)$	The centers of cubes of T_2	$D_2 \times D_2 \times D_2$	8

3.3.2. We now pass to the description of the functions φ_v in $C(\mathbf{I}^3)$. Let $v \in D_o^3$. We have to look for a pyramidal function which is equal to 1 at v and to 0 at all other vertices of T_1, i.e. at points constructed in steps $\Omega(1,1)$, $\Omega(1,2)$, $\Omega(1,3)$. Let $v = (0,0,0)$, say. Then φ_v must vanish at $e_1 = (1,0,0)$, $e_2 = (0,1,0)$, $e_3 = (0,0,1)$ and at $c = (\frac{1}{2},\frac{1}{2},\frac{1}{2})$. The function

$$\psi(x) = \max[0, \ 1 - (|x_1|+|x_2|+|x_3|)], \quad x = (x_1,x_2,x_3) \in \mathbf{R}^3$$

does satisfy this requirement and is piecewise affine; yet, it is not T_1-pyramidal: the simplex spanned by $e_1,e_2,e_3,0$ is the support of ψ restricted to $\mathbf{I}^3$ and it is not in T_1. It can be checked easily that the desired function is

$$\hat{\theta}(x) = \max[0, \ 1 - \max(|x_1|,|x_2|,|x_3|) - \mathrm{med}(|x_1|,|x_2|,|x_3|)]$$

$$= \max[0, \ 1 - (|x_1|+|x_2|+|x_3|) + \min(|x_1|,|x_2|,|x_3|)]$$

where $\mathrm{med}(a_1,a_2,a_3)$ denotes the median of $\{a_1,a_2,a_3\}$. Let us examine the function $\hat\theta$.

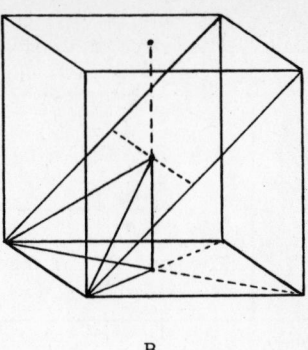

A   B

Figure A above shows the pyramide spanned by the bottom face and the center c. The simplex spanned by $0,e_1,e_2,c$ is the intersection of halfspaces $x_3 \leq x_1$, $x_3 \leq x_2$, $x_1 + x_2 \leq 1$, $x_3 \geq 0$. The plane $x_1 = x_2$ bisects it into two simplexes belonging to T_1 (one of them is shown in picture B, together with the supporting plane $x_1 = x_3$).

Similar inequalities and equations describe other simplexes in T_1 and their faces. The function $\hat\theta$ is affine on each set such that the differences $x_1 - x_2$, $x_1 - x_3$, $x_2 - x_3$, $x_1 + x_2 - 1$ etc. have the same sign.

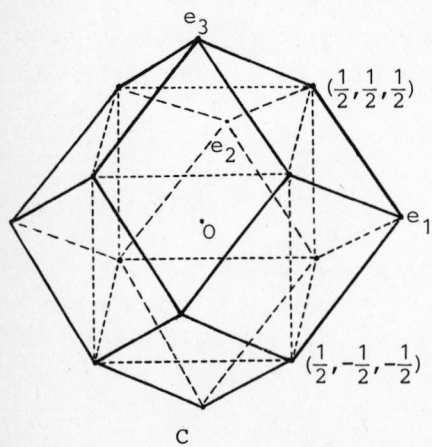

C

The function $\hat\theta$ is actually defined on $\mathbb{R}^3$. Its support (fig.C) is a diamond-face dodecahedron, a symmetric polytope U with 12 congruent faces which is the convex hull of $P \cup Q$, where P is the regular octahedron spanned by 6 points $\pm e_1$, $\pm e_2$, $\pm e_3$ and Q is the cube spanned by 8 points $(\pm\frac{1}{2}, \pm\frac{1}{2}, \pm\frac{1}{2})$.

Translating the function $\hat\theta$ we get the 8 basis functions of step $\Omega(1,1)$:

$$\varphi_v(x) = \hat\theta(x-v), \quad v \in D_o \times D_o \times D_o.$$

3.3.3. The pyramide in picture A above is the support of one of the basis functions of the next step $\Omega(1,2)$; its peak vertex is the center of the bottom face. Rotating and translating this pyramide we get the supports for the functions with vertices at other faces. Denote

$$(2) \qquad \theta^1(x) = \max[0, \; 1 - (2|x_1| + |x_2-x_3| + |x_2+x_3|)],$$

$$(3) \qquad \begin{cases} \theta^2(x_1,x_2,x_3) = \theta^1(x_2,x_3,x_1), \\[2mm] \theta^3(x_1,x_2,x_3) = \theta^1(x_3,x_1,x_2) = \theta^2(x_2,x_3,x_1). \end{cases}$$

It is clear that each of the functions $\varphi_v(x) = \theta^1(x-v)$ for v in $D_o \times D_1 \times D_1$, $\varphi_v(x) = \theta^2(x-v)$ for v in $D_1 \times D_o \times D_1$ and $\varphi_v(x) = \theta^3(x-v)$ for v in $D_1 \times D_1 \times D_o$ is affine on each simplex of T_1 and vanish at each vertex of T_1 different from v. They are the basis functions of step $\Omega(1,2)$. The function with the support shown in picture A is $\varphi_v(x) = \theta^3(x-v)$ where $v = (\frac{1}{2},\frac{1}{2},0)$.

In step $\Omega(1,3)$ we have only one function $\varphi_c(x) = \theta(x-c)$, where c is the center of I^3 and

$$\theta(x) = 1 - 2 \max(|x_1|, |x_2|, |x_3|).$$

The support of φ_c is the whole cube I^3; φ_c is affine on each simplex of T_1, vanishes on the boundary of I^3, and equals 1 at c.

3.3.4. Functions $\hat{\theta}$ and θ are symmetric with respect to x_1,x_2,x_3, but θ^i has a distinguished direction x_i. A significant property of all these functions is that their restrictions from R^3 to R^2 obtained by setting $x_3 = 0$ coincide with previously defined functions of two variables. Specifically,

$$\hat{\theta}_{(3)}(x_1,x_2,0) = \hat{\theta}_{(2)}(x_1,x_2), \qquad \theta_{(3)}(x_1,x_2,0) = \theta_{(2)}(x_1,x_2),$$

$$\theta^1_{(3)}(x_1,x_2,0) = \theta^2_{(3)}(x_1,x_2,0) = \hat{\theta}_{(2)}(2x_1,2x_2),$$

$$\theta^3_{(3)}(x_1,x_2,0) = \theta_{(2)}(x_1,x_2);$$

we added subscripts (2), (3) to distinguish between functions on I^3 and on I^2. Furthermore, setting $x_2 = 0$ we get

$$\hat{\theta}_{(2)}(x_1,0) = 1 - |x_1|, \qquad \theta_{(2)}(x_1,0) = 1 - 2|x_1|,$$

i.e. functions used in the construction of the Faber-Schauder system in $C(\mathbf{I})$. See also 4.1.9.

3.3.5. We shall now outline the construction of the regular pyramidal basis in the case of a d-dimensional cube $\mathbf{I}^d$. In third column, similarly as in 3.3.1, we choose only one generic set of indices.

Step	Description of new vertices	Set of new vertices	Number of new vertices
$\Omega(1,1)$	vertices of $\mathbf{I}^d$	$D_o^d = D_o \times \dots \times D_o$	2^d
$\Omega(1,k)$ $2 \le k \le d-1$	centers of k-dimensional faces	$D_1^k \times D_o^{d-k}$	$\binom{d}{k} \cdot 2^{d-k}$
$\Omega(1,d)$	center of $\mathbf{I}^d$	D_1^d	1
$\Omega(n,1)$ $n \ge 2$	centers of edges of cubes of $T_{n-1}^{(d)}$	$D_{n-1} \times \hat{D}_{n-2}^{d-1}$	$d \cdot 2^{n-2} (2^{n-2}+1)^{d-1}$
$\Omega(n,k)$ $2 \le k \le d-1$	centers of k-dimensional faces of cubes of $T_n^{(d)}$	$D_n^k \times \hat{D}_{n-1}^{d-k}$	$\binom{d}{k} 2^{k(n-1)} (2^{n-1}+1)^{d-k}$
$\Omega(n,d)$	centers of cubes of $T_n^{(d)}$	D_n^d	$2^{d(n-1)}$

$T_n^{(d)}$ denotes the triangulation of $\mathbf{I}^d$ obtained in step $\Omega(n,d)$. Inductively, with respect to d, it can be defined as follows: Take the triangulation $T_n^{(d-1)}$ on each of (d-1)-dimensional faces of $\mathbf{I}^d$ and join each vertex with the center c of $\mathbf{I}^d$. In order to write formulas for functions, denote

(3) $\hat{\theta}(x) = \max[0, \ 1 - \max(|x_1|, \dots, |x_d|) - \mathrm{nxt}(|x_1|, \dots, |x_d|)]$,

where $x = (x_1, \dots, x_d) \in \mathbf{R}^d$ and $\mathrm{nxt}(a_1, \dots, a_d)$ denotes the number next to the largest among $a_1,\ldots,a_d$. In other words, if $q_1,\ldots,q_d$ is a permutation such that $a_{q_1} \leq a_{q_2} \leq \ldots \leq a_{q_d}$, then $\mathrm{nxt}(a_1,\ldots,a_d)=a_{q_{d-1}}$. The function $\hat{\theta}$ equals 1 at 0 and vanishes at each point $e_i = (0,\ldots,1,\ldots,0)$ and at $c = (\frac{1}{2},\ldots,\frac{1}{2})$. The basis functions in step $\Omega(n,1)$ are

$$(4) \qquad \varphi_v(x) = \hat{\theta}(2^{n-1}(x-v)), \qquad v \in D_{n-1} \times \hat{D}_{n-2}^{d-1}.$$

For the step $\Omega(n,d)$ we need the function

$$(5) \qquad \theta(x) = \max(0,\ 1 - 2\max_j |x_j|).$$

Then

$$(6) \qquad \varphi_v(x) = \theta(2^{n-1}(x-v)), \qquad v \in D_n^d.$$

In the step $\Omega(n,k)$, $2 \leq k \leq d-1$ we consider the functions

$$(7) \qquad \theta^{i_1,\ldots,i_k}(x) = \max(0,\ 1 - 2\max_{j=i_1,\ldots,i_k} |x_j| - 2\max_{j \neq i_1,\ldots,i_k} |x_j|),$$

and then

$$(3) \qquad \varphi_v(x) = \theta^{i_1,\ldots,i_k}(2^{n-1}(x-v)) \quad \text{for} \quad v \text{ in } A_1 \times \ldots \times A_d,$$

where $A_{i_m} = D_n$, $A_j = \hat{D}_{n-1}$ for $j \neq i_1,\ldots,i_k$, $m = 1,\ldots,k$.

3.3.6. <u>Example</u>. The picture below shows the cube $\mathbf{I}^4$. The

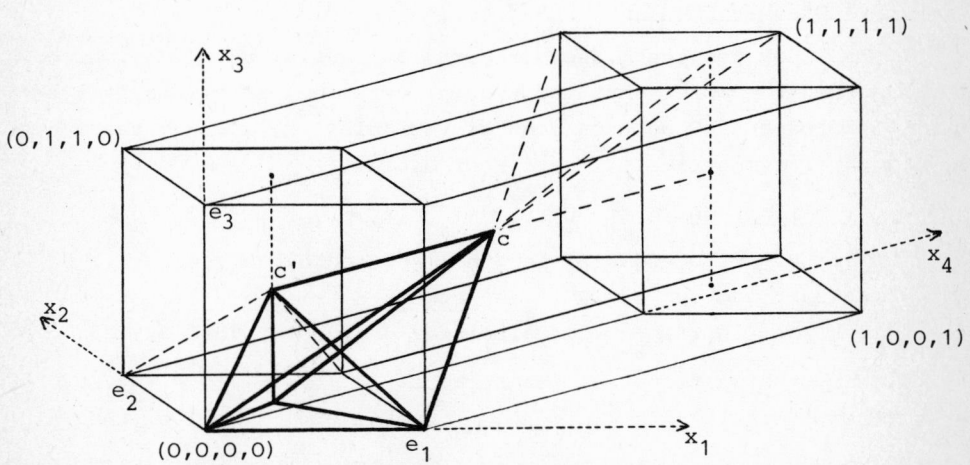

coordinates of sample vertices may help to understand the picture. The 3-dimensional face $x_4 = 0$ is on the left, the face $x_4 = 1$ is on the right. The heavy lines show the simplex spanned on 5 points $(0,0,0,0)$, $e_1 = (1,0,0,0)$, $w = (\frac{1}{2},\frac{1}{2},0,0)$, $c' = (\frac{1}{2},\frac{1}{2},\frac{1}{2},0)$ and $c = (\frac{1}{2},\frac{1}{2},\frac{1}{2},\frac{1}{2})$. The three first points $0, e_1, w$ form a triangle belonging to the first triangulation of the 2-dimensional face $x_1 = 0$, $x_2 = 0$. Adding the fourth point c' we get a simplex of the first triangulation of the 3-dimensional face $x_1 = 0$, $x_2 = 0$, $x_3 = 0$. Adding the fifth point – the center c of $\mathbf{I}^4$ - we get a simplex Δ in the triangulation T_1 of $\mathbf{I}^4$. The function $\hat{\theta}$ defined by (3) in case $d = 4$ is affine on Δ and vanishes at each vertex of Δ except of 0. The function

$$\varphi_w(x) = \theta^{1,2}(x-w) = \max[0,\ 1-2\max(|x_1-\tfrac{1}{2}|,|x_2-\tfrac{1}{2}|) - 2\max(|x_3|,|x_4|)]$$

is affine on Δ and on other simplexes of T_1, it vanishes at points $0, e_1, c', c$, and equals 1 at w.

3.3.7. The diamond-face dodecahedron U shown in fig.C in 3.3.2 has the property that its translations $U + k_1 u_1 + k_2 u_2 + k_3 u_3$ (where $k_i \in \mathbb{Z}$ and u_1, u_2, u_3 are suitable vectors, e.g., $e_1 + e_2$, $e_1 + e_3$, $e_2 + e_3$) fill up the space $\mathbb{R}^3$ without gaps and overlaps (see e.g. H.Steinhaus, Mathematical Snapshots, part VII). This property is related to the possibility of constructing a basis in $C_o(\mathbb{R}^3)$ analogous to that in $C_o(\mathbb{R})$ presented in 2.7.1

3.3.8. Notes. For the inductive description of the regular pyramidal basis, see J.Ryll ['73].

3.4. The squew pyramidal basis in $C(\mathbf{I}^d)$, $d \geq 2$

We shall now construct another pyramidal basis in $C(\mathbf{I}^d)$ which appears somewhat simpler that the regular pyramidal basis and is symmetric with respect to permutations of variables $x_1, \ldots, x_d$ but lacks certains other symmetries of the former basis.

3.4.1. Definition. Let $Q = x_o + a\mathbf{I}^d = \{x_o + \Sigma_{i=1}^{d} b_i e_i : 0 \leq b_i \leq a\}$ be a cube in $\mathbb{R}^d$ with edges of length a, shifted by x_o; $e_1, \ldots, e_d$ denote the unit vectors in $\mathbb{R}^d$. The standard triangulation of Q is the family $T(Q) = \{\Delta_\pi(Q)\}_{\pi \in \Pi_d}$ of simplexes defined as follows: Π_d denotes the set of permutations of numbers $1, \ldots, d$. If $\pi = (j_1, \ldots, j_d)$ is any permutation in Π_d, let $\Delta_\pi(Q)$ be the convex hull of the $d+1$ elements

$$(1) \qquad x_o, \; x_o + u_{j_1}, \; x_o + \sum_{k=1}^{2} u_{j_k}, \ldots, \; x_o + \sum_{k=1}^{d} u_{j_k},$$

where $u_i = ae_i$. Thus, each simplex $\Delta_\pi(Q)$ corresponds to one path from x_o to $x_1 = x_o + \sum_{i=1}^{d} u_i$ via vertices of Q and such that along each segment of the path one coordinate increases and the others are constant. The classical argument shows that

$$(2) \qquad Q = \bigcup_{\pi \in \Pi_d} \Delta_\pi(Q).$$

Indeed, let $x \in Q$. Then $x = x_o + b_1 u_1 + \ldots + b_d u_d$ with $0 \le b_i \le 1$ for $i = 1, \ldots, d$. Permute the coordinates as to get $1 \ge b_{j_1} \ge \ldots \ge b_{j_d} \ge 0$. By Abel's formula,

$$x = x_o + \sum_{i=1}^{d} b_i u_i = c_o x_o + \sum_{k=1}^{d} c_k \left(x_o + \sum_{i=1}^{k} u_{j_i} \right),$$

where $c_o = 1 - b_{j_1}$, $c_k = b_{j_k} - b_{j_{k+1}}$ $(k = 1, \ldots, d-1)$ and $c_d = b_{i_d}$. Thus, $c_i \ge 0$ for $i = 0, \ldots, d$ and $c_o + \ldots + c_d = 1$. This means that $x \in \Delta_\pi(Q)$, where $\pi = (j_1, \ldots, j_d)$. We have shown (2).

If $b_i \ne b_j$ for $i \ne j$, then π is uniquely determined by x; the strict inequalities $b_{j_1} > b_{j_2} > \ldots > b_{j_d}$ are equivalent to $c_i > 0$ for $i = 0, \ldots, d$. Consequently, for different permutations the interiors of simplexes $\Delta_\pi(Q)$ are disjoint. Thus, Q is divided into $d!$ simplexes forming the triangulation $T(Q)$. Obviously,

$$(3) \qquad \mathrm{vert}(T(Q)) = \mathrm{vert}(Q).$$

From the definition it follows that if $x_o = 0$ then $T(Q)$ is symmetric with respect to permutations of coordinates.

Let H_i denote the hyperplane perpendicular to e_i and passing through the point $x_o + u_i/2$ (the center of the edge x_o, $x_o + u_i$). The hyperplanes $H_1, \ldots, H_d$ divide Q into 2^d congruent cubes $\Gamma_k(Q)$, $k = 0, \ldots, 2^d - 1$, of edge $a/2$. If $(\varepsilon_1, \ldots, \varepsilon_d)$ is a sequence of zeros and ones, then

$$\Gamma_k(Q) = x_o + \frac{1}{2} \sum_{i=1}^{d} \varepsilon_i u_i + \frac{a}{2} \mathbf{I}^d \qquad \text{for} \quad k = \sum_{i=0}^{d-1} \varepsilon_i 2^i.$$

3.4.2. <u>Lemma</u>. Let $\{\Delta_\pi(Q)\}_{\pi \in \Pi_d}$ and $\{\Delta_\pi(\Gamma_k(Q))\}_{\pi \in \Pi_d}$ be the standard triangulations of Q and $\Gamma_k(Q)$, respectively. Then for each simplex $\Delta_\pi(\Gamma_k(Q))$ there is a unique simplex $\Delta_{\pi'}(Q)$ such that

$$\Delta_\pi(\Gamma_k(Q)) \subset \Delta_{\pi'}(Q).$$

Proof. Let $\pi = (j_1, \ldots, j_d)$ and let x be in the interior of $\Delta_\pi(\Gamma_k(Q))$. Then $x = x_0 + \frac{1}{2}\Sigma_{i=1}^d \varepsilon_i u_i + \Sigma_{i=1}^d b_i u_i = x_0 + \Sigma_{i=1}^d a_i u_i$, where $\varepsilon_i = 0, 1$, $a_i = \varepsilon_i/2 + b_i$, $0 < a_i < 1$ and

(4) $\qquad \frac{1}{2} > b_{j_1} > b_{j_2} > \ldots > b_{j_d}.$

Denote $M = \{i : \varepsilon_i = 1\}$, $L = \{i : \varepsilon_i = 0\}$. Let π' be the permutation $(j_{r_1}, \ldots, j_{r_m}, j_{q_1}, \ldots, j_{q_n})$ where $r_1 < \ldots < r_m$, $j_{r_j} \in M$, $q_1 < \ldots < q_n$, $j_{q_i} \in L$. Then

$$1 > a_{j_{r_1}} > \ldots > a_{j_{r_m}} > a_{j_{q_1}} > \ldots > a_{j_{q_n}} > 0.$$

Thus, x is in the interior of $\Delta_{\pi'}(Q)$. Now, as long as x is in the interior of $\Delta_\pi(\Gamma_k(Q))$, the sets M, L and the inequalities (4) remain the same. Therefore the permutation π' does not change either. This means that the interior of $\Delta_\pi(\Gamma_k(Q))$ is contained in the interior of $\Delta_{\pi'}(Q)$. $\quad\square$

3.4.3. We shall now consider a sequence $P_0, P_1, \ldots$ of successive partitions of the cube $\mathbf{I}^d$:

$$P_0 = \{\mathbf{I}^d\}, \quad P_1 = \{\Gamma_k(\mathbf{I}^d) : k = 0, \ldots, 2^d - 1\}$$

and $P_n = \{\Gamma_k(Q) : Q \in P_{n-1}, \ k = 0, \ldots, 2^d - 1\}$. Thus, P_n is obtained by dividing each edge of $\mathbf{I}^d$ into 2^n segments of length 2^{-n} and consists of $(2^n)^d$ cubes. Now, let $T_0 = \{\Delta_\pi(\mathbf{I}^d) : \pi \in \Pi_d\}$ and

$$T_n = \{\Delta_\pi(Q) : \pi \in \Pi_d, \ Q \in P_n\} =$$

$$= \{\Delta_\pi(\Gamma_{k_1} \cdots \Gamma_{k_n}(\mathbf{I}^d)) : \pi \in \Pi_d, \ k_1, \ldots, k_n = 0, \ldots, 2^d - 1\}.$$

In other words, T_n is obtained by taking the standard triangulation of each cube in P_n.

In case $d = 2$, the three triangulations $T_0(\mathbf{I}^2)$, $T_1(\mathbf{I}^2)$, $T_2(\mathbf{I}^2)$ are shown below, in 3.4.4.

The set of vertices of T_m is simpler than that of the regular pyramidal basis, namely

$$\text{vert}(T_o) = \text{vert}(\mathbf{I}^d) = (D_o)^d, \quad \text{vert}(T_n) = \text{vert}(P_n) = (\hat{D}_n)^d.$$

Obviously, $\text{vert}(T_n)$ has $(2^n+1)^d$ elements $(n = 0, 1, \ldots)$.

3.4.4. The functions in the squew pyramidal basis are defined as follows. Denote $W = \text{vert}(T_n)$ for $n = 0, 1, \ldots, \quad V_o = W_o, \quad V_n = W_n \backslash W_{n-1}$ for $n \geq 1$. If $v \in V_n$, let φ_v be the T_n-pyramidal function with vertex v (see 3.1.1).

Let $d = 2$. The supports of functions $\varphi_{(0,1)}, \varphi_{(\frac{1}{2},\frac{1}{2})}$ and $\varphi_{(\frac{3}{4},\frac{1}{2})}$ are the shaded areas above. The pattern of successive functions is clear: all φ_v arise from a single function φ be similarity and translation. It is convenient to consider this function as defined on $\mathbf{R}^2$, with support actually contained in $[-1,1]^2$ and with vertex O. Thus $\varphi(x) = \max(0, 1 - \theta(x))$ where $x = (x_1, x_2)$ and

$$\theta(x_1, x_2) = \begin{cases} \max(|x_1|, |x_2|) & \text{if } x_1 x_2 \geq 0, \\ \\ |x_1| + |x_2| & \text{if } x_1 x_2 \leq 0. \end{cases}$$

If $d = 3$, then $\theta(x_1, x_2, x_3)$ is equal to $\max(|x_1|, |x_2|, |x_3|)$ in the octants $x_1, x_2, x_3 \geq 0$ and $x_1, x_2, x_3 \leq 0$ and is equal to
$$\theta_i(x) = |x_i| + \max_{j \neq i} |x_j| \quad \text{if } x_i \geq 0$$
and $x_j \leq 0$ for $i \neq j$ or $x_i \leq 0$ and $x_j \geq 0$ for $i \neq j$ (to show this, we make use of the fact that the planes $x_1 = x_2$, $x_2 = x_3$, $x_1 = x_3$ split $\mathbf{I}^3$ into 6 simplexes of T, see the picture).

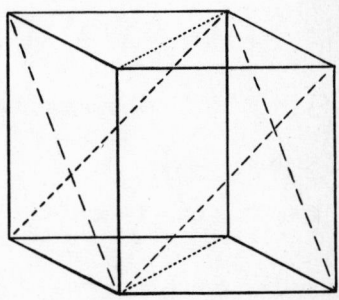

3.4.5. <u>Proposition</u>. The functions φ_v are of the form

(5) $\varphi_v(x) = \varphi(2^n(x-v))$ for v in V_n,

where φ is a function $\mathbf{R}^d$ with $\mathrm{supp}\,\varphi \subset [-1,1]^d$, defined by the formula

(6) $\varphi(x) = \max(0, 1 - \max\limits_{x_i > 0} |x_i| - \max\limits_{x_j < 0} |x_j|)$.

Proof. We consider $[-1,1]^d$ as the union of 2^d cubes of the form $\eta + \mathbf{I}^d$, where $\eta = (\eta_1, \ldots, \eta_d)$ and $\eta_i = -1, 0$. Thus, $\eta + \mathbf{I}^d$ is contained in the cone

$$R_\eta = \{x \in \mathbf{R}^d : x_i \geq 0 \text{ if } \eta_i = 0, \quad x_i \leq 0 \text{ if } \eta_i = -1\}.$$

Moreover, $\eta + \mathbf{I}^d = R_\eta \cap [-1,1]^d$. Let T be the triangulation of $[-1,1]^d$ obtained by taking the standard triangulation of each of 2^d cubes $\eta + \mathbf{I}^d$, $\eta \in \{-1,0\}^d$. The function φ is the T-pyramidal function with peak at 0, extended to $\mathbf{R}^d$ by $\varphi(x) = 0$ for x in $\mathbf{R}^d \smallsetminus [-1,1]^d$. It is clear that it is the desired function φ in (5).

Let us consider φ separately on each of $d!$ simplexes in the standard triangulation of each 2^d cubes $\eta + \mathbf{I}^d$. Let η be fixed and $M = \{i : \eta_i = 0\}$, $L = \{i : \eta_i = -1\}$. Consider the vertex 0 of $\eta + \mathbf{I}^d$. It belongs to a simplex $\Delta = \Delta_\pi(\eta + \mathbf{I}^d)$, where $\pi = (j_1, \ldots, j_d)$, if and only if $j_1, \ldots, j_r \in L$ and $j_{r+1}, \ldots, j_d \in M$. Thus, Δ is the convex hull of

$$\eta, \ \eta + e_{j_1}, \ \eta + e_{j_1} + e_{j_2}, \ldots, \eta + e_{j_1} + \ldots + e_{j_d}.$$

Moreover, $0 = \eta + e_{j_1} + \ldots + e_{j_r}$. A point $x = (x_1, \ldots, x_d)$ belongs to Δ iff

$$1 \geq 1 + x_{j_1} \geq \ldots \geq 1 + x_{j_r} \geq x_{j_{r+1}} \geq \ldots \geq x_{j_d} \geq 0;$$

if $x \in \Delta$, then x has barycentric coordinates $c_0, \ldots, c_d$ in Δ, where

$$c_0 = -x_{j_1}, \quad c_r = 1 + x_{j_r} - x_{j_{r+1}}, \quad c_d = x_{j_d}, \quad c_k = x_{j_k} - x_{j_{k+1}}$$

for $k \neq 0, r, d$. Therefore the $(d-1)$-dimensional face opposite to the vertex 0 is described by $c_r = 0$, i.e., $|x_{j_r}| + x_{j_{r+1}} = 1$. Hence

$\varphi(x) = 1 - |x_{j_r}| + x_{j_{r+1}}$ on Δ. Moreover,

$$|x_{j_r}| = \max_{1 \le k \le r} |x_{j_k}| \quad \text{and} \quad x_{j_{r+1}} = \max_{r < k \le d} x_{j_k}.$$

Thus, we get $\varphi(x) = \max[0, 1 - \max_{i \in L} |x_i| + \max_{i \in M} |x_i|]$, and this is the formula (6). If M or L is empty, the corresponding maximum is understood to be equal to 0. $\square$

3.4.6. The set $V = \cup_{n=0}^{\infty} V_n$ is arranged into a sequence $(v_m)_{m \in N}$ so that the elements of each V_n are ordered lexicographically and if $v_p \in V_n$, $v_q \in V_m$ and $n < m$, then $p < q$. Let $\varphi_m = \varphi_{v_m}$.

It is clear that $(\varphi_m)_{m \in N}$ is a monotone interpolating basis in $C(\mathbf{I}^d)$ with nodes in V. Indeed, $\varphi_m(v_j) = 0$ for $j < m$ and $\|s_m\| \le \|f\|$ for the partial sum s_m of a function f in $C(\mathbf{I}^d)$. Each partial sum interpolates f at previous nodes and hence $\|s_m - f\|$ cannot exceed the oscillation of f on simplexes of successive triangulations and therefore $\|s_m - f\| \longrightarrow 0$.

3.4.7. In order to find formulas for coefficients we shall consider certain operations on dyadic multiindices. Let

$$(7) \qquad D_n^{(d)} = \{\tau = (\tau_1, \dots, \tau_d) \in (\hat{D}_n)^d : \exists_i \tau_i \in D_n\}$$

for $n = 0, 1, 2, \dots$ where $(\hat{D}_n)^d$ denotes the Cartesian power $\hat{D}_n \times \dots \times \hat{D}_n$. Thus,

$$(\tilde{D})^d = (\bigcup_{n=0}^{\infty} D_n)^d = \bigcup_{n=0}^{\infty} D_n^{(d)}.$$

If $n \ne m$, then $D_n^{(d)} \cap D_m^{(d)} = \emptyset$. The operations $\tau \longrightarrow \tau^-$, $\tau \longrightarrow \tau^+$ for $d = 1$ (see 2.1.2) and for $d = 2$ (see 3.2.3) will now be extended to the general case. If $\tau = (\tau_1, \dots, \tau_d) \in D_n^{(d)}$ and $n \ge 1$, let

$$(8) \qquad \tau^- = (\tau_1', \dots, \tau_d'), \qquad \tau^+ = (\tau_1'', \dots, \tau_d'')$$

where $\tau_i' = (\tau_i)^- = \tau_i - 2^{-n}$ and $\tau_i'' = (\tau_i)^+ = \tau_i + 2^{-n}$ for τ_i in D_n and $\tau_i' = \tau_i'' = \tau_i$ for τ_i in $\hat{D}_{n-1}$. Then $\tau = \frac{1}{2}(\tau^- + \tau^+)$.

3.4.8. <u>Lemma.</u> Let $\tau \in D_n^{(d)}$, $n \ge 1$. Then $\tau \in \bar{W}_n$, τ^-, $\tau^+ \in \bar{W}_{n-1}$ and there is a simplex Δ in T_{n-1} such that both τ^- and τ^+ belong to $\text{vert } \Delta$.

Proof. Let $\tau = (\tau_1, \ldots, \tau_d)$ and let $M = \{i : \tau_i \in D_n\}$, $F = \{i : \tau_i = 0\}$, $L = \{i : \tau_i \in \hat{D}_{n-1}\}$, $E = \{i : \tau_i = 1\}$. Then $\tau_i'' = \tau_i' + 2^{-n+1}$ for i in M, $\tau_i' = \tau_i''$ for i in L, $L \cap M = \emptyset$, $L \cup M = \{1, \ldots, d\}$, and $E \cup F \subset L$. Consequently, both τ^- and τ^+ belong to the cube K in P_{n-1}, where

$$K = \{x_0 + \sum_{i=1}^{d} b_i e_i : 0 \le b_i \le 2^{-n+1}\}, \quad x_0 = (x_1^0, \ldots, x_d^0)$$

and $x_0 = \tau_i$ for $i \in F$, $x_i^0 = \tau_i - 2^{-n+1}$ for i in E and $x_i^0 = \tau_i^-$ for $i \in M \cup L$. Let $j_1, \ldots, j_r$ be the elements of E, let $j_{r+1}, \ldots, j_m$ be the elements of M and $j_{m+1}, \ldots, j_d$ - elements of $L \smallsetminus E$. Then $\pi = (j_1, \ldots, j_d)$ is a permutation such that τ^- and τ^+ are in $\Delta_\pi(K)$. $\square$

3.4.9. We shall now compute the coefficients of the expansion $f = \Sigma a_v \varphi_v$ in $C(\mathbf{I}^d)$. By the preceding lemma, if τ is a vertex in V_n, then the function

$$s^{(n-1)} = \sum_{v \in W_{n-1}} a_v \varphi_v$$

is affine on the segment τ^-, τ^+. Consequently, by 3.1.5,

(9) $\quad a_\tau = f(\tau) - s^{(n-1)}(\tau) = f(\tau) - \frac{1}{2}[f(\tau^-) + f(\tau^+)]$.

3.4.10. Exercises. (A) Show that the diagonal of Q joining x_0 and x_1 is contained in each simplex $\Delta_\pi(Q)$. The simplexes $\Delta_\pi(Q)$ have measure $a^d/d!$ each.

(B) What are the symmetries of $\mathbf{I}^d$ which leave the regular triangulation of 3.3.5 invariant and change the triangulation $T(\mathbf{I}^d)$?

(C) Show that all that the simplexes in (2) are congruent. Show that for an analogous triangulation of a cuboid the simplexes need not be congruent.

(D) Show that the triangulation T_n in 3.4.3 has the following property: If $\Delta \in T_n$ and H is a k-dimensional hyperplane in $\mathbf{R}^d$, k=d-1, containing a k-face of Δ, then for any simplex Δ' in T_n, the intersection $H \cap \Delta'$ is either empty or is a face of Δ'. Moreover, for any cube Q in the partition P_n, $Q \cap H$ is the union of k-faces of certain simplexes in T_n.

(E) Construct a basis in $C_0(\mathbf{R}^3)$ decomposing $\mathbf{R}^3$ into unit cubes and using their standard triangulations in a way analogous to that in 2.7.1. (cf. 3.3.7).

3.4.11. <u>Notes</u>. The squew pyramidal bases were considered by R.Bonic, J.Frampton and A.Tromba ['69] and by Z.Ciesielski and K.Gęba (see 3.5.11); the name "squew pyramidal basis" was introduced by J.Ryll ['73].

3.5. Estimation of coefficients for Hölder functions.

3.5.1. The purpose of this section is to prove Z.Ciesielski's results on Faber-Schauder coefficients of functions in the Hölder class

$$\text{Lip}^{\alpha} = \{f \in C(I) : |f(x) - f(x')| \leq c|x - x'|^{\alpha}\}, \quad 0 < \alpha < 1,$$

or more generally, for functions in the class

$$\text{Lip}_{\omega} = \{f \in C(I) : |f(x) - f(x')| \leq c\omega(|x - x'|)\}$$

where the function ω satisfies the conditions (1)-(5) below, and still more generally, for the analogous classes of functions of m variables.

We shall consider functions ω on I satisfying the following conditions, where c_1, c_2, c_3 are positive constants:

(1) $\quad \omega(0) = 0, \quad \omega(\delta) > 0 \quad \text{for} \quad \delta > 0,$

(2) $\quad \omega$ is nondecreasing on I,

(3) $\quad \omega(2\delta) \leq c_1\omega(\delta) \qquad \text{for} \quad 0 \leq \delta \leq \frac{1}{2},$

(4) $\quad \int_0^{\delta} \frac{\omega(t)}{t} dt \leq c_2\omega(\delta) \qquad \text{for} \quad 0 < \delta \leq 1,$

(5) $\quad \delta \int_{\delta}^{1} \frac{\omega(t)}{t^2} \leq c_3\omega(\delta) \qquad \text{for} \quad 0 < \delta \leq 1.$

The reader may verify that the function $\omega(\delta) = \delta^{\alpha}$, $0 < \alpha < 1$, satisfies (1)-(5). For any $u = (u_1, \ldots, u_d) \in R^d$, $u \neq 0$, and for any simplex $\Delta \subset R^d$ define

$$\alpha(u, \Delta) = \sup\{\|x - y\| : x, y \in \Delta, \quad x - y = cu \quad \text{for some} \quad c \quad \text{in} \quad R\},$$

(this supremum is obviously attained) and

$$\alpha(\Delta) = \inf\{\alpha(u,\Delta) : u \in \mathbb{R}^d \setminus \{0\}\}.$$

3.5.2. <u>Theorem</u>. Let $T_1 > T_2 > \ldots$ be a sequence of triangulations of $\mathbf{I}^m$ satisfying the conditions

(6) $\sup\{\text{diam } \Delta : \Delta \in T_n\} \leq a \cdot 2^{-n}$,

(7) $\inf\{\alpha(\Delta) : \Delta \in T_n\} \geq b \cdot 2^{-n}$.

Let $(\varphi_v)_{v \in V}$ be the basis pyramidal with respect to (T_n) (see 3.1.3). Then for every f in $C(\mathbf{I}^m)$ the following conditions are equivalent:

(i) $\omega_f(\delta) = O[\omega(\delta)]$,

(ii) $\displaystyle\sup_{v \in V_n} \| f - S_v(f) \| = O[\omega(2^{-n})]$,

(iii) $\displaystyle\sup_{v \in V_n} |a_v(f)| = O[\omega(2^{-n})]$,

where the constants in (i), (ii), (iii) depend only on ω, c_1, c_2, c_3, a, b, and m.

The implication (i) $\Rightarrow$ (ii) follows from the estimation (6) in 3.1.7 while (ii) $\Rightarrow$ (iii) follows from the formula

$$a_v(f) = f(v) - s^{(n-1)}(v) \quad \text{for} \quad v \text{ in } V_{n-1}$$

(see 3.1.5) and from the obvious equivalence (ii) $\Longleftrightarrow$ (ii'), where

(ii') $\| f - s^{(n-1)} \| = O[\omega(2^{-n})]$.

The implications (ii) $\Rightarrow$ (i) and (iii) $\Rightarrow$ (i) are consequences of Lemmas 3.5.3, 3.5.6. and 3.5.7 below.

It is obvious that the regular pyramidal basis and the squew pyramidal basis in $C(\mathbf{I}^d)$ satisfy (6) and (7) (note that the simplexes in T_n are similar to those in T_1, and hence proportions remain the same).

3.5.3. <u>Lemma</u>. Let $f \in C(\mathbf{I}^m)$, let ω satisfy conditions (1)-(4) and let k be an integer such that $\sup\{|a_v(f)| : v \in V_n\} \leq \omega(2^{-n})$ for $n \geq k$. Then

$$\| f - S_v(f) \| \leq 2c_1 c_2 \omega(2^{-n}) \qquad \text{for} \quad n \geq k.$$

Proof. Let $x \in I^m$ and $v \in V_n$ for some n, $n \geq k$. Then

$$|f(x) - s_v(x)| \leq \sum_{j=n}^{\infty} \sum_{w \in V_j} |a_w| \varphi_w(x) \leq \sum_{j=n}^{\infty} \sup_{w \in V_j} |a_w|$$

$$\leq \sum_{j=n}^{\infty} \omega(2^{-j}) = 2 \sum_{j=n}^{\infty} 2^{-j} \frac{\omega(2^{-j})}{2^{-j+1}} \leq 2 \sum_{j=n}^{\infty} \int_{2^{-j}}^{2^{-j+1}} \frac{\omega(t)}{t} dt$$

$$= 2 \int_{o}^{2^{-n+1}} \frac{\omega(t)}{t} dt \leq 2c_2 \omega(2^{-n+1}) \leq 2c_1 c_2 \omega(2^{-n}). \quad \square$$

3.5.4. <u>Lemma</u>. There exists a constant c_o (depending on a,b,m) such that for each triangulation T of I^m satisfying (6) and (7), for each $n = 1,2,\ldots$ and for each pair x_o, y_o in I^m such that $\| x_o - y_o \| \leq 2^{-n}$ there exist: a sequence $z_o = x_o$, $z_1, \ldots, z_k = y_o$ and simplexes Δ_j in T $(j = 1,\ldots,k)$ such that $k \leq c_o$ and for each j the successive points z_{j-1}, z_j belong to the same simplex Δ_j.

Proof. Let $\| x_o - y_o \| \leq 2^{-n}$. Then there is a cube

$$K = [a_1, b_1] \times \ldots \times [a_m, b_m], \quad \text{where} \quad b_i - a_i = 2^{-n},$$

such that $x_o, y_o \in K$. Denote $T_n^K = \{\Delta \in T_n : \Delta \cap K \neq \emptyset\}$ and

$$\hat{K} = \{z \in I^m : \text{dist}(z, K) \leq a \cdot 2^{-n}\}.$$

By (6), we have $\cup \{\Delta : \Delta \in T_n^K\} \subset \hat{K}$. Obviously, the volume of $\hat{K}$ does not exceed $[(2a+1)2^{-n}]^m$. By (7), the volume of Δ in T_n is not less than $(b2^{-n})^m/m!$. Consequently, the number of simplexes in T_n^K does not exceed $c_o = (2a+1)^m m!/b^m$.

Consider the set Γ of all finite sequences $\Delta_1, \ldots, \Delta_k$ of simplexes in T_n^K such that $x_o \in \Delta_1$ and $\Delta_{j-1} \cap \Delta_j \neq \emptyset$ for $j = 1,\ldots,k$. It is clear that for each Δ in T_n^K there is a sequence (Δ_j) in Γ with Δ as a term. Consequently, there is a sequence $\Delta_1, \ldots, \Delta_k$ in Γ with $y_o \in \Delta_k$. We may assume that $\Delta_i \neq \Delta_j$ for $i < j$ (otherwise we could drop the simplexes $\Delta_{i+1}, \ldots, \Delta_j$). Thus, $k \leq c_o$. Now, choose any z_j in $\Delta_j \cap \Delta_{j+1}$, $j = 1,\ldots,k-1$. $\square$

3.5.5. <u>Lemma</u>. Let ω satisfy (1) and (2). Then the function

$$\Theta(t) = t \int_{t}^{1} \frac{\omega(u)}{u^2} du + \int_{o}^{t} \omega(u) du, \qquad 0 < t < 1,$$

is nondecreasing in I.

Proof. Since ω is monotone, we infer that

$$\theta'(t) = \int_t^1 \frac{\omega(u)}{u^2}du - \omega(t)(\frac{1}{t} - 1) = \int_t^1 \frac{\omega(u) - \omega(t)}{u^2}du \geq 0. \quad \square$$

3.5.6. <u>Lemma</u>. Let $f \in C(I^m)$, let ω satisfy conditions (1) and (2), and let

$$\| f - s^{(n-1)}(f) \| \leq \omega(2^{-n}) \quad \text{for} \quad n = 1, 2, \ldots$$

where $s^{(0)} = 0$. Then there exists $c_4 > 0$ such that

(8) $\qquad \omega_f(\delta) \leq c_4 \delta [2\omega(1) + \int_\delta^1 \frac{\omega(t)}{t^2}dt]$ $\quad$ for $\quad 0 < \delta \leq 1$.

Proof. We shall first estimate $|f(x) - f(y)|$ in the case where x, y belong to the same simplex Δ in T_n. Thus, $\| x - y \| \leq a2^{-n}$. For each $k \leq n$ find Δ_k in T_k such that $\Delta \subset \Delta_k$. By (7), there exist points x_k, y_k in Δ_k such that $x_k - y_k$ is parallel to $x - y$ and $\| x_k - y_k \| \geq b2^{-k}$. Moreover,

$$|\varphi_v(z) - \varphi_v(z')| \leq 1 \quad \text{for} \quad z, z' \in I^m, \quad v \in V.$$

Hence, from the similarity of respective triangles we get

$$|\varphi_v(x) - \varphi_v(y)| = \frac{\| x - y \|}{\| x_k - y_k \|} |\varphi_v(x_k) - \varphi_v(y_k)| \leq \frac{a2^{-n}}{b2^{-k}} \quad \text{for} \quad v \quad \text{in} \quad V_k.$$

Furthermore

$$|a_v| \leq \| f - s^{(n-1)} \| \leq \omega(2^{-n}) \quad \text{for} \quad v \quad \text{in} \quad V_n, \quad n = 1, 2, \ldots$$

Using the above estimations we get

$$|f(x) - f(y)| \leq \sum_{k=1}^{n-1} \sum_{v \in V_k} |a_v| \; |\varphi_v(x) - \varphi_v(y)| + 2\| f - s^{(n-1)} \|$$

$$\leq \sum_{k=1}^{n-1} \omega(2^{-k}) \sum_{v \in \text{vert}(\Delta_k)} |\varphi_v(x) - \varphi_v(y)| + 2\omega(2^{-n})$$

$$\leq \sum_{k=1}^{n-1} \omega(2^{-k}) (m+1)\frac{a}{b} 2^{-n+k} + 2\omega(2^{-n})$$

$$\leq c_5 2^{-n} \sum_{k=1}^{n} 2^{k-1}\omega(2^{-k}) \leq c_5 2^{-n} \sum_{k=1}^{2^n} \omega(\frac{1}{k}).$$

where c_5 depends on a, b, m.

Now, let x, y be any two distinct points of $\mathbf{I}^m$, $\|x - y\| \leq \delta$, $\delta \leq 1$. There exists an integer $n \geq 0$ such that $2^{-n-1} < \|x - y\| \leq 2^{-n}$. By 3.5.4 and 3.5.5, we have

$$|f(x) - f(y)| \leq c_o c_5 2^{-n} \sum_{k=1}^{2^n} \omega\left(\frac{1}{k}\right)$$

$$\leq c_6 2^{-n} 2 \left[\omega(1) + \sum_{k=1}^{2^n - 1} \int_{1/(k+1)}^{1/k} \frac{\omega(t)}{t^2} dt \right]$$

$$\leq 2 c_6 2^{-n} \left[\omega(1) + \int_{2^{-n}}^{1} \frac{\omega(t)}{t^2} dt \right]$$

$$\leq 2 c_6 \|x - y\| \cdot 2 \left[\omega(1) + \int_{\|x-y\|}^{1} \frac{\omega(t)}{t^2} dt \right]$$

$$\leq c_4 \left[\delta \omega(1) + \delta \int_{\delta}^{1} \frac{\omega(t)}{t^2} dt + \int_{\|x-y\|}^{\delta} \omega(t) dt \right]$$

$$\leq c_4 \delta \left[2\omega(1) + \int_{\delta}^{1} \frac{\omega(t)}{t^2} dt \right]. \quad \square$$

3.5.7. <u>Lemma.</u> Let ω satisfy (1) and (5). Then $\omega(\delta) \geq c_9 \delta$.

Proof. If $0 < \delta \leq \frac{1}{2}$, then $c_7 \delta \leq \delta \int_{\delta}^{1} \omega(t) t^{-2} dt \leq c_3 \omega(\delta)$, where $c_7 = \int_{1/2}^{1} \omega(t) t^{-2} dt$. If $\frac{1}{2} \leq \delta \leq 1$, then $\omega(\delta) \geq c_8 \geq c_8 \delta$. $\quad \square$

3.5.8. Let $0 < \alpha < 1$. Let $\mathrm{Lip}_o^{\alpha}(\mathbf{I}^d)$ be the space of Hölder-condition functions of order α on $\mathbf{I}^d$ vanishing at 0, with the norm

$$\|f\|_{\alpha} = \sup\left\{ \frac{|f(x) - f(y)|}{\|x - y\|^{\alpha}} : x, y \in \mathbf{I}^d \right\}.$$

In other words, $\|f\|_{\alpha}$ is the best constant c in the inequality

$$|f(x) - f(y)| \leq c \|x - y\|^{\alpha}.$$

It is easy to verify that $\mathrm{Lip}_o^{\alpha}(\mathbf{I}^d)$ is a nonseparable Banach space.

3.5.9. <u>Corollary.</u> Let $f \in C(\mathbf{I}^d)$, let $(\varphi_v)_{v \in V}$ be the squew pyramidal basis and let $0 < \alpha < 1$. Then the following conditions are equivalent

(a) $f \in \mathrm{Lip}_o^{\alpha}(\mathbf{I}^d)$,

(b) $\|f - s^{(n-1)} f\|_{\infty} \leq c \left(\frac{1}{2^n}\right)^{\alpha}$,

(c) $\qquad |a_v(f)| \le c'(\frac{1}{2^n})^\alpha$ for v in V_n. $\square$

3.5.10. <u>Proposition</u>. Let $0 < \alpha < 1$. Then the space $Lip_o^\alpha(\mathbf{I}^d)$ is isomorphic (as a linear topological space) to l^∞.

Proof. Let $(\varphi_m)_{m=1,2,\ldots}$ be the squew pyramidal basis in $C(\mathbf{I}^d)$ with the function φ_o for $v = 0$ omitted. Let $a = (a_n)$ be in l^∞. Denote

$$(9) \qquad f = \sum_{m=1}^{\infty} b_m a_m \varphi_m = \sum_{n=1}^{\infty} \sum_{v \in V_n} b_v a_v \varphi_v + \sum_{v \in V_o \setminus \{0\}} b_v a_v \varphi_v$$

where $b_m = 2^{-n\alpha} = b_v$ if φ_m is a function φ_v with $v \in V_n$. Since

and $\sum_{v \in V_n} |b_v a_v| \varphi_v(x) \le ||a||_\infty 2^{-n\alpha} \sum_{v \in V_n} \varphi_v(x) \le ||a||_\infty 2^{-n\alpha}$ for x in $\mathbf{I}^d$

$\Sigma 2^{-n\alpha} < \infty$, the series (9) converges in $C(\mathbf{I}^d)$; moreover $f \in C(\mathbf{I}^d)$, $f(0) = 0$ and

$$|| f ||_\infty \le || a ||_\infty \frac{1}{1-2^{-\alpha}}.$$

By 3.5.8, $f \in Lip^\alpha$. Moreover, by Theorem 3.5.2, $\omega_f(\delta) \le c\delta^\alpha$ for some constant c depending on $|| a ||_\infty$ but not on the sequence (a_n) itself. Thus, the operator $u_\alpha(a) = f$ is bounded, linear, and maps l^∞ onto $Lip_o^\alpha(\mathbf{I}^d)$. $\square$

3.5.11. <u>Notes</u>. Theorem 3.5.2 is due to Z.Ciesielski ['60a] who proved it for the Faber-Schauder system in $C(\mathbf{I})$. The isomorphisms of space Lip^α and l^∞ and of spaces

$$lip^\alpha = \{f : f(x_1 - f(x_2) = o(|x_1 - x_2|^\alpha)\} \quad \text{and} \quad C_o(\mathbf{N})$$

were proved in Ciesielski ['60b]. At that time he raised the question whether an analogous characterization of Hölder functions is true for functions of several variables. In 1969 Z.Ciesielski and K.Gęba solved the problem for the squew pyramidal basis but did not publish this as one of main corollaries, namely 3.5.9, was obtained independently for the same basis by Bonic, Frampton and Tromba ['69]. In Bonic ['69], p.69, this theorem is proved for vector-valued Hölder functions $f : \mathbf{I} \longrightarrow \mathbf{F}$. See also Ciesielski ['75].

The above exposition is based on J.Ryll ['73], who generalized Ciesielski's proof to any pyramidal basis satisfying (6) and (7). Ryll in his proof used the same method as Ciesielski did; the only essential differences are: condition (7) and lemma 3.5.4.

Z.Ciesielski ['60b] gave also explicit estimations (from above and from below) for the norm $\| u_\alpha \|$ of the operator $u_\alpha : 1^\infty \to \text{Lip}_o^\alpha(I)$. In particular, he has shown that $\| u_\alpha \| \to \infty$ as $\alpha \to 1$. Thus, the method fails for $\alpha = 1$. Yet, $\text{Lip}_o^1(I)$ is isometrically isomorphic to $L^\infty(I)$ (each f in $\text{Lip}_o^1(I)$ is an indefinite integral of some g in $L^\infty(I)$) and A.Pełczyński ['58] has shown that $L^\infty(I)$ is linearly homeomorphic to 1^∞.

Z.Ciesielski ['63] has found a characterization of Hölder functions for the Franklin system, analogous to 3.5.9.

Bonic, Frampton and Tromba ['69] and Ryll ['78] proved similar theorems for the space $C^{(k+s)}(I^d)$ of all functions f in $C^{(k)}(I^d)$ such that all the partial derivatives of order k satisfy the Hölder condition with the exponent s $(k \geq 1,\ 0 < s < 1)$.

3.6. A construction of a basis in $C_o(X /\!/ Y)$

3.6.1. Suppose that X is a convex polytope of dimension d, $X \subset I^d$, and $T_1 > T_2 > \ldots$ is a sequence of triangulations of X (3.1.3) satisfying $\lim d_n = 0$. Denote $W_n = \text{vert}(T_n)$, $V_1 = W_1$, $V_{n+1} = W_{n+1} \setminus W_n$, $V = \bigcup_{n \in \mathbf{N}} V_n$. Let $(\varphi_v)_{v \in V}$ be the corresponding pyramidal basis in $C(X)$. Let Y be a closed subset of X. We shall construct a basis in the space

$$C_o(X /\!/ Y) = \{f \in C(X) : f(y) = 0 \text{ for } y \text{ in } Y\}.$$

If $v \in W_n$, let φ_v^n be the T_n-pyramidal function with vertex v and

(1) $\qquad Q_n(v) = \{x \in X : \varphi_v^n(x) > 0\}.$

$Q_n(v)$ need not be the set of interior points (relative to X or $\mathbf{R}^d$) of the union of simplexes in T_n having v as a vertex. It is clear that

$$Q_{n+1}(v) \subset Q_n(v) \subset \{x \in X : \varphi_v(x) > 0\} \quad \text{for} \quad v \quad \text{in} \quad W_n.$$

Denote $W_n' = \{v \in W_n : Q_n(v) \cap Y = \emptyset\}$. Then $W_n' \subset W_{n+1}'$. We claim that for each n in $\mathbf{N}$ there is a $k = k(n)$ in $\mathbf{N}$ such that

(2) $\qquad W_n \setminus Y \subset W_{k(n)}'.$

Indeed, Y is closed, $d_k \to 0$ and W_n is finite; consequently there is a k such that $d_k < \text{dist}(W_n \setminus Y, Y)$. Then $W_n \subset W_k'$.

Thus, $\bigcup_{n=1}^{\infty} W_n' = \bigcup_{n=1}^{\infty} W_n \setminus Y$. Denote $V_n' = W_n' \setminus W_{n-1}'$ for $n \geq 2$, $V_1' = W_1'$, $V' = \bigcup_{n=1}^{\infty} V_n'$ (some of the sets V_n' may be empty). Clearly, $V' = \bigcup_{n=1}^{\infty} \overline{W}_n' = V \setminus Y$.

If $v \in V_n'$, define ψ_v as the function equal to 1 at v, equal to 0 at each point of $W_n \setminus \{v\}$, and affine on each simplex of T_n. Obviously,

$$\{x \in X : \psi_v(x) > 0\} = Q_n(v) \quad \text{and} \quad \psi_v \in C_o(X /\!/ Y).$$

The set V' will now be ordered as follows: If $v \in V_n'$, $w \in V_k'$ and $k < n$, then $w < v$; in each V_n' we take the lexicographic order induced by $\mathbf{R}^d$ (this order of V' may be quite different from that induced by V). Thus the elements of V' are arranged into a single sequence $v_1 < v_2 < \ldots$ We recall (see 3.1.4) that

(3)   if $w > v$ and $\psi_v(w) > 0$, then $\{x : \psi_w(x) > 0\} \subset \{x : \psi_v(x) > 0\}$.

3.6.2. <u>Proposition</u>. The sequence (ψ_{v_m}) is a basis in $C_o(X /\!/ Y)$. Moreover, if $v \in V_n'$ then

$$\| f - s_v \| \leq \omega_f(2d_{n-1}) \quad \text{where} \quad s_v = \sum_{w \leq v} a_w \psi_w.$$

Proof. Let $f \in C_o(X /\!/ Y)$. Denote $s^{(0)} = 0$, $s^{(n)} = \sum_{w \in W_n'} a_w \psi_w$ for $n \geq 1$. We define the coefficients a_v as $a_v = f(v) - s^{(n-1)}(v)$ for v in V_n'. Thus, $a_v = f(v)$ for v in V_1'. If $v \in V_n'$, $n > 1$, then

$$s^{(n)}(v) = s^{(n-1)}(v) + \sum_{w \in V_n'} a_w \psi_w(v) = s^{(n-1)}(v) + a_v = f(v);$$

similarly, $s^{(k)}(v) = f(v)$ for $k > n$. Each partial sum s_w, $w \in \overline{W}_n'$, is affine on each simplex of T_n (still, it may be identically 0 on certain simplexes containing points of Y). By (2), the partial sum $s_{k(n)}$ coincides with f at each point of W_n.

We now want to estimate the number $|f(x) - s^{(n)}(x)|$. Denote

$$X_n = \mathbf{U}\{\Delta : \Delta \in T_n, \quad \text{vert}(\Delta) \subset W_n'\}.$$

If $x \in X_n$, then the inequality (5) in 3.1.7 is still valid. If $x \in Y$, then $f(x) = 0 = s^{(n)}(x)$. Thus, suppose that $x \in X \setminus (X_n \cup Y)$. Then there exists a simplex Δ_o in T_n such that $x \in \Delta_o$ and some vertex

v_o of Δ_o does not belong to W_n', i.e., there is a y in $Q_n(v_o) \cap Y$. Therefore $f(y) = 0$ and

$$(4) \qquad \varrho(x,y) \le \varrho(x,v_o) + \varrho(v_o,y) \le 2d_n.$$

If $w \in \text{vert}(\Delta_o)$, then either $w \in W_n'$ and then $s^{(n)}(w) = f(w)$ or else $w \notin W_n'$ and then $s^{(n)}(w) = 0$ (indeed, suppose that $\psi_v(w)$ were positive for some v in W_n'; then $v < w$ and, by (3) the set $\{x : \psi_w(x) > 0\}$ would be contained in $\{x : \psi_v(x) > 0\}$ and hence disjoint with Y; this, however would imply that $w \in W_n'$). Consequently, $s^{(n)}(x)$ is a convex combination of the numbers 0, $f(x_1),\ldots,f(x_j)$, where $\{x_1,\ldots,x_j\}$ is a subset of Δ_o (possibly empty). Thus, either $s^{(n)}(x)$ is between the infimum and the supremum of f on Δ_o and then $|f(x) - s^{(n)}(x)| \le \omega_f(d_n)$ or else

$$|s^{(n)}(x)| < \inf\{|f(z)| : z \in \Delta_o\}$$

and then, by (4), $|f(x) - s^{(n)}(x)| \le |f(x)| = |f(x) - f(y)| \le \omega_f(2d_n)$ for each x in $X \setminus (X_n \cup Y)$. Thus, $\| f - s^{(n)} \| \le \omega_f(2d_n)$.

Now let $U \subset V_n'$ and $s_U = s^{(n-1)} + \Sigma_{v \in U} a_v \psi_v$. Then for each v in W_n', $s_U(v)$ is equal to $s^{(n)}(v)$ if $v \in W_{n-1}' \cup U$ and is equal to $s^{(n-1)}(v)$ if $v \in W_n' \setminus U$. Thus, if $x \in X$, $s_U(x)$ is a convex combination of 0 and values of f on a simplex of T_{n-1}; hence $\| f - s_U \| \le \omega_f(2d_{n-1})$. $\square$

3.6.3. <u>Example</u>. Let Y be the closed subset of I^2 consisting of three components shown in heavy black in the right-hand half of the picture.

▱ the set X_3

▱ the set $I^2 \setminus (X_3 \cup Y_3)$

▱ the set Y_3

⊚   vertices of V_1'

●   vertices of V_2'

·   vertices of V_3'

—   edges of T_3

—   boundary of P_3

There are 5 vertices in V_1 and only one of them, namely $v = (0,1)$ belongs to V_1'; the corresponding function ψ_v is positive on the triangle $x_2 > x_1$. There are 5 vertices in V_2'. Note that the function ψ_w for $w = (0,0)$ is positive only on the set $x_1 + x_2 < \frac{1}{2}$ and is different from the function φ_w for the same vertex in the basis in $C(I^2)$ constructed in 3.2.1 (in fact, $\psi_w(x) = \varphi_w(2x)$ for x in I^2); for the remaining 4 vertices of V_2' the function ψ_v is identical with φ_v. There are 19 vertices in V_3'. For the vertices $(1,1)$ and $(\frac{1}{2},0)$ the functions ψ_v are different from the corresponding functions φ_v; for the remaining 17 vertices we have $\psi_v = \varphi_v$. There are 16 vertices in $W_3 \smallsetminus W_3'$. Only two of them, namely $(\frac{1}{2},\frac{1}{2})$ and $(\frac{5}{8},\frac{3}{8})$ belong to Y. The remaining 14 elements of $W_3 \smallsetminus W_3'$ belong to W_k' for sufficiently large k.

Besides the set X_n defined above we may consider the following sets:

$$Z_n = U\{\Delta : \Delta \in T_n, \ Y \cap \text{int}\, \Delta = \emptyset\}, \quad Y_n = U\{\Delta : \Delta \in T_n, \ \text{vert}(\Delta) \cap W_n' = \emptyset\},$$

and $P_n = U\{\Delta : \Delta \in T_n, \ Y \cap \Delta = \emptyset\}$. Obviously, $X_n \subset X \smallsetminus Y_n \subset Z_n$ and $X_n \subset P_n \subset Z_n$. The set Z_n need not be disjoint with Y; in fact, the center $(\frac{1}{2},\frac{1}{2})$ belongs to both sets. Y need not be contained in Y_n if some v_o in $\text{vert}(T_n)$ is an isolated point of Y. In the example above there are 4 triangles in T_3 which are contained in Y_3 and disjoint with Y; a fifth triangle in T_3 is contained in Y_3 and has the interior disjoint with Y.

3.6.4. <u>Exercises</u>. (A). In case of the above basis in $C_o(X /\!/ Y)$ the inequality $\omega_{s_v}(\delta) \le \omega_f(\delta)$ need not hold; yet $\omega_{s_v}(\delta) \le \omega_f(2\delta)$.

(B) If $\psi_v \ne \varphi_v$ for some v in V' then there exists one common function θ in $\{\hat{\theta},\theta\} \cup \{\theta^{i_1,\ldots,i_k} : 1 \le i_1 < \ldots < i_k \le d\}$ and numbers n,m such that $\varphi_v(x) = \theta(2^m(x-v))$, $\psi_v(x) = \theta(2^n(x-v))$.

(C) Construct a basis in $C_o(I^d /\!/ \partial I^d)$ where ∂I^d denotes the boundary with respect to R^d.

3.6.5. <u>Notes</u>. C.Bessaga ['57] constructed an analogous basis in $C_o(I^\omega /\!/ Y)$ for an arbitrary $Y \subset I^\omega$ (see 4.2.6).

If X is a closed subset of a compact metric space X, then $C_o(X /\!/ Y)$ is isomorphic to $C(Z)$, where Z is a one-point compactification of $X \smallsetminus Y$; consequently, $C_o(X /\!/ Y)$ has a basis.

See also Section 4.1.

3.7. Piecewise multiaffine bases in $C(\mathbf{I}^d)$, $d \geq 2$

3.7.1. Definitions. A function on a cuboid

(1) $\qquad Q = [a_1,b_1] \times \ldots \times [a_d,b_d]$

will be called __multiaffine__ iff it is a finite sum of functions of the
type $h(x) = \Pi_{i=1}^d h_i(x_i)$, where $x = (x_1,\ldots,x_d) \in Q$ and $h_i(t)=c_i t+c_i'$
with $c_i, c_i' \in \mathbf{R}$. A function h will be called __piecewise multiaffine__
on a polytope $P \subset \mathbf{R}^d$ iff P is the union $Q_1 \cup \ldots \cup Q_k$ of cuboids
such that h is multiaffine on each Q_j separately.

In this section we shall deal with bases in $C(\mathbf{I}^d)$ consisting
of piecewise multiaffine functions. The simplest example is the tensor
product of d bases $(\varphi_m^{(1)}),\ldots,(\varphi_m^{(d)})$ of piecewise affine functions
on I, e.g., the tensor product of d copies of the Schauder system
(φ_m) in $C(\mathbf{I})$ or the Franklin system (f_m^F).

The support of $\varphi_{m_1} \otimes \ldots \otimes \varphi_{m_d}$ is a cuboid (1) with $a_i, b_i \in D$.
Infinitely many sets $\text{supp}(\varphi_{m_1} \otimes \ldots \otimes \varphi_{m_d})$ are of the form (1) with
$[a_1,b_1] = [0,1]$. Thus, diameters of these supports do not tend to 0.

3.7.2. We shall now construct another piecewise multiaffine
basis (ψ_m) in $C(\mathbf{I}^d)$ such that the support of each ψ_m is a cube
(or the intersection of $\mathbf{I}^d$ with a cube) and the diameter of $\text{supp}\,\psi_m$
tends to 0. Specifically, let

$$V = (\tilde{D})^d = \bigcup_{n=0}^{\infty} D_n^{(d)} \quad \text{(see 3.4.7).}$$

The functions $(\psi_\tau)_{\tau \in V}$ are defined as follows: if $\tau = (\tau_1,\ldots,\tau_d) \in D_n^{(d)}$
and $x = (x_1,\ldots,x_d) \in \mathbf{R}^d$, then

(2) $\qquad \psi_\tau(x) = \prod_{i=1}^{d} \psi(2^n(x_i - \tau_i)),$

where ψ is a hat function of a single variable defined as

$$\psi(t) = \max(0, 1 - |t|).$$

It is clear that $\text{supp}\,\psi_\tau = \mathbf{I}^d \cap (\tau + [-2^{-n}, 2^{-n}]^d)$ for τ in $D_n^{(d)}$.
Thus, if τ does not belong to the boundary of $\mathbf{I}^d$ (in $\mathbf{R}^d$), i.e.,
if $0 \neq \tau_i \neq 1$ for $i = 1,\ldots,d$, then $\text{supp}\,\psi_\tau$ is simply the cube
$[-2^{-n}, 2^{-n}]^d$ shifted by ψ.

$$D_0 \times D_0 \qquad D_1 \times D_1 \qquad D_0 \times D_1 \qquad D_2 \times D_2 \qquad D_2 \times D_1 \qquad D_2 \times D_0$$

The picture above shows the supports of sample functions ψ_τ in the case where $d = 2$, with vertices chosen from sets $D_j \times D_k$, $j, k = 0, 1, 2$.

3.7.3. <u>Lemma</u>. If $\tau \in D_n^{(d)}$, then ψ_τ vanishes at each point in $D_0^{(d)} \cup \ldots \cup D_{n-1}^{(d)}$ and at each point of $D_n^{(d)} \setminus \{\tau\}$. Moreover, $\psi_\tau(\tau) = 1$. $\quad \square$

3.7.4. We shall need the functionals $\Gamma_\tau : C(\mathbf{I}^d) \longrightarrow \mathbf{R}$ defined as follows: If $\tau \in D_n^{(d)}$, let

$$(3) \qquad \Gamma_\tau f = \frac{1}{2^d} \sum_{\varepsilon \in \Omega^d} f(\tau^\varepsilon)$$

where $\Omega = \{-1, 1\}$, $\tau = (\tau_1, \ldots, \tau_d)$, $\tau^\varepsilon = (\tau_1^{\varepsilon_1, n}, \ldots, \tau_d^{\varepsilon_d, n})$, $\varepsilon = (\varepsilon_1, \ldots, \varepsilon_d)$, $\varepsilon_i = \pm 1$, $\tau_i^{\varepsilon_i, n} = \tau_i + \varepsilon_i 2^{-n}$ for τ_i in D_n, $\tau_i^{\varepsilon_i, n} = \tau_i$ for τ_i in $\hat{D}_{n-1} = D_0 \cup \ldots \cup D_{n-1}$. If $\tau \in (D_n)^d$, i.e., if each coordinate of τ is a dyadic number of order n, then $\Gamma_\tau f$ is the average of values of f at 2^d vertices of the cube $\tau + [-2^{-n}, 2^{-n}]^d$. In general, if $\tau \in D_n^d$ and we denote

$$M = \{i : \tau_i \in D_n\}, \quad L = \{i : \tau_i \notin D_n\}, \quad m = \mathrm{card}\, M,$$

then each number $f(\tau^\varepsilon)$ in (3) is repeated 2^{d-m} times; therefore all points τ^ε with $\varepsilon \in \Omega^d$ belong to an m-dimensional hyperplane $\{x : x_i = \tau_i$ for i in $L\}$ and $\Gamma_\tau f$ is the average of the values of f at 2^m vertices of the m-dimensional cube $\tau + [-2^{-n}, 2^{-n}]^m$.

3.7.5. <u>Lemma</u>. Let h be a multiaffine function on an m-dimensional cuboid Q of the form (1). Then the average of the values of h at 2^m vertices of the cuboid is equal to the value of h at the center z of the cuboid.

Proof. It is enough to prove the lemma for products $h(x) = x_{i_1} \ldots x_{i_k}$. Assume that $h(x) = x_1 \ldots x_k$, say. Let $z = (z_1, \ldots, z_m)$, $t_i = (a_i + b_i)/2$, $\delta_i = b_i - a_i$. Then

$$2^{-m} \sum_{\varepsilon \in \Omega^m} h(z_1 + \varepsilon_1 \delta_1, \ldots, z_m + \varepsilon_m \delta_m) = 2^{-k} \sum_{\varepsilon \in \Omega^k} \prod_{j=1}^{k} (z_j + \varepsilon_j \delta_j)$$

$$= 2^{-k} [(z_k - \delta_k) + (z_k + \delta_k)] \sum_{\varepsilon \in \Omega^{k-1}} \prod_{j=1}^{k-1} (z_j + \varepsilon_j \delta_j)$$

$$= 2^{-k+1} z_k \sum_{\varepsilon \in \Omega^{k-1}} \prod_{j=1}^{k-1} (z_j + \varepsilon_j \delta_j) .$$

Hence, step by step, we get $z_k z_{k-1} \cdots z_1$. $\quad \square$

3.7.6. <u>Lemma</u>. Let h be a multiaffine function on a cuboid Q of the form (1). Then $\sup h$ and $\inf h$ are attained at vertices of Q. $\quad \square$

3.7.7. <u>Proposition</u>. The system $(\psi_\tau)_{\tau \in V}$, ordered analogously to the regular pyramidal basis, is an interpolating basis in $C(\mathbf{I}^d)$ with nodes in V.

Proof. We begin with defining the coefficients. If $f \in C(\mathbf{I}^d)$, let

$$(4) \qquad a_\tau = f(\tau) \qquad \text{for } \tau \text{ in } D_o^{(d)},$$

$$(5) \qquad a_\tau = 2^{-d} \sum_{\varepsilon \in \Omega^d} [f(\tau) - f(\tau^\varepsilon)] = f(\tau) - \Gamma_\tau f \quad \text{for } \tau \text{ in } D_n^{(d)}, \quad n \geq 1.$$

We are going to prove that the series $\sum a_\tau \psi_\tau$ converges to f uniformly. Denote

$$S^{(n)} = \sum_{k=0}^{n} \sum_{\tau \in D_k^{(d)}} a_\tau \psi_\tau \qquad (n = 0, 1, \ldots) .$$

We shall show by induction that

$$(6) \qquad S^{(n)}(v) = f(v) \quad \text{for } v \text{ in } \hat{D}_n^{(d)} = D_o^{(d)} \cup \ldots \cup D_n^{(n)} .$$

If $n = 0$ and $v \in D_o^{(d)}$, then $S^{(0)}(v) = f(v)$ by Lemma 3.7.3. Assume that (6) holds for some n. Let $v \in \hat{D}_{n+1}^{(d)}$. If $v \in \hat{D}_n^{(d)}$, then $\psi_\tau(v) = 0$ for τ in $D_{n+1}^{(d)}$ and hence $S^{(n+1)}(v) = S^{(n)}(v) = f(v)$. If $v \in D_{n+1}^{(d)}$, then all points τ^ε belong to $\hat{D}_{n-1}^{(d)}$ for ε in $\Omega^{(d)}$. Therefore, by (6) and by Lemmas 3.7.3. and 3.7.5,

$$S^{(n+1)}(v) = S^{(n)}(v) + a_v = S^{(n)}(v) + f(v) - \Gamma_v f =$$

$$= S^{(n)}(v) + f(v) - \Gamma_v S^{(n)} = f(v) .$$

Now, let $K = w + [0, 2^{-n}]^d$ be any cube in the partition P_n (see 3.4.3) and let $V_K = \text{vert}(K) \subset \hat{D}_n^{(d)}$. Let $x \in K$, then by Lemma 3.7.6,

$$\inf_{v \in V_K} f(v) = \inf_{v \in V_K} S^{(n)}(v) \leq S^{(n)}(x) \leq \sup_{v \in V_K} S^{(n)}(v) = \sup_{v \in V_K} f(v).$$

Thus, both $S^{(n)}(x)$ and $f(x)$ are two numbers between the infimum and the supremum of f on K. Therefore

(7) $\qquad |f(x) - S^{(n)}(x)| \leq \omega_f(\text{diam } K) = \omega_f(\sqrt{d}2^{-n})$

and, moreover,

(8) $\qquad |a_\tau| \leq \omega_f(\sqrt{d}2^{-n})$ for τ in $D_n^{(d)}$.

Consequently, since $\text{card}(V_K) = 2^d$ and $\psi_\tau(x) = 0$ for τ in $D_n^{(d)} \setminus V_K$,

$$\sum_{\tau \in D_n^{(d)}} |a_\tau| \psi_\tau(x) \leq 2^d \omega_f(\sqrt{d}2^{-n}).$$

Thus, the series $\Sigma a_\tau \psi_\tau$ is uniformly convergent to f. By Proposition 1.3.2, this expansion is unique. $\square$

3.7.8. <u>Exercises</u>. (A) Let $\tau = (\tau_1, \ldots, \tau_d) \in D_n^{(d)}$. Then

$$\psi_\tau \leq \varphi_{\tau_1} \otimes \ldots \otimes \varphi_{\tau_d},$$

where φ_τ are the Faber–Schauder dyadic hat functions. The equality holds if and only if $\tau \in (D_n)^d$, i.e., each τ_i is of the same order n.

(B) Show that functions ψ_m are piecewise harmonic, i.e., $\Delta\psi_m = 0$ on suitable cubes.

(C) Let $\Gamma^h f(x)$, where $x = (x_1, \ldots, x_d)$, denote the average of 2^d values of f at vertices of the cube

(9) $\qquad Q = [x_1 - h, \ x_1 + h] \times \ldots \times [x_d - h, \ x_d + h].$

Show the following: (a) If $f(x) = \Sigma a_{ij} x_i x_j + \Sigma b_i x_i + c$ is a polynomial of degree 2, then $f - \Gamma^h f = -h^2 \Delta f$, where Δ denotes the Laplacian. (b) If $f \in C^{(2)}(G)$, where G is a domain in $\mathbf{R}^d$ containing the cube (9), then $h^{-2}\Gamma^h f(x) \longrightarrow -\Delta f(x)$ and $h \longrightarrow 0$ (Hint: use the Taylor formula). (c) Interpret the formula (5) accordingly.

3.7.9. <u>Notes</u>. The basis $(\psi_\tau)_{\tau \in V}$, which may be called the "cube multiaffine basis", has been constructed by Ryll ['73]. See also 3.8.4.

3.8. Survey of bases in $C^{(k)}(I^d)$ and in A

3.8.1. $C^{(k)}(I^d)$ will denote the space of p-times continuously differentiable real-valued functions on I^d with the norm

(1) $\qquad \| f \|^{(p)} = \sup\{ \| D^\alpha f \| : 0 \leq \alpha \leq p \},$

where $\alpha = (\alpha_1, \ldots, \alpha_d)$, $|\alpha| = |\alpha_1| + \ldots + |\alpha_d|$ and

$$D^\alpha f = \frac{\partial^{|\alpha|} f}{\partial x_1^{\alpha_1} \ldots \partial x_d^{\alpha_d}} .$$

A will denote the space of all functions continuous on the closed disc $\bar{K} = \{z \in C : |z| \leq 1\}$ and analytic in the interior of the disc, with the supremum norm.

The problems of how to construct bases in $C^{(k)}(I^d)$ and in A were open for almost 40 yers. They were raised (for $C^{(1)}(I^2)$ and for A) in the classical treatise Banach ['32], Remarques (for dimension $d = 1$ the problem is trivial, see 2.3.7).

The purpose of this section is to present briefly, without proofs, Ciesielski's construction of a basis in $C^{(1)}(I^2)$ and Bočkariev' construction of a basis in A.

3.8.2. <u>Proposition</u> (Ciesielski ['66]). The indefinite integrals of the Franklin functions augmented with the constant function 1 form a basis in $C(I)$.

We recall that the Haar system is an orthogonal basic sequence in $L^\infty(I)$ and the indefinite integrals of Haar functions (i.e., Faber-Schauder functions φ_n) together with constant 1 form a basis in $C(I)$. The Franklin dyadic functions (f_n^F) form an orthogonal basis in $C(I)$. Denote

$$f_1^C(t) = 1, \qquad f_n^C(t) = \int_0^t f_{n-2}(u)\,du \qquad \text{for } n = 2,3,\ldots, \quad t \in I.$$

The proposition above asserts that the sequence (f_n^C) is a basis in $C(I)$. The coefficients of a function g in $C(I)$ can be expressed as the Stieltjes integrals

$$a_n = \int_0^1 f_n^F(t)\,dg(t), \qquad n \geq 1,$$

and $a_o = g(0)$. Thus

$$g(x) = g(0) + \sum_{n=0}^{\infty} [\int_0^1 f_n^F(t)\,dg(t)]\int_0^x f_n^F(u)\,du$$

and the convergence is uniform for x in $\mathbf{I}$.

3.8.3. <u>Theorem</u> (Ciesielski-Schonefeld). The set of functions

$$h_{n,m}^{(0)} = f_n^C \otimes f_m^C, \quad \text{i.e.,} \quad h_{n,m}(x_1,x_2) = f_n^C(x_1)f_m^C(x_2),$$

with a tensor product order (see below), is a basis in $C^{(1)}(\mathbf{I}^2)$.

Observe that by Theorem 1.4.3, $(h_{n,m})$ is a basis in $C(\mathbf{I}^2)$. Thus, each g in $C(\mathbf{I})$ can be written as

$$(2) \qquad g = \Sigma a_{n,m} h_{n,m}^{(0)}.$$

Moreover, the systems

$$h_{n,m}^{(1)} = f_n^F \otimes f_m^C \quad \text{and} \quad h_{n,m}^{(2)} = f_n^{(C)} \otimes f_m^F$$

are also bases. Yet, the ways they are ordered are not identical. In general, the functions $\varphi_n \otimes \psi_m$ in a tensor product of bases form an infinite matrix:

$$
\begin{array}{llll|l}
\varphi_1 \otimes \psi_1 & \varphi_1 \otimes \psi_2 & \cdots & & \varphi_1 \otimes \psi_n \\
\varphi_2 \otimes \psi_1 & \varphi_2 \otimes \psi_2 & \cdots & & \varphi_2 \otimes \psi_n \\
\vdots & & & & \\
\hline
\varphi_n \otimes \psi_1 & \varphi_n \otimes \psi_2 & \cdots & & \varphi_n \otimes \psi_n
\end{array}
$$

The lines drawn above indicate the new functions. In the basis $h_{n,m}^{(0)}$ these new functions are arranged first horizontally, then vertically (from the top to the bottom):

$$\varphi_n \otimes \psi_1, \ldots, \varphi_n \otimes \psi_{n-1}, \quad \varphi_1 \otimes \psi_n, \ldots, \varphi_{n-1} \otimes \psi_n, \quad \varphi_n \otimes \psi_n.$$

In the basis $h_{n,m}^{(1)}$ these functions are arranged first vertically, then horizontally:

$$\varphi_1 \otimes \psi_n, \ldots, \varphi_{n-1} \otimes \psi_n, \quad \varphi_n \otimes \psi_1, \ldots, \varphi_n \otimes \psi_{n-1}, \quad \varphi_n \otimes \psi_n,$$

where in the basis $h_{n,m}^{(2)}$ the functions are arranged first vertically, then horizontally, but on the sides of an $(n+1) \times n$ rectangle rather than of an $n \times n$ square:

$$\varphi_{n+1} \otimes \psi_1, \ldots, \varphi_{n+1} \otimes \psi_{n-1}, \quad \varphi_1 \otimes \psi_n, \ldots, \varphi_{n+1} \otimes \psi_n.$$

This shift of indices is needed for the crucial property:

$$(3) \qquad \frac{\partial h_{n,m}^{(0)}}{\partial x_1} = h_{n-1,m}^{(1)}, \qquad \frac{\partial h_{n,m}^{(0)}}{\partial x_2} = h_{n,m-1}^{(2)}.$$

Indeed, the derivatives of constants vanish and a row or column of the above matrix disppears.

Identities analogous to (3) are valid for the partial sums. Thus, the convergence of the series $\Sigma h_{n,m}^{(1)}$ and $\Sigma h_{n,m}^{(2)}$ implies the convergence of first partial derivatives of $\Sigma h_{n-1,m}^{(1)}$ and this is what is needed for the convergence of the series (2) in the C^1-norm, i.e., in the norm

$$(4) \qquad \| g \|^{(1)} = \max(\| g \|_\infty, \| \frac{\partial g}{\partial x_1} \|_\infty, \| \frac{\partial g}{\partial x_2} \|_\infty).$$

For details we refer the reader to Ciesielski ['69].

3.8.4. We shall now outline the construction of the Bočkariev basis in A (see S.V.Bočkariev ['74], Z.Wronicz ['79]).

Consider the Franklin functions f_n^F with dyadic breakpoints in I (see 2.5.1) and define new functions as follows:

$$f_n(x) = \begin{cases} \dfrac{1}{\sqrt{2\pi}} f_n^F(\dfrac{x}{\pi}) & \text{if} \quad 0 \leq x \leq \pi, \\[2ex] \dfrac{1}{\sqrt{2\pi}} f_n^F(-\dfrac{x}{\pi}) & \text{if} \quad -\pi \leq x \leq 0 \end{cases}$$

($n = 0, 1, \ldots$). Obviously, the functions f_n are even, 2π-periodic, satisfy the Lipschitz condition, and $\int_{-\pi}^{\pi} f_n(x) f_m(x) \, dx = \delta_{nm}$. The conjugate functions

$$\tilde{f}_n(x) = \frac{1}{2\pi} \int_{-\pi}^{\pi} \frac{f_n(x-t)}{\tan\frac{t}{2}} dt , \quad n \geq 1$$

(see e.g. K.Hoffman ['62], Ch.6) are continuous, odd, and the functions

$$g_n(z) = g_n(re^{ix}) = \frac{1}{2\pi} \int_{-\pi}^{\pi} \frac{f_n(t)+i\tilde{f}_n(t)}{\sqrt{2}} \frac{1-r^2}{1-2r\cos(t-x)+r^2} dt$$

$$= \frac{1}{2\pi} \int_{-\pi}^{\pi} \frac{f_n(t)}{\sqrt{2}} \frac{e^{it}+z}{e^{it}-z} dt ,$$

belong to the space A. The sequence

$$g_0 = \frac{1+i}{2\sqrt{\pi}}, \quad g_1, \ g_2, \dots$$

is called the <u>Bočkariev system</u>. Since f_n is orthogonal to the function 1 for $n \geq 1$, from the above formulas it follows that $g_n(0) = 0$. We shall show that the sequence $(\tilde{f}_n)$ as well as the sequence of boundary values of (g_n) are orthonormal in $L^2(-\pi,\pi)$, and moreover each f_n is orthogonal to each $\tilde{f}_m$. Indeed, if $0 \leq n < m$, then

$$\int_{-\pi}^{\pi} g_n(e^{it}) \overline{g_m(e^{it})} dt$$

$$= \frac{1}{2} \int_{-\pi}^{\pi} (f_n f_m + \tilde{f}_n \tilde{f}_m) dt + \frac{i}{2} \int_{-\pi}^{\pi} (f_m \tilde{f}_n - f_n \tilde{f}_m) dt .$$

The integral in the imaginary part vanishes because the products $f_m \tilde{f}_n$ and $f_n \tilde{f}_m$ are odd functions. Moreover, if $0 \leq n < m$, then by the Cauchy formula,

$$0 = g_n(0) g_m(0) = \frac{1}{2\pi i} \int_{|z|=1} \frac{g_n(z) g_m(z)}{z} dz = \frac{1}{2\pi} \int_{-\pi}^{\pi} g_n(e^{it}) g_m(e^{it}) dt$$

$$= \frac{1}{4\pi} \int_{-\pi}^{\pi} (f_n f_m - \tilde{f}_n \tilde{f}_m) dt + \frac{i}{4\pi} \int_{-\pi}^{\pi} (f_n \tilde{f}_m + f_n \tilde{f}_m) dt$$

and hence $\int_{-\pi}^{\pi} \tilde{f}_n \tilde{f}_m dt = 0$.

Bočkariev has shown that $(g_n)_{n=0,1,\dots}$ is a basis in A considered as a complex Banach space (that is, the coefficients in the expansion $\Sigma a_n g_n$ are to be complex).

It is also possible to construct a basis in A considered as a Banach space over **R** (Wronicz ['79]).

3.8.5. <u>Notes</u>. The construction in 3.9.3 is due to Z.Ciesielski ['69] and (independently) to S.Schonefeld ['69a]. Its extension to the space $C^{(1)}(I^d)$ is immediate.

On the other hand, proofs of existence of bases in $C^{(k)}(I^d)$ require different thechniques. Z.Ciesielski and J.Domsta ['72] found an orthogonal basis in $C^{(k)}(I^d)$, $k = 1, 2, \ldots$ whereas S.Schonefeld ['69b] found an interpolating basis in $C^{(k)}(I^d)$ for $k = 1, \ldots, 4$ and in $C^{(k)}(T^d)$ where $T = \{z \in C : |z| = 1\}$. J.Ryll ['78] defined a system $(\varphi_n^q)_{n \in N}$ such that (φ_n^q) is an interpolating basis in $C(I^d)$, (φ_n^q) is a basis in each space $C^{(k)}(I^d)$ for $k = 1, \ldots, q$ and

$$\text{diam}(\text{supp } \varphi_n^q) \longrightarrow 0 \quad \text{as} \quad n \longrightarrow \infty.$$

If $q = 0$, the above basis is just the multiaffine cube basis considered in 3.7.2. See also G.E.Tkebučava ['77].

Z.Ciesielski and T.Figiel ['82], ['83] constructed, for each k in **N**, a basis (φ_n) in $C(X)$, where X is a smooth compact manifold (with boundary or without), which is simultaneously a basis in all Besov and Sobolev spaces of order not exceeding k; moreover, the biorthogonal system (ψ_n) is a basis in the Besov and Sobolev spaces (with the same exponents) of functions vanishing on the boundary of X.

Z.A.Čanturija ['71] found a counter example to possible extension of 3.8.2: an orthonormal basis in $L^1(I)$ such that the indefinite integrals are not a basis in $C(I)$.

Z.Ciesielski extended Bočkariev method to the case of several complex variables. He constructed bases in uniform closures of complex polynomials on polydiscs and in closures with respect the norm

$$\| w \|_d^{(m)} = \sum_{|p| \leq m} \sup_{|z_i| \leq 1} |D^p w(z_1, \ldots, z_d)|,$$

where $z_i \in C$, $p = (p_1, \ldots, p_d)$, $|p| = p_1 + \ldots + p_d$, w is a complex polynomial, and the differentiation D^p is taken in the complex variable. The results were announced in Ciesielski ['74] but since that time the proofs - involving tedious, technical computations - have been in a manuscript form only.

Chapter 4

FURTHER PROPERTIES OF BASES IN C(X)

In this chapter, several topics are presented. In 4.1 it is shown that in certain situations the existence of a linear extension operator $C(Y) \longrightarrow C(X)$, where $Y \subset X$, is equivalent to the possibility of splitting a basis in $C(X)$ to terms in $C_o(X//Y)$ and terms yielding a basis in $C(Y)$. In 4.2 and 4.3 bases of prism-pyramidal functions are constructed first in $C(I^\omega)$ and then in $C(X)$, where X is an arbitrary closed subset of I^ω. In 4.4 another construction of a basis in an arbitrary space $C(X)$ is presented. In 4.5 a mixed approximation-interpolation problem is considered and applied to the existence of bases consisting of polynomials. Auerbach bases in finite--dimensional subspace of $C(X)$ are discussed in 4.6.

The two final subsections are devoted to proofs of nonexistence of bases in $C(X)$ with additional properties, e.g., bounded and at the same time orthogonal with respect to some measure.

4.1. Relations between bases in $C(X)$, in $C_o(X//Y)$ and linear extension operators

In the following lemmas, which are well known and straight-forward, $k_1 < k_2 < \ldots$ and $m_1 < m_2 < \ldots$ are two sequences such that the sets $K = \{k_n\}_{n \in \mathbf{N}}$ and $M = \{m_n\}_{n \in \mathbf{N}}$ are infinite, disjoint, and $K \cup M = \mathbf{N}$.

4.1.1. <u>Lemma</u>. Let a Banach space F be a direct sum $G \oplus H$ of its subspaces, with corresponding complementary projections $P : F \longrightarrow G$, $I-P : F \longrightarrow H$. Let (ψ_n) be a basis in G, (θ_n) a basis in H, and let (η_n) and (ζ_n) be the corresponding sequences of coefficient functionals in G^* and H^*, respectively. Denote $\varphi_{k_n} = \psi_n$, $\varphi_{m_n} = \theta_n$, $\xi_{k_n} = P^* \eta_n$, $\xi_{m_n} = (I-P)^* \zeta_n$. Then (φ_n) is a basis in F and (ξ_n) is its sequence of coefficient functionals. Moreover

$$(1) \qquad c \leq c_1 \| P \| + c_2 \| I-P \|,$$

here c, c_1, c_2 are the basis constants of (φ_n), (ψ_n) and (θ_n), respectively. $\Box$

4.1.2. <u>Lemma</u>. Let (φ_n) be a basis in a Banach space F such that the convergence of $\Sigma a_n \varphi_n$ implies the convergence of $\Sigma a_{k_n} \varphi_{k_n}$. Then F can be written as a direct sum $F = G \oplus H$, where G is spanned by (φ_{k_n}), H is spanned by (φ_{m_n}), and $P(\Sigma a_n \varphi_n) = \Sigma a_{k_n} \varphi_{k_n}$ is a projection $P : F \longrightarrow G$. □

4.1.3. <u>Definition</u>. Let Y be a closed subset of a compact metric space X. We recall (see #21.1.1) that $\Lambda : C(Y) \longrightarrow C(X)$ is said to be a <u>linear extension operator</u> if it is a bounded linear operator and for each g in $C(Y)$ the function Λg is an extension of g, i.e., $\Lambda g(y) = g(y)$ for y in Y.

Let F be the range of such a Λ, i.e., the subspace of extended functions. Then $F \subset C(X)$ and the operator

$$\Gamma_1 : C(X) \longrightarrow F$$

defined as $\Gamma_1 = \Lambda(\text{rest}_Y f)$ is a projection with $\| \Gamma_1 \| = \| \Lambda \|$, while

$$\Gamma_2 : C(X) \longrightarrow C_0(X /\!/ Y)$$

defined as $\Gamma_2 f = f - \Gamma_1 f$ is the complementary projection, and $\| \Gamma_2 \| = 1 + \| \Lambda \|$. Moreover

$$(2) \qquad \Theta : C(X) \longrightarrow C(Y) \times C_0(X /\!/ Y)$$

defined as $\Theta f = (\text{rest}_Y f, \Gamma_2 f)$ is a linear bijection with $\| \Theta \| = 1 + \| \Lambda \|$, $\| \Theta^{-1} \| \leq 1 + \| \Lambda \|$.

4.1.4. <u>Proposition</u>. Let Λ be a linear extension operator. Let (φ_n) be a basis in $C(Y)$ and (Ψ_n) a basis in $C_0(X /\!/ Y)$. Then the sequence $\tilde{\varphi}_1, \Psi_1, \tilde{\varphi}_2, \Psi_2, \ldots$, where $\tilde{\varphi}_n = \Lambda(\varphi_n)$, is a basis in $C(X)$. Specifically, if $\Theta_{2n-1} = \tilde{\varphi}_n$, $\Theta_{2n} = \Psi_n$ and $f \in C(X)$, then

$$f = \Gamma_1 f + \Gamma_2 f = \Lambda(\Sigma a_n \varphi_n) + \Sigma b_n \Psi_n = \Sigma(a_n \tilde{\varphi}_n + b_n \Psi_n).$$

Conversely, if an expansion $f = \Sigma(a_n \tilde{\varphi}_n + b_n \Psi_n)$ is given, $\Gamma_1 f$ can be obtained by setting all b_n equal to 0 (and keeping a_n unchanged), and similarly we get $\Gamma_2 f$ setting $a_n = 0$. Thus, the direct decomposition $C(X) = F \oplus C_0(X /\!/ Y)$ corresponds to splitting the series $\Sigma c_n \Theta_n$ to two series: $\Sigma a_n \tilde{\varphi}_n$ and $\Sigma b_n \Psi_n$. □

4.1.5. <u>Proposition</u>. Let $\|\Lambda\| = 1$ and let both (φ_n), (ψ_n) be monotone. Then the basis constant c of (θ_n) satisfies the inequality $c \le 3$ and the number 3 is best possible.

Proof. By (1) we have $c \le c_1 \|\Lambda\| + c_2 (1 + \|\Lambda\|) \le 3$. To show that $\|\Lambda\| = 1$, $c_1 = c_2 = 1$ do not imply that $c < 3$, consider the following example. Identify the interval $\mathbf{I}$ with the subset $I_1 = \mathbf{I} \times \{0\}$ of $\mathbf{I}^2$. Let ε be a fixed dydadic number, e.g. $\varepsilon = \frac{1}{2}$. If $g \in C(\mathbf{I})$, let $\Lambda g = \tilde{g}$, where

$$\tilde{g}(x_1, x_2) = \begin{cases} g(x_1) & \text{if } 0 \le x_2 \le 1-\varepsilon, \\ g(x_1) \cdot [\frac{2}{\varepsilon}(1-x_2) - 1] & \text{if } 1-\varepsilon \le x_2 \le 1. \end{cases}$$

Thus, $\Lambda : C(\mathbf{I}) \longrightarrow C(\mathbf{I}^2)$ is a linear extension operator and $\|\Lambda\| = 1$. For each fixed x_1 in $\mathbf{I}$, the function $\tilde{g}$ is affine on the segment joining $(x_1, 1-\varepsilon)$ with $(x_1, 1)$. Moreover, $\tilde{g} \in C(\mathbf{I}^2)$ and $\tilde{g}(x_1, 1) = -g(x_1)$. Let (φ_n) be the Faber-Schauder system in $C(\mathbf{I})$ and let $\tilde{\varphi}_m = \Lambda(\varphi_m)$. Let (Ψ_m) be the basis in $C_o(\mathbf{I}^2 /\!/ I_1)$ constructed by the method described in 3.6.2 for the triangulations (T_n) used in the construction of the regular pyramidal basis in $C(\mathbf{I}^2)$. Then (Ψ_m) consists of all functions (φ_v) constructed in 3.2.1 except of those having peaks in I_1.

We shall show that the basis constant of $\tilde{\varphi}_1, \Psi_1, \tilde{\varphi}_2, \Psi_2, \ldots$ in $C(\mathbf{I}^2)$ is equal to 3. Let $f(x_1, x_2) = x_1$. Then $f \in C(\mathbf{I}^2)$, $\|f\| = 1$ and the expansion of f with respect to the above basis is $f = s^{(1)} + r$, where

$$s^{(1)} = 0 \cdot \tilde{\varphi}_o + 1 \cdot \tilde{\varphi}_1 + 0 \cdot \varphi_{(0,1)} + 2 \cdot \varphi_{(1,1)} + 0 \cdot \varphi_{(\frac{1}{2}, \frac{1}{2})}$$

and r is the sum of terms corresponding to the triangulations T_n, $n \ge 2$. Thus,

$$\|s^{(1)}\| = \|\tilde{\varphi}_1 + 2\varphi_{(1,1)}\| \ge |\tilde{\varphi}_1(1, 1-\varepsilon) + 2\varphi_{(1,1)}(1, 1-\varepsilon)| = 3 - 2\varepsilon. \quad \square$$

4.1.6. <u>Proposition</u>. Let Y be a closed subset of a compact metric space X. Let (θ_n) be a basis in $C(X)$ such that:

(a) the sequence $\varphi_n = \text{rest}_Y \theta_{k_n}$ is a basis in $C(Y)$,

(b) $\theta_{m_n} \in C_o(X /\!/ Y)$ for $n = 1, 2, \ldots$

(c) for each $f = \Sigma a_n \theta_n$ in $C(X)$ the subseries $\Sigma a_{k_n} \theta_{k_n}$ converges in $C(X)$.

Then for each $g = \Sigma b_n \varphi_n$ in $C(Y)$ the series $\Sigma b_n \theta_{k_n}$ converges in $C(X)$ and $\Lambda(\Sigma b_n \varphi_n) = \Sigma b_n \theta_{k_n}$ yields a linear extension operator $\Lambda : C(Y) \longrightarrow C(X)$.

Proof. Let $g \in C(Y)$, $g = \Sigma b_n \varphi_n$. There exists an f in $C(X)$ such that $\text{rest}_Y f = g$. Let $f = \Sigma a_n \theta_n = \Sigma a_{k_n} \theta_{k_n} + \Sigma a_{m_n} \theta_{m_n}$. Then $\text{rest}_Y f = \text{rest}_Y(\Sigma a_{k_n} \theta_{k_n}) = \Sigma a_{k_n} \varphi_n$. Clearly, $a_{k_n} = b_n$. Thus, the operator $\Lambda(g) = \Sigma b_n \theta_{k_n}$ is well defined. Since the operator $Q_m(f) = \Sigma_{n=1}^{m} a_{k_n} \theta_{k_n}$ converge for each f in $C(X)$, we have $\|\Lambda\| \leq \sup \|Q_m\| < \infty$. $\square$

4.1.7. <u>Proposition</u>. Let Y be a closed subset of a compact metric space X. Let (θ_n) be a basis in $C(X)$ satisfying (a) and such that

(d) for each $g = \Sigma b_n \varphi_n$ in $C(Y)$ the series $\Sigma b_n \theta_{k_n}$ converges in $C(X)$,

(e) if $f = \Sigma a_n \theta_n$ is in $C(X)$ and $g = \text{rest}_Y f$, then $g = \Sigma a_{k_n} \varphi_n$.

Then $\Lambda(g) = \Sigma b_n \theta_{k_n}$ is a linear extension operator. Moreover, conditions (b) and (c) also hold. $\square$

4.1.8. We shall now consider two polyhedra $Y \subset X$, with $\dim X = d$, $\dim Y = d' < d$. We assume that X and Y are unions of members of a simplicial complex C and of its subcomplex C', respectively. Furthermore $T_1 > T_2 > \dots$, W_n, V_n, V, φ_v are triangulations, vertices and pyramidal functions for the space X, whereas $T_1' > T_2' > \dots$, W_n', V_n', V' and φ_v' are analogous triangulations, vertices and functions for Y; (φ_m) and (φ_m') are the corresponding sequences of functions φ_v and φ_v' respectively. Furthermore, we assume that

(i) $T_n' = \{\Delta \cap Y : \Delta \in T_n, \ \dim(\Delta \cap Y) = d'\}$;

(ii) $W_n' = W_n \cap Y$, $V_n' = V_n \cap Y$, $V' = V \cap Y$;

(iii) if $\Delta \in T_n$ and $\Delta \cap Y \neq \emptyset$, then $\Delta \cap Y$ is a common face of Δ and Δ' where Δ' is a simplex in T'; conversely each face of each Δ' in T_n' is of the form $\Delta \cap Y$ with Δ in T_n;

(iv) if $v \in V \setminus Y$, then φ_v vanishes on Y;

(v) if $v \in V'$, then $\varphi_v' = \text{rest}_Y \varphi_v$.

A typical example is $X = \mathbf{I}^d$, Y a k-dimensional face of $\mathbf{I}^d$, $(k < d)$, T_n, T_n'-triangulations for the regular pyramidal bases $(\varphi_v)_{v \in V}$ in $C(X)$ and $(\varphi_v')_{v \in V'}$ in $C(Y)$ (see 3.3.4). Another example is $X = \mathbf{I}^d$, $Y = \partial(\mathbf{I}^d)$ = the boundary of $\mathbf{I}^d$ (in $\mathbf{R}^d$) (note that if v, w belong to different $(d-1)$-dimensional faces of $\mathbf{I}^d$ and $v \neq w$, then supports of φ_v and φ_w are non-overlapping).

4.1.9. <u>Proposition</u>. Let $X, Y, (\varphi_v)_{v \in V}$ and $(\varphi_v')_{v \in V'}$ satisfy (i) - (v) and let $\lim d_n = 0$ (see 3.1.3). For each g in $C(Y)$ let $g = \Sigma_{v \in V'} a_v \varphi_v'$ be the expansion of g. Then the series

$$(3) \qquad \Lambda f = \sum_{v \in V'} a_v \varphi_v$$

converges in $C(X)$. The map $g \longrightarrow \Lambda g$ is a linear extension operator $\Lambda : C(Y) \longrightarrow C(X)$, $\|\Lambda\| = 1$, and $\Lambda \geq 0$.

Proof. Let $g \in C(Y)$. Denote $s_m' = \Sigma_{j=1}^m a_j \varphi_j'$, $s_m = \Sigma_{j=1}^m a_j \varphi_{k_j}$, where φ_{k_j} is the function in $(\varphi_v)_{v \in V}$ corresponding to the vertex v_j of φ_j' (note that the series (3) has no terms $a_v \varphi_v$ with v in $V \smallsetminus Y$). We may assume that $\|g\| = 1$. Consequently, $\|s_m'\| \leq 1$ for $m = 1, 2, \ldots$ We shall prove that $\|s_m\| \leq 1$. If $m = 1$, the inequality is obvious. Suppose it is true for some m. The graph of the function s_m in $\mathbf{R}^{d+1}$ is a piecewise affine surface between the hyperplanes $x_{d+1} = -1$ and $x_{d+1} = 1$. Adding a new term $a_{m+1} \varphi_{k_{m+1}}$ means that we get a new peak (at v_{m+1}), which is also between these hyperplanes and is affinely joined with a finite number of points on the surface. Hence $\|s_{m+1}\| \leq 1$. If $g \geq 0$, then we consider the hyperplanes $x_{d+1} = 0$, $x_{d+1} = 1$ and infer that $s_m \geq 0$ for $m = 1, 2, \ldots$

In order to prove that the series (3) is convergent we apply the above estimation to the difference $s_p - s_q$. Let $\varepsilon > 0$. Then $\|s_p' - s_p'\| < \varepsilon$ for p, q sufficiently large, and hence $\|s_p - s_q\| \leq \|s_p' - s_q'\| < \varepsilon$. Thus, the conditions (a), (b), (d) and (e) are satisfied. $\square$

4.1.10. <u>Proposition</u>. Let Y be a closed subset of a compact space X. Let $(\varphi_n)_{n \in \mathbf{N}}$ be an interpolating basis in $C(X)$ with nodes $(t_n)_{n \in \mathbf{N}}$ (see 1.3.1). Let $K = \{k_n\}_{n \in \mathbf{N}}$ and $M = \{m_n\}_{n \in \mathbf{N}}$ be as in 4.1.1 Let $t_{k_n} \in Y$, $t_{m_n} \in X \smallsetminus Y$ and $rest_Y \varphi_{m_n} = 0$ for n in $\mathbf{N}$.

Denote $\psi_n = \text{rest}_Y \varphi_{k_n}$ and $u_n = t_{k_n}$. Then $(\psi_n)_{n \in \mathbb{N}}$ is an interpolating basis in $C(Y)$ with nodes (u_n).

Proof. Let $g \in C(Y)$. There exists an f in $C(X)$ such that $\text{rest}_Y f = g$. Let $f = \Sigma a_n \varphi_n$. Then

$$g = \text{rest}_Y (\sum_{n \in \mathbb{N}} a_n \varphi_n) = \sum_{n \in K} a_n \text{rest}_Y \varphi_n = \sum_{n \in \mathbb{N}} a_{k_n} \psi_n.$$

Since (φ_n) is interpolating, $\varphi_n(t_n) \neq 0$ and $\varphi_n(t_j) = 0$ for $n > j$. Consequently, $\psi_n(u_n) \neq 0$ and $\psi_n(u_j) = 0$ for $n > j$. This implies (see 1.3.2) that (ψ_n) is a basis in $C(Y)$. $\square$

4.1.11. Exercises. (A) Show that conditions (a), (b), (d) imply (c) and (e).

(B) Construct a basis $(\hat{\varphi}_m)$ in $C(\mathbf{I}^2)$ combining 4.1.1 with (2) for $x = \mathbf{I}^2$, $Y = \{0\} \times \mathbf{I}$, and making use of the basis (ψ_k) in $C_0(\mathbf{I}^2 /\!/ Y)$, constructed from the regular pyramidal basis (φ_m) in the way described in 3.6.1. Compare the bases $(\hat{\varphi}_m)$ and (φ_m).

4.2. A prism-pyramidal basis in $C(\mathbf{I}^\omega)$

4.2.1. We shall now consider the Hilbert cube $\mathbf{I}^\omega$ consisting of all sequences $x = (x_1, x_2, \ldots)$ such that $x_n \in \mathbf{I}$, with the product topology. It will be convenient to use the distance function

$$(1) \qquad \rho(x,y) = \sum_{n=1}^{\infty} 2^{-n} |x_n - y_n|.$$

We are going to construct a basis in $C(\mathbf{I}^\omega)$. Yet, the method of pyramidal bases does not work here. E.g., let $0 = (0,0,\ldots)$ be a vertex in $\mathbf{I}^\omega$. Conceivably, one may think of a function θ on $\mathbf{I}^\omega$, which is piecewise (in some sense) affine on $\mathbf{I}^\omega$, equals 1 at the vertex 0 and equals 0 at points of some (generalized) triangulation. Analogy suggest that θ should vanish at unit vectors $e_n = (0, \ldots, 0, 1, 0 \ldots)$; however, the sequence (e_n) tends to 0 (in the product topology of $\mathbf{I}^\omega$) and therefore such a function θ would be discontinuous.

We shall use prisms rather than simplexes. The given function f in $C(\mathbf{I}^\omega)$ will be approximated by partial sums of the series, on sets of the form

$$I_0^n = \{x \in \mathbf{I}^\omega : x_j = 0 \text{ for } j > n\}.$$

In order to simplify the notation we shall often identify $\mathbf{I}^n$ with I_o^n. We shall also use the fact the $\bigcup_{n=1}^{\infty} I_o^n$ is dense in $\mathbf{I}^\omega$; specifically, $\rho(x, I_o^n) \leq 2^{-n}$ for each x in $\mathbf{I}^\omega$. In other words, the lengths of perpendicular edges of successive cubes $I_o^1 \subset I_o^2 \subset I_o^3 \subset \ldots$ tend to 0.

4.2.2. <u>Definitions</u>. The basis functions ψ_v will first be defined on finite-dimensional cubes and then extended to $\mathbf{I}^\omega$ by composing then with projections

$$\pi_n^\omega : \mathbf{I}^\omega \longrightarrow \mathbf{I}^n, \quad \text{where} \quad \pi_n^\omega(x) = (x_1, \ldots, x_n).$$

Thus, each function ψ_v will depend only on a finite number of variables.

By a <u>prism-pyramidal function</u> on $\mathbf{I}^n$ we shall mean a function ψ of the form $\psi(x_1, \ldots, x_n) = \Theta(x_{m_1}, \ldots, x_{m_k})$ where Θ is a pyramidal function on $\mathbf{I}^k$ and $1 \leq m_1 < \ldots < m_k \leq n$, $k \leq n$.

Let $T_n(\mathbf{I}^k)$ denote the n-th triangulation of $\mathbf{I}^k$ for the regular pyramidal basis. If $\Delta \in T_n(\mathbf{I}^k)$ we shall use the symbol $\Delta \times \mathbf{I}^\omega$ to denote the set of all κ in $\mathbf{I}^\omega$ such that $\pi_n^\omega(x) \in \Delta$. Moreover, let

$$T_n^k = \{\Delta \times \mathbf{I}^\omega : \Delta \in T_n(\mathbf{I}^k)\}, \quad W_n^k = \{x \in I_o^n : \pi_n^\omega(x) \in \text{vert } T_n(\mathbf{I}^k)\}.$$

4.2.3. We first consider the triangulation $T_1(\mathbf{I})$ and two functions Θ_o, Θ_1 in $C(\mathbf{I})$ defined by $\Theta_o(t) = 1$, $\Theta_1(t) = t$ for t in $\mathbf{I}$. Composing them with the projection π_1^ω we get two functions ψ_v ($v \in W_1^1$) in $C(\mathbf{I}^\omega)$, namely $\psi_{(0,0,\ldots)}$ is the constant function 1, while $\varphi_{e_1}(x) = x_1$, where $e_1 = (1,0,0,\ldots)$. Thus,

$$s^{(1)}(x) = b_o \psi_o(x) + b_{e_1} \psi_{e_1}(x) = f(0) + [f(1) - f(0)]x_1$$

(cf. 2.4.7). Now, we pass to $T_2(\mathbf{I}^2)$. The set W_2^2 has 13 elements; φ_v has been already defined for two of them, marked with heavy dots in Fig.A below. Among the remaining 11 points there are 3 points

$(0,1)$, $(1,1)$, $(\frac{1}{2},\frac{1}{2})$ belonging to vert $T_1(I^2)$. Yet, no further pyramidal functions will be constructed for $T_1(I^2)$. In other words, these three points are now regarded as points of the second triangulation and the pyramidal functions with peaks at those 3 points have to vanish at all other vertices of $T_2(I^2)$. Consequently, the new functions Θ_v for these 3 points (see Fig. B and C) are different from the respective functions φ_v in the regular pyramidal basis in $C(I^2)$. For the remaining 8 vertices, Θ_v coincide with the respective functions defined in 3.2.1. As before, we define $\psi_v = \Theta_v \circ \pi_2^\omega$.

For each of the above mentioned 3 points $v_3 = e_2$, $v_4 = e_1 + e_2$, $v_5 = \frac{1}{2}(e_1 + e_2)$, the corresponding coefficient b_{v_j} is the difference $f(v_j) - s^{(1)}(v_j)$. Since $s^{(1)}$ does not depend on the second coordinate, we have $s^{(1)}(v_3) = f(0)$, $s^{(1)}(v_4) = f(e_1)$, $s^{(1)}(v_5) = \frac{1}{2}[f(0) + f(e_1)]$. Thus, the partial sum $s_5 = \Sigma_{j=1}^{5} b_{v_j} \psi_{v_j}$ is affine on each set in T_1^2 and $s_5(v_j) = f(v_j)$ for $j = 1,\ldots,5$. Therefore, though functions Θ_{v_j} and coefficients b_{v_j} are different from the previous functions φ_{v_j} in the regular pyramidal basis in $C(I^2)$ and from previous coefficients a_{v_j}, the sum s_5 is identical with the analogous sum $\Sigma a_{v_j} \varphi_{v_j} \circ \pi_2^\omega$. In other words, the effect of replacing the pyramidal functions $\varphi_{v_1}, \varphi_{v_2}$ by prism-pyramidal functions $\Theta_{v_1}, \Theta_{v_2}$ has been compensated by a different choice of coefficients. Consequently, for the remaining vertices $v_6,\ldots,v_{13}$ in W_2^2 the construction of the basis functions is the same as in 3.2.1 and we have $\Theta_{v_j} = \varphi_{v_j}$ and $b_{v_j} = a_{v_j}$ for $j = 6,\ldots,13$.

4.2.4. Suppose now that the functions ψ_v ($v \in W_n^n$) have been defined so that for each f in $C(I^\omega)$ the partial sum

$$s^{(n)} = \Sigma_{v \in W_n^n} a_v \psi_v$$

does not depend on variables $x_{n+1}, x_{n+2},\ldots$, is equal to f at each point of W_n^n and is affine on each prism in T_n^n.

Let Θ_v be the function on I^{n+1} pyramidal with respect to the triangulation $T_n(I^{n+1})$, with the peak at v, where $v = (\tau_1,\ldots,\tau_n,\tau_{n+1},0,0,\ldots)$ is in W_n^{n+1} and $\tau_{n+1} \neq 0$. Denote $b_v = f(v) - s^{(n)}(v)$ and

$$\pi_n^{n+1}(x_1,\ldots,x_n,x_{n+1},0,0,\ldots) = (x_1,\ldots,x_n,0,0,\ldots).$$

If $\pi_n^{n+1}(v) \in W_n^n$, then $b_v = f(v) - s^{(n)}(v) = f(v) - s^{(n)}(\pi_n^{n+1}v) = f(v) - f(\pi_n^{n+1}v)$ and $|b_v| \leq \omega_f(2^{-n})$. If $\pi_n^{n+1}(v) \notin W_n^n$, then

$b_v = f(v) - \frac{1}{2}[f(v') + f(v'')]$ for some vertices v', v'' in W_n^n such that $\pi_n^{n+1}v = \frac{1}{2}(v' + v'')$ and $s^{(n)}$ is affine on the segment v', v''; hence

$$(2) \qquad |b_v| \leq \omega_f(2^{-n} + d_n), \quad \text{where} \quad d_n = \sup\{\operatorname{diam}\Delta : \Delta \in T_n(\mathbf{I}^n)\}.$$

Note that $\operatorname{diam}\Delta$ refers to the distance (1). Thus, the partial sum $\tilde{s} = s^{(n)} + \Sigma b_v \psi_v$, where the summation is extended over all v in $W_n^{n+1} \setminus W_n^n$, is affine on each prism in T_n^{n+1} and $\tilde{s}(w) = f(w)$ for all w in W_n^{n+1}. Moreover, $\| b_v(\cdot) \| \leq 2$, where $\| b_v(\cdot) \|$ denotes the variation of the corresponding measure b_v. For each v in $W_{n+1}^{n+1} \setminus W_n^{n+1}$ we define $\Theta_v = \varphi_v$, $b_v = a_v$ (where φ_v and a_v are as in 3.3.5) and $\psi_v = \Theta_v \cdot \pi_{n+1}^\omega$. We also define $V = \cup V_n$, where $V_n = W_{n+1}^{n+1} \setminus W_n^n$.

4.2.3. <u>Proposition</u>. The sequence $(\psi_v)_{v \in V}$ is a monotone interpolating basis in $C(\mathbf{I}^\omega)$.

Proof. We are going to show that the series $\Sigma b_v \varphi_v$ is uniformly convergent to f on $\mathbf{I}^\omega$. The estimation (5) in 3.1.7 is valid here, i.e.,

$$|f(x) - s^{(n)}(x)| \leq \omega_f(d_n) \quad \text{for } x \text{ in } I_o^n.$$

Let $x \in \mathbf{I}^\omega \setminus I_o^n$ and let $x' = (x_1,\ldots,x_n,0,0,\ldots)$. Then $\rho(x,x') \leq 2^{-n}$ and $x' \in \Delta$ for some Δ in $T_n(\mathbf{I}^n)$. By the Darboux property there exists an x'' in Δ such that $s^{(n)}(x') = f(x'')$. Hence

$$|f(x) - s^{(n)}(x)| = |f(x) - f(x'')| \leq \omega_f(2^{-n} + d_n).$$

Now, let U be any subset of $W_{n+1}^{n+1} \setminus W_n^n$. Clearly, $r_U = \Sigma_{w \in U} b_v \varphi_v$ is affine on each prims in T_{n+1}^{n+1}; if $w \in W_{n+1}^{n+1}$ then $r_U(w)$ is equal either to b_w or to 0. Consequently, by (2),

$$\| r_U \| \leq \max\{|b_v| : v \in U\} \leq \omega_f(2^{-n} + d_n).$$

Hence $\| f - (s^{(n)} + r_U) \| \leq 2\omega_f(2^{-n} + d_n)$. This guarantees that the series $\Sigma b_v \psi_v$ is uniformly convergent to f regardless of the order in each set $W_{n+1}^{n+1} \setminus W_n^n$. By 1.3.2, $(\psi_v)_{v \in V}$ is a basis in $C(\mathbf{I}^\omega)$. $\square$

4.2.4. <u>Proposition</u>. Let Y be a closed subset of $\mathbf{I}^d$, $1 \leq d \leq \omega$. Let $(\varphi_n)_{n \in \mathbf{N}}$ be the regular pyramidal basis in $C(\mathbf{I}^d)$, $1 \leq d < \omega$ [the prism-pyramidal basis in $C(\mathbf{I}^\omega)$, respectively], interpolating with nodes (t_n). Let $k_1 < k_2 < \ldots$ be the sequence of all integers such that $\varphi_{k_n}(y_n) > 0$ for some y_n in Y and let $m_1 < m_2 < \ldots$ be the sequence of the remaining numbers in $\mathbf{N}$. Thus, $\mathrm{rest}_Y \varphi_{m_n} = 0$ for n in $\mathbf{N}$. Denote $u_n = t_{k_n}$, $Y_1 = Y \cup \{u_n\}_{n \in \mathbf{N}}$ and $\psi_n = \mathrm{rest}_{Y_1} \varphi_{k_n}$. Then Y_1 is a closed subset of X and $(\psi_n)_{n \in \mathbf{N}}$ is an interpolating basis in $C(Y_1)$ with nodes (u_n).

Proof. For each $\varepsilon > 0$ the set $\{n \in \mathbf{N} : \mathrm{diam}(\mathrm{supp}\ \varphi_n) \geq \varepsilon\}$ is finite. Consequently, the set $\{n \in \mathbf{N} : \varrho(u_n, Y) \geq \varepsilon\}$ is also finite. Therefore Y_1 is closed. Each φ_{m_n} vanishes on Y_1 (indeed, if $\varphi_{m_n}(t_{k_j})$ were positive, then $m_n < k_j$ and $\mathrm{supp}\ \varphi_{m_n} \subset \mathrm{supp}\ \varphi_{k_j}$ by 3.1.4, and hence φ_{k_j} would vanish on Y). Thus, by 4.1.10, (ψ_n) is a basis in $C(Y_1)$. $\square$

4.2.5. <u>Exercises</u>. (A) Let $\varphi_0(y) = 1-y$, $\varphi_1(y) = y$ for y in $\mathbf{I}$, $\tilde{\varphi}_0 = \varphi_0 \circ \pi_1^\omega$, $\tilde{\varphi}_1 = \varphi_1 \circ \pi_1^\omega$, $\tilde{\tilde{\varphi}}_0(x) = \varphi_0(x_2)$, $\tilde{\tilde{\varphi}}_1(x) = \varphi_1(x_2)$. Show that the functions $\tilde{\varphi}_0, \tilde{\varphi}_1, \tilde{\tilde{\varphi}}_0, \tilde{\tilde{\varphi}}_1$ cannot be extended to a basis.

(B) Let Y be a closed subset of $\mathbf{I}^\omega$. Construct a basis in $C_0(\mathbf{I}^\omega /\!/ Y)$ analogous to that in 3.6.1.

4.2.6. <u>Notes</u>. The above proposition (for $d = \omega$) is implicitly contained in F.S.Vaher ['55] and is the key to her proof that for any uncountable compact space Y the space $C(Y)$ has a basis. Indeed, Y is homeomorphic to a subset Y_0 of $\mathbf{I}^\omega$ and there exists a closed subset Y_1 of $\mathbf{I}^\omega$ such that $Y_1 \setminus Y_0$ is countable and $C(Y_1)$ has a basis. Now, $C(Y)$ and $C(Y_1)$ are isomorphic (as linear topological spaces; this is essentially due to K.Borsuk ['33]). Therefore $C(Y)$ also has a basis. Yet, this result follows also from a theorem of A.A. Milutin ['66] that $C(Y)$ is isomorphic to $C(\mathbf{I})$. Much stronger results, due to V.I.Gurariĭ ['66] and to E.Michael and A.Pełczyński ['66], are presented in next subsections. Vaher's construction is now only of historical interest. A basis in $C_0(\mathbf{I}^\omega /\!/ Y)$ was constructed by C.Bessaga ['57].

4.3. Gurariĭ's construction of an interpolating basis in C(X).

4.3.1. <u>Theorem</u> (Gurariĭ). Let X be a compact metric space and let (t_n) be a sequence dense in X $(t_i \neq t_j$ for $i \neq j)$. Let $\varepsilon > 0$. Then there exists an interpolating basis (ψ_n) in $C(X)$ with nodes (t_n) and with a basis constant less than $1 + \varepsilon$.

Proof. We may assume that X is infinite. Also, by Urysohn theorem, we may assume that X is a closed subset of the unit ball of a separable Hilbert space H and the linear span of X is dense in H. Let (ε_n) be a sequence of positive numbers such that $\varepsilon_n \downarrow 0$, $\varepsilon_1 < \frac{1}{2}$, $(1-\varepsilon_1)^{-1} < 1 + \varepsilon$.

If E is a subspace of H and $\pi_E : H \longrightarrow E$ is the orthogonal projection, then for each set $Y \subset E$ we shall denote

$$\tilde{Y} = \pi_E^{\leftarrow}(Y) = \{x \in H : \pi_E x \in Y\} = \{y + z : y \in Y, z \in E^{\perp}\}.$$

If $g : Y \longrightarrow \mathbb{R}$ is any function, $\tilde{g}$ will denote its extension $g \circ \pi_E$ to $\tilde{Y}$, i.e., $\tilde{g}(y + z) = g(y)$ for y in Y, z in $E^{\perp}$.

We shall define by induction the following sequences:

1^O a sequence of positive numbers (δ_n), $\delta_n \downarrow 0$,

2^O a sequence $q_1 < q_2 < \dots$ of natural numbers,

3° a sequence of finite-dimensional subspace $E_1 \subset E_2 \subset \dots$ of H,

4° a sequence of cubes $P_n \subset E_n$, $n = 1, 2, \dots$,

5° a sequence of triangulations T_n of P_n, $n = 1, 2, \dots$,

6° a sequence of sets $Y_n \subset P_n$ such that $\tilde{Y}_{n+1} \subset \tilde{Y}_n$ and $\cap_{n=1}^{\infty} \tilde{Y}_n = X$,

7° a sequence of functions $\varphi_n : Y_n \longrightarrow R$, $n = 1, 2, \dots$.

We first choose $\delta_1 = \frac{1}{2}$ and q_1 such that $(t_i)_{i=1,\dots,q_1}$ is a δ_1-net in X. E_1 is the linear span of $t_1, \dots, t_{q_1}$ in H; hence $d_1 \leq q_1$, where $d_1 = \dim(E_1)$. P_1 is a cube in E_1 containing $\pi_{E_1}(X)$, $\dim(P_1) = d_1$. T_1 is a triangulation of P_1 into d_1-dimensional simplexes such that $\operatorname{diam} \Delta < \delta_1$ for Δ in T_1 and each t_m $(m = 1, \dots, q_n)$ is a vertex of at least one simplex in T_1. Let $(\Delta_j^1)_{j=1,\dots,r_1}$ be those simplexes in T_1 which satisfy the condition

(1) $\quad \tilde{\Delta}_j^1 \cap X \neq \emptyset$, i.e., $\Delta_j^1 \cap \pi_{E_1}(X) \neq \emptyset$ for $j = 1, \dots, r_1$.

Thus, $\{t_1, \dots, t_{q_1}\} \subset W_1 = \bigcup_{j=1}^{r_1} \operatorname{vert}(\Delta_j^1)$; let $y_1^1, \dots, y_{p_1}^1$ be a sequence of the remaining vertices of simplexes Δ_j^1, $j = 1, \ldots, r_1$. Thus

$$\{y_1^1, \ldots, y_{p_1}^1\} = W_1 \setminus \{t_1, \ldots, t_{q_1}\} \quad \text{and} \quad y_i^1 \neq y_k^1 \quad \text{for} \quad i \neq k$$

Let $Y_1 = \cup_{j=1}^{r_1} \Delta_j^1$. We shall show that

$$(2) \qquad \forall_j \ \forall_{x \in \Delta_j^1} \ \min_{m=1,\ldots,q_1} \| x - t_m \| < 2\delta_1.$$

Indeed, let $x \in \Delta_j^1$. By (1), there is an x' in $\tilde{\Delta}_j^1 \cap X$. Since $\{t_1, \ldots, t_{q_1}\}$ is a δ_1-net in X, there exists an m such that $1 \leq m \leq q_1$ and $\| x' - t_m \| < \delta_1$. From the definition of E_1 it follows that $t_m \in E_1$. Let $x'' = \pi_{E_1} x'$. Then $x'' - t_m$ is orthogonal to $x' - x''$ and hence $\| x'' - t_m \| \leq \| x' - t_m \| < \delta_1$. Taking into account that $x, x'' \in \Delta_j^1$ and $\mathrm{diam}(\Delta_j^1) < \delta_1$ we get (2).

For each y_j^1 $(j = 1, \ldots, p_1)$ there is a t_{k_j} among $t_1, \ldots, t_{q_1}$ such that

$$\| y_j^1 - t_{k_j} \| = \inf_{m=1,\ldots,q_1} \| y_j^1 - t_m \| ;$$

if there is more than one such t_{k_j}, we choose one of them and denote it by $\sigma(y_j^1)$. In this way we get a map

$$\sigma : \{y_1^1, \ldots, y_{p_1}^1\} \longrightarrow \{t_1, \ldots, t_{q_1}\}.$$

The functions $\varphi_1, \ldots, \varphi_{q_1}$ are defined as the unique functions in $C(Y_1)$ satisfying the following conditions:

a) φ_m is affine on each simplex Δ_j^1 $(j = 1, \ldots, r_1)$,

b) $\varphi_m(t_m) = 1$ $(m = 1, \ldots, q_1)$,

c) $\varphi_m(y_j^1) = 1$ if $\sigma(y_j^1) = t_m$,

d) $\varphi_m(v) = 0$ for any other vertex v in W_1.

Assume now that we have already defined positive numbers $\delta_1 > \ldots > \delta_n$, indices $q_1 < \ldots < q_n$, subspaces $E_1 \subset \ldots \subset E_n$, cubes $P_1, \ldots, P_n$, triangulations $T_1, \ldots, T_n$, compact sets $Y_1, \ldots, Y_n$ and functions $\varphi_1, \ldots, \varphi_{q_1}$ in $C(Y_1), \ldots, \varphi_{q_{n-1}+1}, \ldots, \varphi_{q_n}$ in $C(Y_n)$; assume also that for $k = 1, \ldots, n$ the following conditions hold: $P_k \subset E_k$, $\dim(P_k) = \dim(E_k) = d_k$; T_k is a triangulation of P_k such that the set $T_k' = \{\Delta \cap E_{k-1} : \Delta \in T_k\}$ is a subtriangulation of T_{k-1} and $\Delta \cap E_{k-1}$ is a face of Δ for Δ in T_k, hence $\text{vert}(T_{k-1}) \subset \text{vert}(T_k)$; each t_m, where $q_{k-1} < m \leq q_k$, is in $\text{vert}(T_k)$; $\text{diam}(\Delta) < \delta_k$ for Δ in T_k; $\Delta_1^k, \ldots, \Delta_{r_k}^k$ are those simplexes in T_k for which $\tilde{\Delta}_j^k \cap X \neq \emptyset$ $(j = 1, \ldots, r_k)$;

$$W_k = \bigcup_{j=1}^{r_k} \text{vert}(\Delta_j^k) \quad \text{and} \quad Y_k = \bigcup_{j=1}^{r_k} \Delta_j^k; \quad X \subset \tilde{Y}_k \quad \text{and} \quad Y_k \subset \tilde{Y}_{k+1};$$

$y_1^k, \ldots, y_{p_k}^k$ are those vertices in W_k which do not belong to $\{t_1, \ldots, t_{q_k}\}$; $\varphi_m(t_i) = 0$ for $m = 2, \ldots, q_n$ and $i < m$.

If $q_{j-1} < m \leq q_j$, define $L(k,m)$ as the linear span in $C(\tilde{Y}_j)$ of restrictions of functions $\tilde{\varphi}_k, \ldots, \tilde{\varphi}_m$ to the set $\tilde{Y}_j$; $K(k,m)$ denotes the unit ball in $L(k,m)$. We choose $\delta_{n+1} > 0$ so that for each g in $K(1, q_n)$ and for each $z, z' \in Y_n$

(3) $\| z - z' \| < 2\delta_{n+1}$ implies $|g(z) - g(z')| < \frac{1}{4}\varepsilon_n$.

Such a δ_{n+1} exists because for each bounded subset G of a finite-dimensional subspace of $C(Y_n)$ the moduli of continuity $\omega_g(\delta)$, $g \in G$, do not exceed $\text{const} \cdot [\omega_{\varphi_1}(\delta) + \ldots + \omega_{\varphi_{q_n}}(\delta)]$ and the extension from Y_n to $\tilde{Y}_n$ and from φ_j to $\tilde{\varphi}_j$ does not change the moduli of continuity.

In turn, we choose q_{n+1} so that $\{t_{q_n+1}, \ldots, t_{q_{n+1}}\}$ be a δ_{n+1}-net in $X \setminus \{t_1, \ldots, t_{q_n}\}$. Denote $E_{n+1} = \text{span}(t_1, \ldots, t_{q_{n+1}})$, $\dim(E_{n+1}) = d_{n+1}$. Choose a d_{n+1}-dimensional cube P_{n+1} so that $\pi_{E_{n+1}}(X) \subset P_{n+1}$ and $P_{n+1} \cap E_n = P_n$. Choose a triangulation T_{n+1} of P_{n+1} so that each t_m $(q_n < m \leq q_{n+1})$ is a vertex of some Δ in T_{n+1}, $\text{diam}(\Delta) < \delta_{n+1}$ for Δ in T_{n+1}, $\{\Delta \cap E_n : \Delta \in T_{n+1}\}$ is a subtriangulation of T_n, and $\text{vert}(T_n) \subset \text{vert}(T_{n+1})$. Denote by $\Delta_1^{n+1}, \ldots, \Delta_{r_{n+1}}^{n+1}$ the simplexes in T_{n+1} which intersect $\pi_{E_{n+1}}(X)$ and let

$$Y_{n+1} = \bigcup_{j=1}^{r_{n+1}} \Delta_j^{n+1}.$$

Then $X \subset \tilde{Y}_{n+1} \subset \tilde{Y}_n$. Similarly as in the proof of (2), we can show that

(4) $x \in Y_{n+1} \setminus \{t_1, \ldots, t_{q_n}\}$ implies $\inf_{q_n < m \leq q_{n+1}} \| x - t_m \| < 2\delta_{n+1}$.

Let $y_1^{n+1}, \ldots, y_{p_{n+1}}^{n+1}$ $(y_i^{n+1} \neq y_j^{n+1}$ for $i \neq j)$ be all the vertices in

$$W_{n+1} \setminus \{t_1, \ldots, t_{p_{n+1}}\}, \quad \text{where} \quad W_{n+1} = \bigcup_{j=1}^{r_{n+1}} \text{vert}(\Delta_j^{n+1}).$$

For each y_i^{n+1} choose $\sigma(y_i^{n+1})$ in the set $\{t_{q_n+1}, \ldots, t_{q_{n+1}}\}$ so that

(5) $\qquad \|y_i^{n+1} - \sigma(y_i^{n+1})\| = \inf\{\|y_i^{n+1} - t_k\| : q_n < k \leq q_{n+1}\}.$

Thus, by (4) we get

(6) $\qquad \|y_i^{n+1} - \sigma(y_i^{n+1})\| < 2\delta_{n+1}$ for $i = 1, \ldots, p_{n+1}.$

It is easy to see that $\bigcap_{n=1}^{\infty} \tilde{Y}_n = X$. Indeed, let $x \in \bigcap_{n=1}^{\infty} \tilde{Y}_n$, i.e., $\pi_{E_n}(x) \in Y_n$ for $n = 1, 2, \ldots$ and $\pi_{E_n}(x) \in \Delta_j^n$ for some j. Since $\text{diam}(\Delta_j^n) < \delta_n$ and $\Delta_j^n \cap \pi_{E_n}(X) \neq \emptyset$, there is an x_n in X such that $\|\pi_{E_n} x - \pi_{E_n} x_n\| < \delta_n.$
Since X is compact, $x_{n_k} \longrightarrow x'$, $x' \in X$. We recall that $\text{span}(X)$ is dense in H; therefore $\pi_{E_n}(x) \longrightarrow x$ as $n \longrightarrow \infty$ and $\pi_{E_{n_k}}(x_{n_k}) \longrightarrow x'$. Consequently, $x = x'$ and $x \in X$.

If $q_n < m \leq q_{n+1}$ we define functions φ_m as the unique functions in $C(Y_{n+1})$ satisfying the following conditions:

a) φ_m is affine on each simplex Δ_j^{n+1} $(j = 1, \ldots, r_{n+1})$,

b) $\varphi_m(t_m) = 1,$

c) $\varphi_m(y_i^{n+1}) = 1$ if $\sigma(y_i^{n+1}) = t_m,$

d) $\varphi_m(v) = 0$ for any other vertex v in $W_{n+1}.$

Denote $c = 1 - \varepsilon_1$, where ε_1 was given at the beginning. Thus, $c > \frac{1}{2}$. We shall prove that for each n and each k such that $q_{n-1} < k \leq q_n$, the following inequality holds:

(7) $\qquad \max_{1 \leq m \leq k} |g(t_m)| \geq c \sup_{x \in \tilde{Y}_n} |g(x)|$ for g in $L(1,k).$

We shall proceed by induction with respect to n. Suppose that (7) is satisfied for $k = 1, \ldots, q_{n-1}$ (if $n = 1$, we mean that $q_0 = 0$; thus, some assumptions involving q_n are void for $n = 1$). Let $q_{n-1} < k \leq q_n$ and $g \in L(1,k)$, $\|g\| = \sup\{|g(x)| : x \in \tilde{Y}_n\} = 1.$ Then

$g = g_1 + g_2$, where $g_1 = \text{rest}_{\tilde{Y}_n} g_1'$ for some g_1' in $L(1, q_{n-1})$ and $g_2 \in L(q_{n-1}+1, k)$ (if $n = 1$, then $g_1 = 0$). We distinguish two cases:

1. $\|g_1\|_{\tilde{Y}_n} \geq 1$. Then also $\|g_1\|_{\tilde{Y}_{n-1}} \geq 1$. By the induction hypothesis, there is an index m_0 such that $m_0 \leq q_{n-1}$ and $|g_1(t_{m_0})| \geq c$. From the condition d) above it follows that $g_2(t_{m_0}) = 0$. Hence

$$\max_{1 < m < k} |g(t_m)| \geq |g(t_{m_0})| = |g_1(t_{m_0})| \geq c.$$

2. $\|g_1\|_{\tilde{Y}_n} < 1$. Then $\|g_2\|_{\tilde{Y}_n} > 0$. Since the function $|g|$ is affine on each simplex Δ_j^n $(j = 1, \ldots, r_n)$, it attains its supremum at a vertex v in $\{t_m\}_{m=1, \ldots, q_n} \cup \{y_i^n\}_{i=1, \ldots, p_n}$; thus $|g(v)| = 1$. It is clear that $v \neq t_m$ for $m = k+1, \ldots, q_n$ (note that $g_2(t_m) = 0$ and hence $|g(t_m)| < 1$). Analogously, $v \neq y_i^n$ if $\sigma(y_i^n) = t_m$ and $k < m \leq q_n$. If $v = t_m$ with $1 \leq m \leq k$, we get (7). Thus, it remains to consider the case where $v = y_m^n$, $\sigma(y_m^n) = t_{m_0}$, $q_{n-1} \leq m_0 \leq k$. From the condition c) above it follows that $g_2(t_{m_0}) = g_2(v)$. By the induction hypothesis,

$$\|g_1\|_{\tilde{Y}_{n-1}} \leq \frac{1}{c} \max_{1 < m < q_{n-1}} |g_1(t_m)| \leq \frac{1}{c} \|g_1\|_{\tilde{Y}_n} < \frac{1}{c} < 2.$$

Consequently, $\frac{1}{2} g_1 \in K(1, q_{n-1})$. Moreover, by (6),

$$\|v - t_{m_0}\| = \|y_i^n - \sigma(y_i^n)\| < 2\delta_n.$$

Hence, by (3), $|g_1(v) - g_1(t_{m_0})| < \varepsilon_{n-1}$. Therefore

$$|g(t_{m_0})| = |g(v) + [g(t_{m_0}) - g(v)]| = |g(v) + [g_1(t_{m_0}) - g_1(v)]|$$

$$\geq |g(v)| - \varepsilon_{n-1} \geq 1 - \varepsilon_1 = c.$$

Thus, (7) has been proved.

Denote $\psi_m = \text{rest}_X \tilde{\varphi}_m$, $m = 1, 2, \ldots$ Since $\tilde{Y}_n \supset X$ for $n = 1, 2, \ldots$ we infer that

$$(8) \qquad \max_{1 < k < m} |s_m(t_k)| \geq c \sup_{x \in X} |s_m(x)| = c \|s_m\|, \quad \text{where} \quad s_m = \sum_{j=1}^{m} a_j \psi_j.$$

Consequently, by Proposition 1.3.5, (ψ_m) is a basic sequence with basis constant $1/c$. Moreover, $1/c = 1/(1-\varepsilon_1) < 1 + \varepsilon$.

Since $\psi_m(t_k) = 0$ for $k < m$ and $\psi_m(t_m) = 1$, the coefficients of the expansion $\sum a_m \psi_m$ can be computed in the way shown in 1.3.3.

It remains to be shown that $\text{span}(\psi_1, \psi_2, \ldots)$ is dense in $C(X)$. Let $f \in C(X)$, $\|f\| = 1$, $\varrho > 0$. There is an n such that

(9) $\qquad \omega_f(4\delta_n) < \frac{\varrho}{4}$ and $\varepsilon_{n-1} < \varrho$.

There is a unique function of the type $h = \sum_{m=1}^{q_n} b_m \tilde{\varphi}_m$ in $C(\tilde{Y}_n)$ which interpolates f at nodes $t_1, \ldots, t_{q_n}$. Let $x \in X \setminus \{t_1, \ldots, t_{q_n}\}$. We claim that

(10) $\qquad |f(x) - h(x)| < \varrho$.

There is an j_o such that $1 \le j_o \le r_n$ and $x \in \tilde{\Delta}_{j_o}^n$. The function h can be decomposed as $h = h_1 + h_2$, where $h_1 \in L(1, q_{n-1})$, $h_2 \in L(q_{n-1} + 1, q_n)$. Obviously, $h_2(t_k) = 0$ and $|f(t_k)| \le 1$ for $k = 1, \ldots, q_{n-1}$. Consequently,

$$|h_1(t_k)| = |h(t_k)| = |f(t_k)| \le 1$$

and hence, by (8),

$$\|h_1\| \le \frac{1}{c} \max |h_1(t_k)| \le \frac{1}{c} < 2.$$

There exists a k_o such that

(11) $\qquad \|x - t_{k_o}\| < \delta_n$ and $q_{n-1} < k_o \le q_n$.

Thus, $t_{k_o} \in E_n$ and $\|\pi_{E_n} x - t_o\| < \delta_n$. Since $\text{diam}(\Delta_{j_o}^n) < \delta_n$ and $\pi_{E_n} x \in \Delta_{j_o}^n$, we have

(12) $\qquad \|z_k - t_{k_o}\| < 2\delta_n$, $k = 1, \ldots, d_n + 1$,

where $z_1, \ldots, z_{d_n+1}$ are the vertices of $\Delta_{j_o}^n$ and $d_n = \dim(\Delta_{j_o}^n)$. Combining (9) with (11) we get

(13) $\qquad |f(x) - f(t_{k_o})| < \frac{\varrho}{4}$.

In order to estimate $|f(t_{k_o}) - h(z_k)|$, $k = 1, \ldots, d_n + 1$, we distinguish two cases:

1. z_k is one of the points t_m $(m = 1,\ldots,q_n)$. Then $h(z_k) = f(z_k)$. Inequalities (12) together with (9) imply that

$$|h(z_k) - f(t_{k_o})| = |f(z_k) - f(t_{k_o})| \leq \frac{\varrho}{4}.$$

2. $z_k = y_i^n$ for some i $(1 \leq i \leq p_n)$. From (6) it follows that

$$(14) \qquad \| \sigma(z_k) - t_{k_o} \| \leq \| \sigma(z_k) - z_k \| + \| z_k - t_{k_o} \| \leq 4\delta_n.$$

Hence, since $\sigma(z_k) = t_m$ for some m with $1 \leq m \leq q_n$, we get from (9)

$$(15) \qquad |h(\sigma(z_k)) - f(t_{k_o})| = |f(\sigma(z_k)) - f(t_{k_o})| \leq \frac{\varrho}{4}.$$

Moreover, $h_2(\sigma(z_k)) = h_2(z_k)$ and $\| z_k - \sigma(z_k) \| < 2\delta_n$. Clearly, $\frac{1}{2}h_1 \in K(1,q_{n-1})$ and we can apply (3):

$$(16) \qquad |h(z_k) - h(\sigma(z_k))| = |h_1(z_k) - h_1(\sigma(z_k))| \leq \frac{2\varepsilon_{n-1}}{4} < \frac{\varrho}{2}.$$

Combining (16) and (15) we obtain

$$(17) \qquad |f(t_{k_o}) - h(z_k)| < \frac{3}{4}\varrho, \qquad k = 1,\ldots,d_n+1.$$

Thus, we have proved the estimation (17) in either case. Here $f(t_{k_o})$ is a constant; since (17) is valid for all vertices z_k of $\Delta_{j_o}^n$ and h is affine on $\Delta_{j_o}^n$, the inequality must be valid on the whole simplex. In particular,

$$(18) \qquad |f(t_{k_o}) - h(x)| < \frac{3}{4}\varrho.$$

Inequalities (18) and (13) give (10). Thus, the sums $\Sigma a_n\psi_n$ are dense in $C(X)$. Consequently, (ψ_n) is a basis. We have already shown that it is interpolating with nodes (t_n). $\Box$

4.3.2. <u>Corollary</u>. Let X, (t_n) and (ψ_n) be as above. Then the Banach-Mazur distance between $F_n = \text{span}(\psi_1,\ldots,\psi_n)$ and l_n^∞ does not exceed $1/c$.

Proof. This follows from (8) and 1.3.6. $\Box$

4.3.3. <u>Example</u>. Let $X = \{0,\frac{1}{2}\} \cup \{\frac{1}{2} + \frac{1}{n} : n = 1,2,\ldots\}$. Then $X \subset I$

and we may think of X as a subset of the cube I^ω. Let $t_1 = 0$, $t_n = 1/2 + 1/n$, $n = 2,3,\ldots$ Then $\{t\}_{n=1,2,\ldots}$ is dense in X and by the preceding theorem there is an interpolating basis in $C(X)$ with nodes (t_n). Yet, suppose that we the construction of the system (φ_n) above replacing the condition $\varphi_m(\sigma(y_i^n)) = 1$ by $\varphi_m(\sigma(y_i^n)) = 0$. Then (ψ_n) need not be a basis.

Indeed, suppose that δ_n and q_n are chosen. Obviously, each E_n is of dimension 1 and the triangulation T_n is a partition

$$0 = a_o^{(n)} < a_1^{(n)} < \ldots < a_{i_n}^{(n)} < t_{q_n} < t_{q_n - 1} < \ldots < t_2 = 1$$

of the interval I, where $a_{i_n}^{(n)} < \frac{1}{2}$. If $f(x) = 1$ for x in X and $(a_{i_n}^{(n)})$ tends to $\frac{1}{2}$ quickly enough, then the values of each partial sum s_m at $\frac{1}{2}$ will be smaller than $\frac{1}{2}$, say, and will not tend to 1.

4.3.4. Example. We shall show a situation where the construction in 4.3.1 does not give a monotone basis. Suppose that $X \subset I$ and for some interval $[a,b]$, the intersection $X \cap [a,b]$ consists of points $v_1 < v_3 < v_2$ and $v_3 + \tau n^{-1}$, $n = 1,2,\ldots$ where $v_3 - v_1 = v_2 - v_3 = \frac{1}{2}\delta_n$, $\tau = \frac{1}{8}\delta_n$ (δ_n is as in 4.3.1), and the function f in $C(X)$ satisfies the conditions: $\|f\| = 1$, $f(v_1) = \frac{1}{2}$, $f(v_2) = 0$, $f(v_3) = f(v_3 + \tau n^{-1}) = 1$ (see the figure below). Suppose that previous partitions have been

chosen so that the m-th partial sum s_m is affine on the interval $[v_1, v_2]$. Then $s_m(v_3) = \frac{1}{4}$. If t_{m+1} is one of the numbers $v_3 + \tau n^{-1}$, then the corresponding coefficient a_{m+1} is equal to

$$a_{m+1} = f(t_{m+1}) - s_m(t_{m+1}) = \frac{3}{4} + \frac{1}{16n}.$$

Furthermore, suppose that v_3 is none of elements $t_1, t_2, \ldots$ and $\sigma(v_3) = t_{m+1}$. Then $\psi_{m+1}(v_3) = \psi_{m+1}(t_{m+1}) = 1$ and

$$s_{m+1}(v_3) = s_m(v_3) + a_{m+1}\psi_{m+1}(v_3) > \frac{1}{4} + \frac{3}{4} = 1.$$

Thus, $\|s_{m+1}\| > 1 = \|f\|$.

4.3.5. <u>Notes</u>. Theorem 4.3.1 was proved by Gurariĭ ['66a]. In a footnote he has mentioned that is possible to construct a basis in C(X) with basis constant 1 if we drop the assumption that the basis be interpolating at given nodes. A proof of this fact, due to E.Michael and A.Pełczyński ['66], is presented in the next section. In 1966 Gurariĭ raised the question whether in each space C(X) there is an interpolating basis with nodes given a priori having basis constant 1. The problem seems to be still open.

It should be pointed out that in 4.3.1 we construct an interpolating basis for the given <u>sequence</u> of nodes, that is, the nodes are ordered in advance.

J.Ryll ['78], ['80] has constructed for each p in $\mathbb{N}$, for each X belonging to a certain class of compact connected subsets of $\mathbb{R}^d$ and for each dense sequence (t_n) of nodes, an interpolating basis in C(X) which at the same time is a basis in each space $C^{(m)}(X)$, $m = 1, \ldots, p$.

4.4. Bases in C(X) constructed with peaked partitions of unity.

The purpose of this section is to prove the existence of monotone bases in all separable spaces C(X) and $C_o(X)$, using a significant technique of peaked partitions of unity. The arguments below are valid for both scalar fields: $\mathbb{R}$ and $\mathbb{C}$.

4.4.1. The space l_n^∞ of all n-tuples $(a_1, \ldots, a_n)$ of scalars is the same as the space $l^\infty\{1, \ldots, n\}$ of all bounded functions on the set $\{1, \ldots, n\}$. The unit vectors in l_n^∞ will be denoted by $e_i^{(n)}$, $i \leq n$. Thus, $e_i^{(n)}(j) = \delta_{ij}$.

If $\Gamma : l_m^\infty \longrightarrow l_n^\infty$ is a linear operator, let $u_j = \Gamma(e_i^{(m)})$, $j \leq m$. Then

$$\Gamma a.k = \Gamma(\sum_{j \leq m} a_j e_j^{(m)})(k) = \sum_{j \leq m} a_j u_j(k)$$

for $a = (a_1, \ldots, a_m)$ in l_m^∞, $k \leq n$. Clearly, $u_j(k)$ are entries of the matrix determined by Γ. The norm of Γ is equal to

$$(1) \qquad \| \Gamma \| = \max_{k \leq n} \sum_{j \leq m} |u_j(k)|.$$

This is a special case of a general formula for the norm of a bounded linear operator $\Gamma : C(X) \longrightarrow C(Y)$ (see #18.8.2), but it can also be proved directly:

$$\| \Gamma \| = \sup_{\| a \|=1} \| \Gamma(a) \| = \sup_{\| a \|=1} \sup_{k \le n} | \Gamma a.k |$$

$$= \sup_{k \le n} \sup_{\| a \|=1} | \Gamma a.k | = \sup_{k \le n} \sum_{j \le m} | u_j(k) |.$$

4.4.2. Lemma. Γ is an isometric embedding if and only if the following conditions hold:

(2) $\qquad \sum_{j \le m} | u_j(k) | \le 1 \qquad$ for $\quad k \le n,$

(3) $\qquad \| u_j \| = 1 \qquad$ for $\quad j \le m,$

(4) $\qquad$ if $\; | u_j(k) | = 1 \;$ and $\; i \ne j, \;$ then $\; u_i(k) = 0 \;$ for $\; j \le m, \; i \le m, \; k \le n.$

Proof. Let Γ be an isometric embedding. Then (2) and (3) are obvious. If $| u_j(k) | = 1$ and $i \ne j$, then there exists a $\;c\;$ such that $\; | c | = 1 \;$ and

$$1 = \| e_j^{(m)} + c e_i^{(m)} \| = \| u_j + c u_i \| \ge | u_j(k) + c u_i(k) | = 1 + | u_i(k) |.$$

Therefore $u_i(k) = 0.$ Assume now that (2), (3), (4) hold. Note that (1) and (2) imply the inequality $\| \Gamma \| \le 1.$ To prove that $\| \Gamma \| \ge 1,$ consider an $\; a = \Sigma a_j e_j^{(m)}.$ By (3), there are $k_1, \ldots, k_m$ in $\{1, \ldots, n\}$ such that $\; | u_j(k_j) | = 1 \;$ and hence, by (4), $\; u_j(k_i) = 0 \;$ for $\; i \ne j.$ Thus,

$$\| \Gamma a \| = \max_{k \le n} | \sum_{j \le m} a_j u_j(k) | \ge \max_{i \le m} | \sum_{j \le m} a_j u_j(k_i) | = \max_{i \le m} | a_i | = \| a \|. \quad \square$$

4.4.3. Lemma. Suppose that $m < n,$ $G \subset l_n^\infty$ and $\Gamma : l_m^\infty \longrightarrow G$ is a linear isometrical bijection. Then there is a subspace E of l_n^∞ such that $G \subset E$ and E is isometrically isomorphic to $l_{m+1}^\infty.$

Proof. The sequences $u_j = \Gamma(e_j^{(m)})$ $(j = 1, \ldots, m)$ satisfy conditions (2), (3), (4). Since $m < n,$ there are more points in $\{1, \ldots, n\}$ than sequences $u_1, \ldots, u_m.$ Consequently, either there exist j, k', k'' such that $j \le m,$ $k' \le n,$ $k'' \le n,$ $k' \ne k'',$ $| u_j(k') | = 1,$ $| u_j(k'') | = 1,$ or there exists a k such that $k \le n$ and $| u_i(k) | < 1$ for each $i \le m.$ In either case, there exists a k_o such that each u_j attains its norm at a point in $\{1, \ldots, n\} \setminus \{k_o\}.$ Let

(5) $\qquad h_j = u_j - u_j(k_o) e_{k_o}^{(n)} \qquad$ for $\; j \le m \;$ and $\; h_{m+1} = e_{k_o}^{(n)},$

and let E be the span of $h_1, \ldots, h_{m+1}$. Then $G \subseteq E \subseteq \mathrm{cl}_n^\infty$, h_{m+1} vanishes at points where $h_1, \ldots, h_m$ attain their norms, and all h_j ($j \leq m$) vanish at k_o. Thus, by Lemma 4.4.2, $\Lambda(a) = \Sigma_{j=1}^{m+1} a_j h_j$ yields a linear isometry. $\square$

4.4.4. <u>Definitions</u>. Let X be compact. By a <u>partition of unity</u> in $C(X)$ we mean here a finite set $\{p_1, \ldots, p_n\}$ of nonnegative functions in $C(X)$ such that

$$p_1 + \ldots + p_n = 1_X$$

(one may also consider infinite, locally finite partitions of unity, #20.4.1). A <u>peaked partition of unity</u> is a partition of unity $\{p_j\}_{j \leq n}$ such that $\| p_j \| = 1$ for $j \leq n$. If this is the case, then there exist $x_1, \ldots, x_n$ in X such that $p_j(x_j) = 1$ for $j \leq n$ and hence $p_j(x_i) = 0$ for $i \neq j$. Points $x_1, \ldots, x_n$ with these property are called <u>peak points</u> of the functions $p_1, \ldots, p_n$. A subspace G of $C(X)$ is called a <u>peaked partition subspace</u> iff it is spanned by a peaked partition of unity.

Note that $\{p_1, \ldots, p_n\}$ is a peaked partition of unity if and only if the function $p = (p_1, \ldots, p_n)$ maps X into the simplex spanned by the unit vectors $v_1, \ldots, v_n$ in $\mathbb{R}^n$ and $p(X)$ contains all the vertices.

4.4.5. <u>Proposition</u>. Let $p_1, \ldots, p_n$ and $x_1, \ldots, x_n$ be as above and let $E = \mathrm{span}\{p_i\}_{i \leq n}$. If $a = (a_i)_{i \leq n}$ is in l_n^∞, let $g = \Sigma_{i=1}^n a_i p_i$. Then

$$(6) \qquad a_i = g(x_i) \qquad \text{for} \quad i \leq n,$$

i.e., g is a function in E having values $a_1, \ldots, a_n$ at nodes $x_1, \ldots, x_n$. Moreover,

$$(7) \qquad \| \sum_{i=1}^n a_i p_i \| = \max_{i \leq n} |a_i| = \| a \| .$$

Consequently, the map $\Lambda : l_n^\infty \longrightarrow E$ defined as $\Lambda(a) = \Sigma a_i p_i$ is a linear isometry and the map

$$(8) \qquad \pi(f) = \sum_{i=1}^n f(x_i) p_i \qquad \text{for} \quad f \quad \text{in} \quad C(X)$$

is a projection from $C(X)$ onto E with $\| \pi \| = 1$.

Proof. Equality (6) follows from $p_i(x_j) = \delta_{ij}$. Hence

$$\| a \| \; = \; \max_{i < n} |a_i| \; = \; \max_{i < n} |g(x_i)| \; \leq \; \| g \| \; .$$

On the other hand,

$$\| g \| \; = \; \sup_{x \in X} | \sum_{i < n} a_i p_i(x)| \; \leq \; \| a \| \; \sup_{x \in X} \sum_{i < n} p_i(x) \; = \; \| a \| \; . \quad \square$$

4.4.6. <u>Proposition</u>. Let E be a subspace of $C(X)$ isometrically isomorphic to some l_n^∞ and let $1_X \in E$. Then E is a peaked partition subspace.

Proof. The images $u_1,\dots,u_n$ in E of the unit vectors $e_j^{(n)}$, $j \leq n$, form a basis in E, $\|u_j\| = 1$ for $j \leq n$ and there are $x_1,\dots,x_n$ such that $|u_j(x_j)| = 1$. Let $p_j = c_j u_j$ where $c_j = \overline{u_j(x_j)}$. Then, by Lemma 4.4.2, $p_j(x_k) = \delta_{ik}$ and $(p_j)_{j \leq n}$ is a basis in E. Therefore there are scalars $(a_j)_{j \leq n}$ such that $1_X = \Sigma a_k p_k$. Consequently,

$$a_k \; = \; \sum_{j < n} a_j p_j(x_k) = 1 \quad \text{and} \quad \sum_{j < n} p_j(x) = 1.$$

Moreover, for arbitrary scalars $b_1,\dots,b_n$ we have

$$\| \sum_{j < n} b_j p_j \| \; = \; \| \sum_{j < n} b_j c_j u_j \| \; = \; \max_{j < n} |b_j c_j| \; = \; \max_{j < n} |b_j|$$

and hence $\Sigma |p_j(x)| \leq 1$ for x in X. Since $\Sigma p_j = 1$, we infer that $p_1,\dots,p_n$ are nonnegative. $\quad \square$

4.4.7. <u>Definitions</u>. Let $\varepsilon > 0$. A partition of unity $\{p_i\}_{i \leq n}$ is said to be ε-subordinated to a cover $\mathcal{U}$ of X iff there are $U_1,\dots,U_n$ in $\mathcal{U}$ such that

$$\sum_{i \in I(X)} p_i(x) \leq \varepsilon \quad \text{for} \quad x \text{ in } X, \quad \text{where } I(x) = \{j : x \notin U_j\}.$$

The partition $\{p_i\}_{i \leq n}$ is <u>subordinated</u> to $\mathcal{U}$ iff it is 0-subordinated, i.e., $p_i(x) = 0$ for x in $X \setminus U_i$, $i \leq n$.

4.4.8. <u>Lemma</u>. Let $\{U_j\}_{j \leq n}$ be an open cover of X and let $x_i \in U_i$, $x_i \neq x_j$ for $i \neq j$. Then there is a peaked partition of unity $\{p_j\}_{j \leq n}$ such that p_i vanishes on $X \setminus U_i$ ($i \leq n$) and $p_i(x_j) = \delta_{ij}$.

Proof. Denote $V_i = U_j \setminus \{x_j : j \neq i\}$. Then $\{V_j\}_{j \leq n}$ is also an open cover of X. There exists a partition of unity $\{p_j\}_{j \leq n}$ subordinated to $\{V_j\}_{j \leq n}$ (see #20.4.2). Obviously, $p_i(x_j) = 0$ for $i \neq j$. Thus $p_i(x_i) = 1$ for $i \leq n$. $\quad \square$

4.4.9. <u>Lemma</u>. Let $\{p_i\}_{i \leq n}$ be a partition of unity in $C(X)$ such that $p_i(x_i) = 1$ for some $x_1, \ldots, x_n$ in X. Let $\mathcal{O}$ be any open cover of X and let $0 < \varepsilon < 1$. Then there exist: a number $m > n$, a partition of unity $\{q_j\}_{j \leq m}$ in $C(X)$ and points $z_1, \ldots, z_m$ in X such that $\{q_j\}_{j \leq m}$ is ε-subordinated to $\mathcal{O}$, $q_j(z_j) = 1$ for $j \leq m$, $z_i = x_i$ for $i \leq n$ and

(9) $\operatorname{span}\{p_i\}_{i \leq n} \subset \operatorname{span}\{q_j\}_{j \leq m}.$

Proof.   Denote $\lambda = 1 - \varepsilon$. For each z in X, choose U_z in $\mathcal{O}$ so that $z \in U_z$. Let

(10) $V_i(z) = U_z \cap \{x \in X : p_i(x) > \lambda p_i(z)\}$ if $p_i(z) > 0$

and $V_i(z) = U_z$ if $p_i(z) = 0$. The sets $V(z) = \cap_{i=1}^{n} V_i(z)$, where $z \in X$, form an open cover of X. Therefore there exist $z_1, \ldots, z_m$ such that $\cup_{j=1}^{m} V(z_j) = X$. Obviously, we may assume that $m > n$ and $z_i = x_i$ for $i \leq n$. By Lemma 4.4.8, there is a partition of unity $\{g_j\}_{j \leq m}$ such that g_j vanishes outside $V(z_j)$ for $j \leq m$ and $g_j(z_k) = \delta_{jk}$. We shall adjust this partition to make sure that the new one be subordinated to $\mathcal{O}$ and satisfy (9). First, we define auxiliary functions

(11) $h_i = \sum_{j=1}^{m} p_i(z_j) g_j$ for $i \leq n$.

Thus, each h_i is a function in $\operatorname{span}\{g_j\}_{j \leq m}$ interpolating p_i at nodes z_j, $i \leq n$, $j \leq m$ (cf. formula (8)). If $g_j(x) > 0$, then $x \in V(z_j)$ and, by (10), $p_i(x) \geq \lambda p_i(z_j)$ for $i \leq n$. Consequently,

(12) $p_i(x) = \sum_{j=1}^{m} p_i(x) g_j(x) \geq \lambda \sum_{j=1}^{m} p_i(z_j) g_j(x) = \lambda h_i(x)$

for x in X. Let $\gamma_i(x) = \sup\{c : c \leq 1, \ p_i(x) \geq c h_i(x)\}$, $i \leq n$. Then by (12)

(13) $\gamma(x) \geq \lambda$, where $\gamma(x) = \min_{i \leq n} \gamma_i(x).$

Since $h_i(z_k) = p_i(z_k)$ for $i \leq n$, $k \leq m$, we have $\gamma(z_k) = 1$. Moreover,

(14) $p_i(x) \geq \gamma(x) h_i(x)$ for $i \leq n$.

Let now i be fixed, $i \leq n$. We claim that $\gamma_i : X \longrightarrow \mathbb{R}$ is continuous. The continuity on the open set $W_i = \{x \in X : h_i(x) > 0\}$ is obvious, bacause

$$\gamma_i(x) = \min[1, p_i(x)/h_i(x)] \quad \text{for} \quad x \quad \text{in} \quad W_i.$$

Let us show that γ_i is continuous at each point y_0 such that $h_i(y_0) = 0$. Clearly, $\gamma_i(y_0) = 1$. If $p_i(y_0) > 0$, then $p_i(x) > h_i(x)$ for x in a neigborhood W' of y_0 and hence $\gamma_i(x) = 1$ for x in W'. Thus, assume that $p_i(y_0) = 0$ Denote

$$J_i = \{j : j \leq m, \quad p_i(z_j) > 0\}$$

and $Y_i = \{x \in X : p_i(x) < \lambda p_i(z_j) \text{ for all } j \text{ in } J_i\}$. Obviously, Y_i is open and $y_0 \in Y_i$. Let $x \in Y_i$. Then $h_i(x) = 0$. Indeed, by (11) it is enough to show that for each $j \leq m$ either $p_i(z_j) = 0$ or $g_j(x) = 0$. If $p_i(z_j) > 0$ for some $j \leq m$, then $j \in J_i$ and from the assumption $x \in Y_i$ it follows that $x \notin V_i(z_j)$; **consequently,** $x \notin V(z_j)$ and $q_j(x) = 0$. Therefore h_i vanishes on Y_i and hence it is continuous at y_0. Thus, each γ_i is continuous on X and so is γ.

We are now able to define: $q_j(x) = \gamma(x)g_j(x)$ for $j = n+1, \ldots, m$ and

$$(15) \quad q_j(x) = \gamma(x)g_j(x) + [p_j(x) - \gamma(x)h_j(x)] \quad \text{for} \quad j \leq n.$$

Obviously, $q_j \in C(X)$ for $j \leq m$ and $q_j \geq \gamma g_j \geq 0$ by (14). If $x = z_j$, the expression in the bracket in (15) vanishes; this ensures the property $q_j(z_j) = \gamma(z_j)g_j(z_j) = 1$ for $j \leq m$. Moreover, $p_i(z_j)$ $= p_i(x_j) = \delta_{ij}$ for $i \leq n$, $j \leq n$. Consequently, by (11) and (15), for $i \leq n$ we get

$$(16) \quad \sum_{i=1}^{m} p_i(z_j)q_j = \gamma \sum_{j=1}^{m} p_i(z_j)g_j + p_i - \gamma h_i = p_i.$$

Thus, $p_i \in \text{span}\{q_j\}_{j \leq m}$. Moreover, by $\Sigma p_i = 1_X$ and (16),

$$\sum_{j=1}^{m} q_j = \sum_{j=1}^{m} [\sum_{i=1}^{n} p_i(z_j)]q_j = \sum_{i=1}^{n} \sum_{j=1}^{m} p_i(z_j)q_j = \sum_{i=1}^{n} p_i = 1_X.$$

This means that $\{q_j\}_{j \leq m}$ is a partition of unity.

It remains to be shown that $\{q_j\}_{j \leq m}$ is ε-subordinated to the cover $\mathcal{U}$. For each j find U_j in $\mathcal{U}$ such that $V(z_j) \subset U_j$. Let

$x \in X$ and $I(x) = \{j : j \leq m, \ x \notin U_j\}$. Since g_j vanishes outside $V(z_j)$, by (13) we get

$$\sum_{j \in I(x)} q_j(x) = 1 - \sum_{j \notin I(x)} q_j(x) \leq 1 - \gamma(x) \sum_{j \notin I(x)} g_j(x)$$

$$= 1 - \gamma(x) \sum_{j \leq m} g_j(x) = 1 - \gamma(x) \leq 1 - \lambda = \varepsilon. \quad \Box$$

4.4.10. Lemma. Let $f \in C(X)$, $f \neq 0$, $\varepsilon > 0$. Then there exist: a $\delta > 0$ and an open cover $\mathcal{U}$ of X such that for each partition of unity $\{q_j\}_{j \leq m}$ δ-subordinated to $\mathcal{U}$ we have

(17) $\mathrm{dist}(f, \mathrm{span}\{q_j\}_{j \leq m}) \leq \varepsilon$.

Proof. Let $\delta = \varepsilon/4 \| f \|$. Consider the family $\mathcal{U}$ of all sets

$$U(a) = \{x \in X : |f(x) - a| < \varepsilon/2\}, \qquad |a| \leq \| f \|.$$

Clearly, $\mathcal{U}$ is an open cover of X. Suppose that $\{q_j\}_{j \leq m}$ is δ-subordinated to $\mathcal{U}$. Then there exist scalars $a_1, \ldots, a_m$ such that $|a_j| \leq \| f \|$ and for all x in X

$$\sum_{j \in I(x)} q_j(x) < \delta \quad \text{where} \quad I(x) = \{j : j \leq m, \ x \notin U(a_j)\}.$$

We claim that $g = \sum_{j=1}^{m} a_j q_j$ satisfies the inequality $\| f - g \| \leq \varepsilon$. Indeed, let $x \in X$. Then

$$|f(x) - g(x)| = \Big| \sum_{j \leq m} [f(x) - a_j] q_j(x) \Big| \leq \sum_{j \leq m} |f(x) - a_j| q_j(x)$$

$$= \Big(\sum_{j \in I(x)} + \sum_{j \notin I(x)} \Big) |f(x) - a_j| q_j(x)$$

$$\leq \sup_{j \notin I(x)} |f(x) - a_j| + \delta \sup_{j \leq m} |f(x) - a_j| \leq \frac{\varepsilon}{2} + 2\delta \| f \| = \varepsilon. \quad \Box$$

4.4.11. Lemma (see Kelley ['55], Ch.5, Th.26). Let $\{U_i\}_{i \leq n} =$ be an open cover of a compact metric space X. Then there exists a number $r > 0$ such that for each x in X the open ball $K(x, r)$ is contained in some member of $\mathcal{U}$ (i.e., the cover $\{K(x, r)\}_{x \in X}$ is a refinement of $\{U_i\}_{i \leq n}$).

Proof. Let $f_i(x) = \mathrm{dist}(x, X \setminus U_i)$ and $f(x) = \max\{f_i(x) : i \leq n\}$. Clearly f_i and f are continuous. If $x \in X$, then x belongs to some U_i and hence $f(x) \geq f_i(x) > 0$.
Let $r = \frac{1}{2} \inf \{f(x) : x \in X\}$. Since X is compact, r is positive.

If $x \in X$, then there is an i such that $f_i(x) > r$; this means that $K(x,r) \subset U_i$. $\square$

4.4.12. <u>Theorem</u>. Let X be an infinite compact metric space and let $\tilde{E}_k$ be a k-dimensional peaked partition subspace of $C(X)$. Then there exists an increasing sequence $E_1 \subset E_2 \subset \ldots$ of peaked partition subspaces of $C(X)$ such that $E_k = \tilde{E}_k$, $\dim E_n = n$ for n in $\mathbb{N}$ and $\cup E_n$ is dense in $C(X)$. Moreover, there exist projections $P_n : C(X) \longrightarrow E_n$ such that $\| P_n \| = 1$.

Proof. There exists a sequence $(\mathcal{U}_m)_{m \in \mathbb{N}}$ of finite open covers of X such that $U \in \mathcal{U}_m$ implies $\text{diam}(U) < 1/m$. Let G_o be the sub-space spanned by 1_X and let $G_1 = \tilde{E}_k$. We now apply Lemma 4.4.9 inductively to get peaked partitions of unity $\{q_j^{(m)}\}_{j \le k_m}$ such that $G_m = \text{span}\{q_j^{(m)}\}_{j \le k_m}$ is 1/m-subordinated to $\mathcal{U}_m$ for $m \ge 2$, $G_m \subset G_{m+1}$, and $q_i^{(m)}(y_j^{(m)}) = \delta_{ij}$ for some $y_j^{(m)}$ in X such that

$$\{y_j^{(m)} : j = 1,\ldots,k_m\} \subset \{y_j^{(m+1)} : j = 1,\ldots,k_{m+1}\}.$$

We shall show that $\cup G_m$ is dense in $C(X)$. Let $f \in C(X)$ and $\varepsilon > 0$. There exist: a $\delta > 0$ and an open cover $\mathcal{U}$ satisfying the condition imposed in Lemma 4.4.10. By Lemma 4.4.11, there exists an m such that $m^{-1} < \delta$ and each $\mathcal{U}_i$ $(i \ge m)$ is a refinement of $\mathcal{U}$. Consequently, $\{q_j^{(m)}\}_{j \le k_m}$ is δ-subordinated to $\mathcal{U}$ and hence $\text{dist}(f,G_m) \le \varepsilon$.

By Lemma 4.4.3, there exist subspaces $E_1 \subset E_2 \subset \ldots$ such that $E_1 = G_o$, $E_{k_m} = G_m$ (we recall that $k_m = \dim(G_m)$) and each E_n is isometrically isomorphic to 1_n^∞. By Proposition 4.4.6, each E_n is a peaked partition subspace of $C(X)$.

Finally, the projections P_n are given by formula (8). $\square$

4.4.13. <u>Theorem</u>. Let X be an infinite compact metric space. Then there exist: a countable dense set $T = \{t_n\}_{n \in \mathbb{N}}$ and a normalized monotone interpolating basis (e_n) with nodes (t_n). Moreover, such a basis can be chosen so that $e_1 = 1_X$.

Proof. The existence of a normalized monotone basis (e_n) follows from Theorem 4.4.12 and Proposition 1.1.7. We recall that e_{n+1} is an element of E_{n+1} such that $\| e_{n+1} \| = 1$ and $P_n(e_{n+1})=0$. Thus, in order to show that (e_n) is interpolating we have to examine the way E_n and P_n are constructed.

Choose any r in $\mathbb{N}$. Let $(p_i^{(r)})_{i \le r}$ be a peaked partition of unity, $t_1, \ldots, t_r \in X$ and $p_i^{(r)}(t_j) = \delta_{ij}$. Denote $e_i = p_i^{(r)}$ for $i \leq r$. Note that if we choose $r = 1$, then $e_1 = 1_X$.

Suppose that the functions $e_1, \ldots, e_n$ have been constructed, $n \geq r$, $\|e_i\| = 1$, $e_i(t_j) = 0$ for $i > j$, $(p_j^{(n)})_{j \leq n}$ is a peaked partition of unity in $E_n = \text{span}(e_j)_{j \leq n}$, $t_1, \ldots, t_n \in X$, $p_j^{(n)}(t_i) = 0$ for $i \neq j$, $E_n \subset G_m$, where G_m is as in the proof of 4.4.12, and m is the smallest number such that $n < k_m$, $G_m = \text{span}\{q_j^{(m)}\}_{j \leq k_m}$, $q_j^{(m)}(t_i^{(m)}) = \delta_{ij}$, $t_j^{(m)} \in X$ for $j \leq k_m$, and $\{t_i\}_{i \leq n} \subset \{t_j^{(m)}\}_{j \leq k_m}$. The operators

$$\Gamma_n : E_n \longrightarrow l_n^\infty \quad \text{and} \quad \Lambda_m : G_m \longrightarrow l_{k_m}^\infty,$$

where $\Gamma_n(f) = (f(t_1), \ldots, f(t_n))$, $\Lambda_m(f) = (f(t_1^{(m)}), \ldots, f(t_{k_m}^{(m)}))$ (i.e., restrictions of f to the sets $\{t_1, \ldots, t_n\}$ and $\{t_1^{(m)}, \ldots, t_{k_m}^{(m)}\}$) are linear isometrical bijections and

$$\Gamma_n^{-1}(a_1, \ldots, a_n) = \sum_{i=1}^n a_i p_i^{(n)}, \quad \Lambda_m^{-1}(b_1, \ldots, b_{k_m}) = \sum_{j=1}^{k_m} b_j q_j^{(m)}.$$

Each sequence $\Lambda_m(e_i)$, where $i \leq n$, attains its $l_{k_m}^\infty$-norm at a point in $\{t_j\}_{j \leq n}$. Let $t_{j_\nu}^{(m)}$ $(1 \leq \nu \leq k_m - n)$ be all the points in $\{t_j^{(m)}\}_{j \leq k_m}$ different from all t_j $(j \leq n)$. In virtue of (5) let

$$(18) \quad e_{n+\nu} = \Lambda_m^{-1}(e_{j_\nu}^{(k_m)}) = q_{j_\nu}^{(m)}, \quad \nu \leq k_m - n.$$

Then, by (8),

$$P_{n+1}(f) = \sum_{i \leq n} f(t_i) p_i^{(n)} \quad \text{for} \quad f \quad \text{in} \quad C(X).$$

Since $q_{j_1}^{(m)}(t_i) = 0$ for $i \leq n$, we get $P_{n+1}(q_{j_1}^{(m)}) = 0$. In this way we infer that $e_{n+1} = q_{j_1}^{(m)}$ satisfies the desired conditions. It is clear from (5) that if we continue this induction process, the next functions $e_{n+1}, \ldots, e_{k_m}$ constructed step by step will be identical with the functions (18) and $e_n(t_j) = 0$ for $j < n \leq k_m$. $\square$

4.4.14. **Corollary.** Let X be a locally compact space such that the one-point compactification γX is metrizable. Then $C_o(X)$ has a monotone normalized interpolating basis (e_n).

Proof. By Theorem 4.4.13, there is a basis $(1_X, e_1, e_2, \ldots)$ in

$C(\gamma X)$. Therefore $(e_1, e_2, \ldots)$ is a basis in $C_O(X)$. $\square$

4.4.15. <u>Exercises</u>. (A) Analyse the meaning of Lemma 4.4.2 in the case where $m = n$.

(B) Formulate Lemma 4.4.3 in geometrical terms, as a statement about intersecting an n-dimensional parallelopiped by an m-dimensional hyperplane.

(C) Let $X = [0, 2\pi]$, $p_1(x) = \frac{1}{2}(1 + \sin x)$, $p_2(x) = \frac{1}{2}(1 - \sin x)$, $q_1(x) = \frac{1}{2}(1 + \cos x)$, $q_2(x) = \frac{1}{2}(1 - \cos x)$ for x in X. Show that $\{p_1, p_2\}$ and $\{q_1, q_2\}$ are peaked partitions of unity such that no peaked partition subspace of $C(X)$ contains all four functions p_1, p_2, q_1, q_2. (Hint: the subspace of $C(X)$ spanned by $\sin x$, $\cos x$ is Euclidean and cannot be isometrically embedded into any 1_n^∞.)

(D) Let T be the triangle in $\mathbb{R}^3$ spanned by the unit vectors v_1, v_2, v_3 and let $p : I \longrightarrow T$ be a continuous function such that $p(t_i) = v_i$, where $1/2 < t_1 < t_2 < t_3 = 1$ and p restricted to $[0, 1/2]$ is a homeomorphism onto a semicircle $A \subset T$. Let $U_1 = [0, 1/2)$, $U_2 = (1/3, 1]$. Show that $\{p_i\}_{i \leq n}$ is a peaked partition of unity such that no peaked partition of unity $\{q_j\}_{j \leq m}$ subordinated to the open cover $\{U_1, U_2\}$ satisfies

$$\text{span}\{p_i\}_{i \leq n} \subset \text{span}\{q_j\}_{j \leq m}.$$

Hint: Show that $p = \Sigma_{j=1}^m p(z_j) q_j$ where $q_j(z_j) = 1$; if $\{q_j\}_{j \leq m}$ were subordinated (not merely ε-subordinated) to $\{U_1, U_2\}$, then $p(X \setminus U_2)$ would be a sub-arc of A contained in the convex hull of the finite set $\{p(z_j) : q_j$ vanishes on $X \setminus U_1\}$.

(E) Let G_m, $q_j^{(m)}$ and $y_j^{(m)}$ be as in the prooof of 4.4.12. Show that the set $\cup_{m=1}^\infty \{y_j^{(m)} : j \leq k_m\}$ is dense in X.

(F) Let $E_1 \subset E_2 \subset \ldots$ be peaked partition subspaces of $C(I)$ such that $\cup E_n$ is dense in $C(I)$. Show that for each interval $(a, b) \subset I$ there exists an f in some E_n, not having a continuous derivative on (a, b). (Hint: if $\{p_i\}_{i \leq n}$ is a partition of unity and $p_i(x_j) = \delta_{ij}$, then each p_i has a maximum or minimum at each x_j.)

(G) If we assume additionally in Lemma 4.4.10 that the partition $\{q_j\}_{j \leq m}$ satisfies $q_j(z_j) = 1$, we can take $g = \Sigma f(z_j) q_j$, i.e., a function g in $\text{span}\{q_j\}_{j \leq m}$ not only approximating f but also interpolating f at $z_1, \ldots, z_m$.

(H) Formulate a variant of Theorem 4.4.12 for finite X. What are E_n ?

(I) Modify the proof of Theorem 4.4.13 to show that for each countable set $T = \{t_n\}_{n \in \mathbb{N}}$ there exist: a set $Y = \{y_m\}_{m \in \mathbb{N}}$ such that $T \subset Y$ and a monotone basis in $C(X)$ with nodes (y_m).

4.4.16. Notes. Results 4.4.1 - 4.4.12 and Exercises (B)-(G) are due to E.Michael and A.Pełczyński ['67]; see also Pełczyński ['64]. The existence of a monotone basis in each separable space $C(X)$, asserted by Gurariĭ ['66a] (see 4.3.5), was proved independently by Michael and Pełczyński ['66]. They also proved that a monotone basis exists in each separable π_1^∞-space, i.e., predual of L-space, and this includes the spaces $C_o(X)$ and $C_\sigma(X)$, where

$$C_\sigma(X) = \{f \in C(X) : f \bullet \sigma = -f\}$$

for some involutory homeomorphism $\sigma : X \longrightarrow X$ (see Day ['73], Lacey ['74]).

4.5. Approximation with nodes in $C(X)$.

4.5.1. The purpose of this section is to show that there are bases and peaked partitions of unity in $C(I)$ consisting of algebraic (or trigonometric) polynomials.

In this section X is a compact space, E is a finite-dimensional subspace of $C(X)$ and $T = \{t_1, \ldots, t_m\} \subset X$. We shall consider the condition:

(*) for every sequence $s = (s_1, \ldots, s_m)$ in $\mathbb{R}^m$ there is an h in E

such that

(1) $h(t_k) = s_k$ for $k = 1, \ldots, m$,

(2) $\|h\| = \sup_k |s_k|$.

2.5.2. Proposition. Let E satisfy (*) and let $f \in C(X)$. Denote

$$E_f^T = \{h \in E : h(t_i) = f(t_i) \quad \text{for} \quad i = 1, \ldots, m\}.$$

Then

(3) $\operatorname{dist}(f, E_f^T) \leq 2 \operatorname{dist}(f, E).$

Thus, the best approximation with interpolation at given nodes can at most double the distance from the subspace.

Proof. There is a g in E such that $\text{dist}(f,E) = \|f-g\|$. In turn, there is an h' in E such that $h'(t_k) = f(t_k) - g(t_k)$ for $k = 1,\ldots,m$ and $\|h'\| \leq \|f-g\|$. Denote $h = h' + g$. Then $h \in E$, $h(t_k) = f(t_k)$ and

$$\text{dist}(f,E_f^T) \leq \|f-h\| \leq \|f-g\| + \|h'\| \leq 2\,\text{dist}(f,E). \quad \square$$

4.5.3. <u>Proposition</u>. If E satisfies (*) for each sequence $s_k = \pm 1$, then it satisfies it for all sequences $(s_1,\ldots,s_m)$ in $\mathbf{R}^m$.

Proof. Without loss of generality we can assume that $\sup|s_k| = 1$. Let

$$K = \{(h(s_1),\ldots,h(s_m)) : h \in E\}.$$

Then $K \subset \mathbf{R}^m$, K is convex and every vertex of $[-1,1]^m$ belongs to K. Hence $[-1,1]^m \subset K$. $\square$

4.5.4. <u>Proposition</u>. Let $E_0 \subset E_1 \subset E_2 \subset \ldots$ be finite-dimensional subspaces of $C(X)$ such that

(**)   if $f \in E_j$ and $g \in E_k$, then $fg \in E_{j+k}$.

Suppose that the some m there are $\varphi_1,\ldots,\varphi_m$ in E_m such that $\varphi_k(t_j) = \delta_{jk}$. Then there exists an n (depending on T) such that E_n satisfies the condition (*).

Proof. Let $s_1,\ldots,s_m$ be given and let $\sup|s_k| = 1$. Denote $c = \Sigma\|\varphi_k\|$ and

$$h_b = \sum_{k=1}^{m} b_k\varphi_k, \quad \text{where} \quad b = (b_1,\ldots,b_m) \in \mathbf{R}^m.$$

Then $\|h_b\| \leq c \sup|b_k|$ and $h_b(t_k) = b_k$. Let $\Theta_n = 2^{n-1}T_n$ where T_n denotes the n-th Tchebycheff polynomial on $[-1,1]$. Thus,

$$\Theta_n(u) = \cos n(\arccos u) = u^n - \binom{n}{2}u^{n-2}(1-u^2) + \ldots$$

Find an even natural number n_0 such that

$$\left|\cos\left(\frac{1}{2} - \frac{1}{n_0}\right)\pi\right| < \frac{1}{c}.$$

Consider the polynomial Θ_{n_o} restricted to the interval $(-1/c, 1/c)$.

If $y_k = \cos(k\pi/n_o)$ for $k = n_o/2 - 1$ and $k = n_o/2$, then $\Theta_{n_o}(y_k) = (-1)^k$

and $|y_k| < \frac{1}{c}$. Thus Θ_{n_o} is a polynomial such that $-1 \le \Theta_{n_o}(u) \le 1$

for $u \in [-1, 1]$ and $\Theta_{n_o}(y_k) = \pm 1$. Consequently, by the Darboux property,

there are $b_1, \ldots, b_m$ in $(-1/c, 1/c)$ such that $\Theta_{n_o}(b_k) = s_k$. Consider
the function

$$h(x) = \Theta_{n_o}(h_b(x)), \quad x \in X.$$

Then $h(t_k) = \Theta_{n_o}(h_b(t_k)) = \Theta_{n_o}(b_k) = s_k$. Since $\|h_b\| \le c \sup |b_k| \le 1$,

we get $\|h\| \le 1$. Moreover, there is an n such that $h \in E_n$;
a routine inspection shows that n depends only on $t_1, \ldots, t_m$ and
$\varphi_1, \ldots, \varphi_m$ but not on $s_1, \ldots, s_m$. $\square$

4.5.5. <u>Corollary</u>. Let $\varepsilon > 0$. For each sequence $t_o = a$, $t_1 = b$,
$t_2, t_3, \ldots$ dense in $[a, b]$ [respectively, for each sequence $t_o = 0$,
$t_1, t_2, \ldots$ dense in $[0, 2\pi]$] there exists an interpolating basis in
$C[a, b]$ [respectively in the space of continuous 2π-periodic functions]
with nodes (t_n) and with basis constant less than $1 + \varepsilon$, consisting
of algebraic [trigonometric] polynomials.

Indeed, the Lagrange interpolating polynomials and trigonometric
polynomials satisfy the assumptions of 4.5.4 (see e.g. Natanson ['49],
p.505). It is enough to apply the Krein-Milman-Rutman theorem (1.1.8). $\square$

4.5.6. <u>Proposition</u>. Let $E_o \subset E_1 \subset \ldots$ be finite-dimensional
subspaces of $C(X)$ satisfying (∗∗). Suppose that for each pair u, v
of points in T there exist: a k and g_{uv} in E_k such that

$$g_{uv}(u) = 1, \quad g_{uv}(v) = 0, \quad \|g_{uv}\| \le 1.$$

Then there exist: an n and $h_1, \ldots, h_m$ in E_n such that $(h_k)_{k \le n}$
is a peaked partition of unity with peaks $(t_k)_{k \le n}$. Consequently, for
every $s = (s_1, \ldots, s_m)$ in $\mathbb{R}^m$ the function

$$h = s_1 h_1 + \ldots + s_m h_m$$

satisfies (1) and (2).

Proof. Since the square of g_{uv} belongs to E_{2k}, we may assume
that $g_{uv} \ge 0$. We are looking for functions $f_1, \ldots, f_m$ such that
$0 \le f_k \le 1$, $f_k(t_k) = 1$ for $k = 1, \ldots, m$ and $f_k(t_j) = 0$ for $j > k$.
To get this we define $f_m = 1$, $f_{m-1} = g_{t_{m-1}, t_m}$ and

$$(4) \qquad f_{m-k} = \prod_{j=0}^{k-1} g_{t_{m-k}, t_{m-j}} \, .$$

In turn, we define $h_1 = f_1$, $h_2 = f_2(1-f_1)$ and

$$(5) \qquad h_k = f_k \prod_{j=1}^{k-1} (1-f_j), \qquad k = 2, \ldots, m.$$

Then $0 \le h_k \le 1$ and $h_k(t_j) = \delta_{jk}$ for $k, j = 1, \ldots, m$. Moreover,

$$h_m + h_{m-1} = \prod_{j=2}^{m-2} (1-f_j),$$

$$h_m + h_{m-1} + h_{m-2} = \prod_{j=1}^{m-3} (1-f_j),$$

etc. Hence $h_m + h_{m-1} + \ldots + h_1 = 1$. $\quad\square$

4.5.7. <u>Exercises</u>. (A) Compare formulas (4) and (5) with the formulas for Haar functions $\tilde{\chi}_m$ on 2^ω (see 2.2.2) and for the coefficients for an interpolating basis (1.3.3).

(B) Let $X = [a,b]$ and let E_n be the space of algebraic polynomials of degree n. Show that estimation for the degree n in 4.5.6 is worse than that in 4.5.4 (i.e., in 4.5.6 n tends more quickly to ∞).

4.5.8. <u>Notes</u>. The above results, including (B), have been proved for ordinary polynomials by W.Wolibner ['51], S.Paszkowski ['55], ['56], ['57] and by S.Mazur (presented on 24 March 1959 at a colloquium of the Polish Mathematical Society in Poznań, unpublished).

K.M.Saĭdukov ['57] constructed an orthonormal basis in $C(I)$ consisting of algebraic polynomials. On the other hand, V.I.Gurariĭ ['63] has shown that no monotone basis in $C(I)$ consists of functions analytic in $(0,1)$. This is related to the following result: Let $n_k > 0$ be a sequence such that $\Sigma 1/n_k < \infty$. Then, by the classical theorem of Müntz, the set of polynomials $\Sigma a_k t^{n_k}$ is not dense in $C(I)$ and its closure F consists of functions analytic in $(0,1)$ (see L.Schwartz ['43]); Gurariĭ has shown that F has no monotone basis.

There are many papers estimating (from above or from below) the degrees of algebraic or trigonometric polynomials forming a basis in $C(I)$. The first result in this line is due to G.Faber ['14], p.202, who showed that for a trigonometric polynomial basis (φ_n) the degree of φ_n cannot be equal to n. Z.A.Čanturija ['67] strengthened this by proving that in fact these degrees have to grow quicker than $n+o(\log n)$; M.I.Kadec ['74] proved that a sequence of polynomials p_n

such that $\deg p_n$ is not bigger than $n+o(\log^2 n)$ cannot be a basis in $C(I)$. Čanturija ['73] (see also ['72]) proved that if ω_n is an increasing sequence such that $\Sigma 1/n\omega_n < \infty$, then in $C(I)$ there exists an orthogonal basis of trigonometric polynomials and an orthogonal basis of algebraic polynomials, of degrees (in either case) not exceeding $n^{3/2}\omega(n)$ for $n > n_o$.

4.6. Auerbach bases

4.6.1. <u>Definitions</u>. Let F be a Banach space. An <u>Auerbach basis</u> in F is a basis (e_m) such that $\| e_m \| = 1$ and $\| e_m^* \| = 1$ for each m, where (e_m^*) is the sequence biorthogonal to (e_m).

4.6.2. <u>Proposition</u>. Each finite-dimensional Banach space has an Auerbach basis.

Proof. Let $h_1,\ldots,h_n$ be any basis in F. If $f_1,\ldots,f_n$ are in F and $f_i = \Sigma_{k=1}^n a_{ik}h_k$, let $V(f_1,\ldots,f_n)$ be the determinant $\det(a_{ik})$. The set $Q = \{(f_1,\ldots,f_n) : \| f_i \| \le 1\}$ is compact; therefore V attains its supremum at an element $(e_1,\ldots,e_n)$ of Q. Denote

$$(1) \qquad e_i^*(f) = \frac{V(e_1,\ldots,e_{i-1},f,e_{i+1},\ldots,e_n)}{V(e_1,\ldots,e_n)}, \quad i = 1,\ldots,n.$$

Then obviously $e_i^*(e_i) = 1$ and $e_i^*(e_j) = 0$ for $i \ne j$. Moreover, $e_i^* \in F^*$ and $|e_i^*(f)| \le 1$ for each f with $\| f \| = 1$; hence $\| e_i^* \| = 1$. $\square$

4.6.3. <u>Example</u>. Let X be a compact subset of $\mathbb{R}$ or $\mathbb{C}$ and let E_n be the space of all (real or complex, respectively) polynomials of degree $\le n-1$, with the norm $\| f \|_X = \sup\{|f(x)| : x \in X\}$. There exists a subset $\{\xi_1,\ldots,\xi_n\}$ of X maximal in the sense of M.Fekete (Über die Verteilung der Wurzeln bei gewissen algebraischen Gleichungen mit ganzzahligen Koeffizienten, Math. Zeitschrift 17(1923), 228-249) i.e., such that

$$(2) \qquad |V(\xi_1,\ldots,\xi_n)| = \sup_{(x_j) \in X^n} |V(x_1,\ldots,x_n)|,$$

where $V(x_1,\ldots,x_n) = \prod_{i<j} |x_i - x_j|$ denotes the Vandermond determinant for nodes $x_1,\ldots,x_n$. Then the Lagrange interpolating polynomials

$$e_j(x) = \frac{V(\xi_1,\ldots,\xi_{j-1},x,\xi_{j+1},\ldots,\xi_n)}{V(\xi_1,\ldots,\xi_n)} = \prod_{\substack{i=1 \\ i \ne j}}^{n} \frac{x - \xi_j}{\xi_i - \xi_j}$$

form an Auerbach basis in E_n; indeed, $\| e_j \|_X = 1$ and

(3)
$$\left\| \sum_{j=1}^{n} a_j e_j \right\|_X \ge \left| \sum_{j=1}^{n} a_j e_j (\xi_k) \right| = |a_k|.$$

Note that argument here is dual to that in the preceding proposition. In 4.6.2, the function e_j^* is in F^* while the variables of V are in F; here the functions e_j are in E_n while the variables in V are in X and may be considered as functionals in E_n^*.

4.6.4. Let us now consider a more general situation: Assume that X is any compact space and E_n is an n-dimensional subspace with a basis $h_1, \ldots, h_n$. If $c_1, \ldots, c_n$ are scalars and $x_1, \ldots, x_n$ are in X, we may look for coefficients $a_1, \ldots, a_n$ such that

(4)
$$\sum_{i=1}^{n} a_i h_i (x_j) = c_j \qquad \text{for} \quad j = 1, \ldots, n,$$

i.e., $f(x_j) = c_j$ where $f = \Sigma a_i h_i$. It is well known that in the case where the determinant

(5)
$$V(x_1, \ldots, x_n) = \det[h_i (x_j)]_{i,j \le n}$$

is different from 0, applying Cramer formulas to sequences $(c_j)_{j \le n}$ and $(\delta_{jk})_{j \le n}$, $k = 1, \ldots, n$, yields

$$f(x) = \sum_{j=1}^{n} c_j g_j (x), \qquad \text{where} \quad g_j (x) = \frac{V(x_1, \ldots, x_{j-1}, x, x_{j+1}, \ldots, x_n)}{V(x_1, \ldots, x_n)}.$$

Obviously, $g_j \in E_n$ for $j \le n$ and $g_j (x_k) = \delta_{jk}$.

4.6.5. <u>Proposition</u>. Let $(\xi_1, \ldots, \xi_n)$ be a sequence in X^n satisfying condition (2) with V given by (5). Then

$$V(\xi_1, \ldots, \xi_n) \ne 0$$

and the functions

(6)
$$e_j (x) = \frac{V(\xi_1, \ldots, \xi_{j-1}, x, \xi_{j+1}, \ldots, \xi_n)}{V(\xi_1, \ldots, \xi_n)}, \qquad j = 1, \ldots, n,$$

form an Auerbach basis in E_n.

Proof. We shall show by induction that V considered as a function on X^n is not identically 0. If $n = 1$, then obviously $V(x_1) = |h_1 (x_1)| \ne 0$. Suppose that $V(x_1, \ldots, x_k) \ne 0$ for some k and $k < n$. Then there are functions b_i such that

$$V(x_1,\ldots,x_k,x) = \sum_{i=1}^{k} b_i(x_1,\ldots,x_k) h_i(x) + V(x_1,\ldots,x_k) h_{k+1}(x).$$

Since $(h_i)_{i \leq k+1}$ are linearly independent, there exists an x_{k+1} in X such that $V(x_1,\ldots,x_k,x_{k+1}) \neq 0$.

Consequently, $V(\xi_1,\ldots,\xi_n) \neq 0$ and the functions (6) are well defined. It is clear that $\|e_j\| = 1$, $e_j(\xi_k) = \delta_{jk}$ and (3) is valid in this case; hence $\|e_j^*\| = 1$ for $j \leq n$. $\square$

4.6.6. <u>Notes</u>. Proposition 4.6.2 is due to Herman Auberbach, a dozent at the University of Lwów, killed by Gestapo in the summer of 1943. The above proof of 4.6.2 is from Lindenstrauss and Tzafiri ['77], p.16.

4.6.3 was observed by C.Bessaga and J.Siciak in 1956. The generalization 4.6.5 was suggested to the author by J.Siciak.

4.7. Biorthogonal systems and kernel representation of partial sums.

4.7.1. Let X be a set, B – a σ-field of subsets of X and let μ be a nonnegative σ-finite measure on B. If B and μ are understood, we shall write $L^p(X)$ instead of $L^p(X,B,\mu)$. $L^p(I)$ will refer to the Lebesgue measure.

In the sequel we assume that sequences $f_n \in L^1(X)$, $g_n \in L^\infty(X)$ $(n = 1,2,\ldots)$ are biorthogonal (see 1.1.9). Thus $\int_X f_n g_m d\mu = \delta_{nm}$ and we get two sequences of linear operators

$$(1) \qquad S_n : L^1(X) \longrightarrow L^1(X) \quad \text{and} \quad T_n : L^\infty(X) \longrightarrow L^\infty(X)$$

where

$$(2) \qquad (S_n f)(x) = \sum_{k=1}^{n} a_k f_k(x), \qquad a_k = \int_X f g_k d\mu,$$

and

$$(3) \qquad (T_n g)(x) = \sum_{k=1}^{n} b_k g_k(x), \qquad b_k = \int_X g f_k d\mu.$$

Substituting the formulas for coefficients we get

$$(4) \qquad S_n f.x = \int_X K_n(x,t) f(t) d\mu(t),$$

where

$$(5) \qquad K_n(x,t) = \sum_{k=1}^{n} f_k(x) g_k(t),$$

and

$$(6) \qquad T_n g \cdot x = \int_X g(t) K_n(t,x) \, d\mu(t).$$

The operator T_n is adjoint to S_n and $\| T_n \| = \| S_n \|$.

Certain considerations about bases in $C(X)$ can be reduced to the case of biorthogonal systems considered above. Specifically:

4.7.2. <u>Proposition</u>. Let X be a compact space and let (g_n) be a basic sequence in $C(X)$, $\| g_n \|_\infty = 1$ for $n = 1, 2, \ldots$ Then there exist: a nonnegative Radon measure μ on X and a sequence of Borel--measurable functions f_n on X such that $f_n \in L_1(X, B, \mu)$ and the sequences (g_n) and (f_n) are biorthogonal. Moreover, $\sup \| f_n \|_1 < \infty$.

Proof. Let G be the closure of the linear span of (g_n) in $C(X)$ and let (g_n^*) be the sequence in G^* biorthogonal to (g_n). Then $\sup \| g_n^* \| < \infty$. By the Hahn-Banach theorem, each g_n^* can be extended to a linear functional μ_n on $C(X)$ with $\| g_n^* \| = \| \mu_n \|$. Denote

$$\mu = \sum_{n=1}^{\infty} 2^{-n} |\mu_n|.$$

It is clear that $\| \mu \| \le \sup \| \mu_n \| < \infty$ and the Radon-Nikodym derivatives f_n of μ_n with respect to μ, $n = 1, 2, \ldots$ have the desired properties. □

We shall need estimations of the norms of the operators (1). The proposition below shows a well-known way of computing the norms of operators $L_1 \longrightarrow L_1$ represented by kernels (see Kantorovič, Vulih, Pinsker ['50], p.306 , Dunford and Schwartz ['58], VI.8.9 and V.9.54).

4.7.3. <u>Proposition</u>. Let $\mu(X) < \infty$. Let K be a product-measurable function on $X \times X$ such that

$$(7) \qquad c = \operatorname*{ess\,sup}_{t \in X} \int_X |K(x,t)| \, d\mu(x) < \infty.$$

Then the integral

$$(8) \qquad \Phi f(x) = \int_X K(x,t) f(t) \, d\mu(t)$$

exists a.e. and represents a bounded linear operator

(9) $\Phi : L^1(X) \longrightarrow L^1(X)$

and $\|\Phi\| \leq c$.

Proof. Assume (7). Then, by the Tonneli theorem,

$$\|\Phi f\|_1 = \int_X |\int_X K(x,t)f(t)d\mu(t)|d\mu(x) \leq \int_{X \times X} |K(x,t)||f(t)|d\mu(x)d\mu(t) =$$

$$= \int_X |f(t)|[\int_X |K(x,t)|d\mu(x)]d\mu(t) \leq c\|f\|_1 .$$

Hence Φ maps $L^1(X)$ into $L^1(X)$ and $\|\Phi\| \leq c$. □

4.7.4. <u>Proposition</u>. Let X be a bounded domain in $\mathbb{R}^m$ with $\lambda(\partial X)=0$ for the Lebesgue measure λ and let K be a product-measurable function on $X \times X$ such that for each f in $L^1(X)$ the integral (8) exists and represents a bounded linear operator (9). Then the number c in (7) is finite and $\|\Phi\| = c$.

Proof. Since $1_X \in L^1(X)$, the integral $\int_X |K(x,t)|dt$ is finite for x in $X \setminus \tilde{E}$ where $\tilde{E}$ is a set such that $\lambda(\tilde{E})=0$. Let f in (8) be the characteristic function of a cube $Q \subset X$ with edges of length a. Then $\|f\|_1 = a^m$ and the inequality $\|\Phi f\|_1 \leq \|\Phi\| \|f\|_1$ gives

$$(10) \quad \int_X |\frac{1}{a^m} \int_Q K(x,t)dt|dx \leq \|\Phi\| .$$

By a theorem on differentiation of indefinite integrals (see e.g. Dunford and Schwartz ['58], III.12.7), if $x \notin \tilde{E}$, then there exists a set E_x such that $\lambda(E_x)=0$ and

$$(11) \quad \lim_{a \to 0} \frac{1}{a^m} \int_{t_1}^{t_1+a} \ldots \int_{t_m}^{t_m+a} K(x,s)ds = K(x,t) \quad \text{for t in } X \setminus E_x.$$

The set E of pairs (x,t) such that (11) does not hold is product-measurable and has the property that

$$\int_X [\int_X \chi_E(x,t)dt]dx = 0.$$

Therefore $\int_X [\int_X \chi_E(x,t)dx]dt = 0$ and there exists a set $\hat{E} \subset X$ such that $\lambda(\hat{E}) = 0$ and for every t in $X \setminus \hat{E}$ the convergence (11) holds for almost all x. Hence, by Fatou lemma and by (10),

$$\int_X |K(x,t)|dx \leq \liminf_{n \to \infty} \int_X |\frac{1}{a^m} \int_t^{t+ae} K(x,s)ds|dx \leq \|\Phi\|$$

for $t \in X \setminus \hat{E}$, where $e = (1,\ldots,1) \in \mathbf{R}^m$. Thus, we get $c \leq \| \Phi \|$. $\square$

4.7.5. **Lemma.** Let $\Phi : L^1(X) \longrightarrow L^1(X)$ be a linear operator of the form (8), satisfying (7). Then the adjoint operator $\Phi^* : L^\infty(X) \longrightarrow L^\infty(X)$ is given by

$$(12) \qquad (\Phi^* g)(x) = \int_X g(t) K(t,x) d\mu(t),$$

$$(13) \qquad \| \Phi^* \| = \| \Phi \| \leq \text{ess sup}_{t \in X} \int_X |K(x,t)| d\mu(x).$$

The verification of this classical property is immediate. $\square$

Note that the kernel $K(x,t)$ in (8) is replaced by $K(t,x)$ in (12). In order to stress the switch of variables it is advisable to write the function g on the left of the kernel.

4.7.6. **Corollary.** Let X be a bounded domain in $\mathbf{R}^n$. Then the norms of the operators S_n and T_n in (1) are equal to

$$(14) \qquad \| S_n \| = \text{ess sup}_{t \in X} \int_X |K_n(x,t)| dx = \| T_n \|. \quad \square$$

4.7.7. **Corollary.** Let (f_n) in $L^1(I^m)$ and (g_n) in $L^\infty(I^m)$ be a biorthogonal system and let $\text{span}\{f_n\}_{n=1}^\infty$ be dense in $L^1(I^m)$. Then (f_n) is a basis in $L^1(I^m)$ if and only if

$$(15) \qquad \sup_n \text{ess sup}_{t \in I^m} \int_{I^m} |K_n(x,t)| dx < \infty. \quad \square$$

4.7.8. **Proposition.** Let X be a compact and let μ be a finite positive Borel measure on X. Let K be a function on $X \times X$ such that for each x in X the function $t \longrightarrow K(t,x)$ is in $L^1(X,\mu)$ and the function $h : X \longrightarrow L^1(X,\mu)$ defined as $h(x) = K(\cdot,x)$ is continuous. Then the formula

$$(16) \qquad \Psi g.x = \int_X g(t) K(t,x) d\mu(t) \qquad \text{for} \quad g \quad \text{in} \quad C(X)$$

gives a bounded linear operator $\Psi : C(X) \longrightarrow C(X)$ and

$$(17) \qquad \| \Psi \| = \sup_{x \in X} \int_X |K(x,t)| d\mu(t).$$

Proof. We have assumed that for every x in X and $\varepsilon > 0$ there exists a neighborhood U of x such that $y \in U$ implies

$$\int_X |K(t,x) - K(t,y)| d\mu(t) < \varepsilon.$$

Hence the function (16) is continuous on X (in fact; it is enough that $g \in L^\infty(X,\mu)$). If x in X is fixed, then (16) is a linear functional on $C(X)$ and its norm is equal to $\int_X |K(t,x)| d\mu(t)$. Thus

$$\|\Psi\| = \sup_{\|g\|=1} \sup_{x \in X} |\Psi g.x| = \sup_{x \in X} \sup_{\|g\|=1} |\Psi g.x| = \sup_{x \in X} \int |K(t,x)| d\mu(t).$$

The last number is finite (in virtue of compactness of X and continuity of h). $\square$

4.7.9. <u>Corollary</u>. Let X,μ be as above. Let (f_n) in $L^1(X)$ and (g_n) in $C(X)$ be a biorthogonal system such that $\text{span}\{g_n\}_{n \in \mathbb{N}}$ is dense in $C(X)$. Then (g_n) is basis in $C(X)$ if and only if

$$(18) \qquad \sup_n \sup_{x \in X} \int_X |K_n(t,x)| dt < \infty.$$

Indeed, the kernel $K_n(x,t) = \sum_{k=1}^n f_k(x) g_k(t)$ satisfies the assumptions of the preceding proposition. $\square$

4.7.10. We now assume that X is a compact metric space and μ is a probabilistic Radon measure on X such that the carrier of μ is X; this guarantees that $\|f\|_1 = \int_X |f| d\mu$ is a norm on $C(X)$.

Let $(\varphi_n)_{n \in \mathbb{N}}$ be an orthonormal system in $C(X,\mathbb{C})$. It may be regarded as a special case of a biorthogonal system of continuous functions with $f_n = \varphi_n$, $g_n = \overline{\varphi}_n$. For each fixed x in X the partial sum

$$(19) \qquad S_n(f,x) = \sum_{k=1}^n c_k(f) \varphi_k(x), \qquad \text{where} \qquad c_k(f) = \int_X f \overline{\varphi}_k d\mu$$

$(f \in C(X), \ x \in X, \ n \in \mathbb{N})$ is a linear functionsl on $C(X)$. From (4), (5), (6) we get

$$(20) \qquad S_n(f,x) = \int_X D_n(x,t) f(t) dt,$$

where $D_n(x,t) = \sum_{k=1}^n \varphi_k(x) \overline{\varphi_k(t)}$ is called the <u>Dirichlet kernel</u> with respect to $(\varphi_n)_{n \in \mathbb{N}}$. Suppose that the set $E = \text{span}\{\varphi_n\}_{n \in \mathbb{N}}$ is dense in $(C(X), \| \ \|_\infty)$ and hence, by the regularity of μ, E is dense in $L^2(X,\mu)$. If $g \in E$, then obviously $\|S_n g - g\|_\infty \longrightarrow 0$.

Let $x \in X$. Then the norm of the functional (20) on $C(X)$ is equal to

$$(21) \qquad L_n^\varphi(x) = \int_X |D_n(x,t)| d\mu(t).$$

Functions $x \longrightarrow L_n^{\varphi}(x)$, $n = 1,2,\ldots$ are called <u>Lebesgue functions</u> of the system $(\varphi_n)_{n\in N}$. By the Banach-Steinhaus theorem, for a given point x the boundedness of $L_n^{\varphi}(x)$ is a necessary and sufficient condition in order that functionals (19) converge to $f(x)$.

If we are interested in the uniform convergence of the orthonormal series, we consider the operators

$$S_n : C(X) \longrightarrow C(X)$$

given by (19). By 4.7.8,

$$(22) \qquad \| S_n \| = \sup_{x\in X} L_n^{\varphi}(x).$$

Therefore the condition $\sup_n \| S_n \| < \infty$ is necessary and sufficient in order that $\| S_n f - f \|_\infty \longrightarrow 0$ for each f in $C(X)$.

4.7.11. <u>Exercises</u>. (A) What is the measure μ and functions f_n in 4.7.2 in the case of the Faber-Schauder system in $C(I)$?

(B) Let X and μ be as in 4.7.8 and let K be continuous on $X \times X$. Show that the assumption on continuity $X \longrightarrow L^1(X)$ is fulfilled (Hint: #7.7.5).

4.8. Non-existence of bases with certain properties.

4.8.1. We have discussed examples of bases in $C(X)$ which were at the same time orthogonal with respect to a measure: the Haar system (see 2.2), the Franklin system (2.6), and their tensor products (3.6). None of them remains uniformly bounded if it is L^2-normalized. On the other hand, the well-known uniformly bounded orthogonal systems: trigonometric and Walsh, are not basic sequences in L^∞. This observation is a special case of a celebrated theorem of A.M.Olevskiĭ ['66a], ['75]): A uniformly bounded orthonormal system cannot be a basis in $C(I)$.

In this subsection we shall present various results in this line.

4.8.2. <u>Definition</u>. Let $A \subset L^1(X,B,\mu)$. The set A is <u>equiintegrable</u> if

$$(1) \qquad \underset{\varepsilon>0}{\forall} \ \underset{\delta>0}{\exists} \ \underset{E\in B}{\forall} \ \underset{f\in A}{\forall} \ [\mu(E) < \delta \implies \int_E |f| d\mu < \varepsilon].$$

This condition means the uniform μ-continuity of measures $\nu_f(E) = \int_E f d\mu$. If $\mu(X) < \infty$ and A is bounded in $L^1(X)$, condition (1) is equivalent to the uniform countable additivity of these measures, and is a necessary and sufficient condition in order that the set A be weakly sequentially compact in $L^1(X,B,\mu)$ (see Dunford and Schwartz ['58], IV.8.8).

4.8.3. <u>Theorem</u> (M.I.Kadec and A.Pełczyński ['62]). Let $A \subset L^1(X,B,\mu)$. If A fail to satisfy (1), then there exists a sequence of elements of A equivalent to the unit-vector basis of l^1.

4.8.4. <u>Theorem</u> (S.J.Szarek ['79]). Let μ be a probabilistic measure on a field B of subsets of X and let $f_n \in L^1(X)$, $g_n \in L^\infty(X)$ $(n = 1, 2, \ldots)$ be a biorthogonal system such that

(3) $\sup\limits_{n} \| g_n \|_\infty < \infty$

and

(4)   the functions f_n are equiintegrable.

Then

(5) $\sup\limits_{n} \int\limits_{X \times X} \int |K_n(x,t)| \, d\mu(x) \, d\mu(t) = \infty,$

where the kernel K_n is as in 4.7.1, i.e.

(6) $K_n(x,t) = \sum\limits_{k=1}^{n} f_k(x) g_k(t).$

For the proofs of 4.8.3 and 4.8.4 we refer the reader to the original papers.

We shall show that the above theorem of Szarek is a powerful result leading easily to many significant properties of bases.

4.8.5. <u>Corollary</u> (Szarek ['79]). No L^1-normalized basis in $L^1(I^m)$ consists of equiintegrable functions; moreover, if μ is an atomless probability measure, then every L^1-normalized basis (f_n) in $L^1(X,\mu)$ contains a subsequence (f_{n_k}) equivalent to the unit--vector basis of l^1.

Proof. Suppose that (f_n) is a basis in $L^1(I^m)$ and $\| f_n \|_1 = 1$. Then, by 1.1.3, the biorthogonal system (g_n) in $L^\infty(I^m)$ satisfies (3). On the other hand, by 4.7.6 we get

(7) $\sup\limits_{n} \operatorname{ess\,sup}\limits_{t \in I^m} \int\limits_{I^m} |K_n(x,t)| \, dx < \infty.$

If (f_n) were equiintegrable, then (5) would contradict (7). The remaining statement follows from 4.8.3 and form the well-known isomorphism $L^1(X,\mu) \equiv L^1(I)$. $\square$

4.8.6. Corollary (S.Kwapień and S.J.Szarek ['79], Szarek ['79]). There is no normalized order-bounded basis in $L^1(X,\mu)$, where μ is an atomless probability measure on X.

Indeed, any order-bounded set A (i.e., such that all f in A satisfy $|f| \leq h$ for some h in $L^1(I^m)$) is equiintegrable. $\square$

4.8.7. Corollary (S.Kwapień and S.J.Szarek ['79]). Let (g_n) be a normalized basis in C(X). Then the biorthogonal sequence of measures (μ_n) is not order-bounded in M(X).

The argument is similar to those in 4.8.5 and 4.8.6, based on 4.7.2 and 4.7.9, with variables x,t switched in (7). $\square$

4.8.8. Corollary (S.J.Szarek ['79]). Let (g_n) be a normalized basis in C(X) with a biorthogonal sequence (μ_n) in M(X). Then there exists $n_1 < n_2 < \ldots$ such that every sequence (c_k) of numbers tending to 0 is the sequence of coefficients b_{n_k} for some $g = \Sigma b_n g_n$ in C(X), i.e., the operator

$$(8) \qquad g \longrightarrow T(g) = (\mu_{n_k}(g))_{k \in \mathbb{N}}$$

transforms C(X) onto $C_0(\mathbb{N})$.

Proof. By 4.7.2, there exist: a finite positive measure μ in M(X) and functions f_n in $L^1(X,\mu)$ such that $d\mu_n = f_n d\mu$. By 4.8.4 and 4.7.9, the functions f_n are not equiintegrable and hence, by the Kadec-Pełczyński criterion (4.8.3), a subsequence (f_{n_k}) is equivalent to the unit-vector basis (e_n) in l^1. The operator adjoint to the operator

$$(9) \qquad T : C(X) \longrightarrow C_0(\mathbb{N})$$

given by (8) is the operator $T^* : l^1 \longrightarrow M(X)$ given by

$$T^* a.g = \sum_{k=1}^{\infty} a_k \mu_{n_k}(g) \quad \text{for} \quad a = (a_n) \text{ in } l^1, \text{ g in C(X)}.$$

The equivalence of sequences (f_{n_k}) and (e_n) means that T^* is an isomorphic embedding. Consequently, the operator (9) is a surjection. $\square$

4.8.9. Corollary (Olevskii theorem). A uniformly bounded orthonormal system cannot be a basis in C(X).

Proof. Let (φ_n) be a basis in $C(X)$ orthonormal with respect to some measure $\mu \geq 0$ and let $c_n = \|\varphi_n\|_\infty$. Obviously, $c_n \geq 1$. Suppose that $c_n \leq c$ for some constant c. Then $g_n = c_n^{-1}\varphi_n$ would satisfy the assumptions of 4.8.8 and for some subsequence (g_{n_k}) the operator (8) would map $C(X)$ onto $C_o(N)$. This, however, is impossible, since (φ_n) is an orthonormal system and the range of the operator (8) is l^2. □

4.8.10. **Definitions**. Let $1 \leq p < \infty$. A seminormalized sequence (f_n) is called p-Besselian if there exists a constant $c > 0$ such that for any finite sequence (a_n) of scalars

$$(11) \quad \|\sum_n a_n f_n\| \geq c(\sum_n |a_n|^p)^{1/p}.$$

(f_n) is called p-Hilbertian if

$$(12) \quad \|\sum_n a_n f_n\| \leq c(\sum_n |a_n|^p)^{1/p}.$$

Analogously we define the two concepts for $p = \infty$ replacing the l^p-norm by l^∞-norm.

4.8.11. **Corollary** (Szarek ['79], ['80]). Let $1 \leq p < \infty$. Then there is no p-Besselian seminormalized basis in $C(X)$.

Indeed, if (g_n) were such a basis, then for each $g = \sum b_n g_n$ in $C(X)$ and each subsequence (g_{n_k}) we would have

$$(\sum_{k=1}^\infty |b_{n_k}|^p)^{1/p} \leq (\sum_{n=1}^\infty |b_n|^p)^{1/p} \leq \frac{1}{c}\|g\|$$

and this would contradict 4.8.8. □

4.8.12. **Corollary** (Szarek ['79], ['80]). Let $1 < q \leq \infty$. Then there is no q-Hilbertian basis in $L^1(I^m)$.

Indeed, if (f_n) were a q-Hilbertian basis, by 4.8.5 we would have a constant c' and a subsequence (f_{n_k}) such that

$$c(\sum |a_k|^q)^{1/q} \geq \|\sum a_k f_{n_k}\| \geq c'\sum|a_k|$$

for any scalars (a_n). □

4.8.13. **Exercises**. (A) Find explicitly a subsequence of L^1-normalized Haar functions which is equivalent to the unit-vector basis in l^1.

(B) Find explicitly a subsequence χ_{n_k} of the L^∞-normalized

Haar system such that for every (c_n) in $C_o(\mathbb{N})$ the series $\Sigma a_k \chi_{n_k}$ is convergent in L .

(C) Let $\psi_1 = \chi_1$, $\psi_2 = \chi_2$, $\psi_{2^n+k} = \chi_k r_{n+1}$ where (χ_k) and (r_n) are the L^2-normalized Haar and Rademacher systems in $C(I)$, $n = 1, 2, \ldots$ $k = 1, \ldots, 2^n$. Show that:

(a) (ψ_m) is orthonormal on $[0,1]$,

(b) The functions ψ_m $(2^n < m \leq 2^{m+1})$ are linear combinations of Haar functions χ_{2^n+k}, $k = 1, \ldots, 2^n$, and vice versa. Moreover

$$\sum_{k=1}^{m} \psi_{2^n+k}(x) \psi_{2^n+k}(t) = r_{n+1}(x) r_{n+1}(t) \sum_{k=1}^{m} \chi_k(x) \chi_k(t).$$

(c) The Lebesgue functions: $L^\psi_{2^n}$ for (ψ_m) and $L^\chi_{2^n}$ for (χ_m) coincide; for the remaining indices, $L^\psi_{2^n+k}(x) \leq L^\psi_{2^n}(x) L^\chi_k(x) \leq 2$.

(d) If $f \in C(I)$, then the expansion of f with respect to (ψ_m) converges uniformly to f.

(e) $(\psi_m)_{m \in \mathbb{N}}$ contains a uniformly bounded subsequence (ψ_{m_k}).

(f) If $(\psi_m)_{m \in \mathbb{N}}$ is L^∞-normalized, then the sequence $n_1 < n_2 < \ldots$ of indices in 4.8.8 is disjoint with (m_k) except of a finite number of terms.

4.8.14. <u>Notes</u>. In Olevskiǐ ['75] and Szarek ['79], ['80], 4.8.9 appears to be formulated and proved for $X = [0,1]$.

For the basis (ψ_n) in Exercise (C) see Olevskiǐ ['66b] and ['75], p.22.

P.Wojtaszczyk ['73] proved that $C(I)$ has a basis which is p-Hilbertian for all $p < \infty$ and $L^1(I)$ has a basis which is p-Besselian for all $p > 1$.

J.Lindenstrauss and A.Pełczyński ['71] proved that every basis (e_n) in $C(X)$ has the following, surprising property: If $C(X)$ is isometrically isomorphic to a subspace of a Banach space G with a basis (g_n), then there exists a basis (h_n) in G with

$$h_n = \sum_{j=k_n+1}^{k_{n+1}} a_j g_j, \quad k_1 < k_2 < \ldots$$

equivalent to the basis (e_n), i.e., such that

$$c^{-1} \left\| \sum_{i=1}^{m} a_i e_i \right\| \leq \left\| \sum_{i=1}^{m} a_i h_i \right\| \leq c \left\| \sum_{i=1}^{m} a_i e_i \right\|$$

for some $c > 0$; if (e_n) is monotone, such (h_n) can be found with $c < 1 + \varepsilon$.

H.E.Warren ['71] constructed a basis of polygonal functions in $C(I)$ such that $\|e_n\| = 1$ and $e_n \longrightarrow 0$ weakly. P.Wojtaszczyk ['73] proved the existence of such a basis in each separable infinite-dimensional space $C_0(X)$ (and, more generally, in each Banach space with a basis, containing a subspace isomorphic to $C_0(\mathbf{N})$).

BIBLIOGRAPHY

F.G.Arutjunjan ['72] Bases of the spaces $L_1[0,1]$ and $C[0,1]$(Russian), Mat.Zametki 11(1972), 241 - 249 ; Math.Notes 11(1972), 152 - 157.

L.A.Balašov and A.I.Rubinšteǐn ['71] Series with respect to the Walsh system and their generalizations,in : Itogi Nauki, Mat. Analiz 1970, Moscow 1971, 147 - 202.

S.Banach ['32] Théorie des opérations linéaires, Monogr. Mat., Warszawa - Lwów, 1932.

K.G.Beauchamp ['75] Walsh functions and their applications, Academic Press, London - New York, 1975.

C.Bessaga ['57] Bases in certain spaces of continuous functions,Bull. Acad.Polon. Sci.Cl.III, 5(1957), 11 - 14.

C.Bessaga and A.Pełczyński ['60] Banach spaces of continuous functions IV, Studia Math.19(1960), 53 - 62.

S.V.Bočkarev ['74] Existence of a basis in the space of functions analytic in a disc and some properties of the Franklin system (Russian), Mat.Sbornik 95(1974), 3 - 18.

R.A.Bonic ['69] Linear functional analysis, Gordon and Breach, New York - London - Paris 1969.

R.Bonic, J.Frampton and A.Tromba ['69], Λ - manifolds, J.Functional Analysis 3(1969),310 - 320.

C.de Boor ['78] A practical guide to splines, Springer - Verlag, New York - Heidelberg - Berlin, 1978.

K.Borsuk ['33], Über Isomorphie der Funktionalräume, Bull. Internat. Acad.Polon.Sci. 1933, 1 - 10.

Z.A.Čanturija ['67], A generalization of G.Faber's theorem on polynomial bases (Russian), Mat.Zametki 2(1967), 187 - 190.

— ['71] On bases in $C(0,1)$ and $L(0,1)$, Bull.Acad.Polon.Sci., Seř.Math. Astronom.Phys. 19(1971), 1013 - 1021.

— ['72] Bases in the space of continuous functions (Russian), Mat.Sbornik 88(1972), 589 - 608.

— ['73] Orthogonal polynomial bases (Russian), Dokl.Akad.Nauk SSSR 208(1973), 785 - 786.

Z.Ciesielski ['59] On Haar functions and on the Schauder basis of the space $C<0,1>$, Bull.Acad.Polon. Sci.Ser. math. astr.phys. 7(1959), 227 - 232.

— ['60a] Some properties of Schauder basis of the space $C<0,1>$, ibidem, 3(1960), 141 - 144.

— ['60b] On the isomorphisms of the spaces H_α and m, ibidem, 4(1960), 217 - 222.

— ['61] Hölder conditions for realizations of Gaussian processes, Trans.Amer.Math.Soc. 99(1961), 403 - 413.

— ['63] Properties of the orthonormal Franklin system, Studia Math. 23(1963), 141 - 157.

— ['66] Properties of the orthonormal Franklin system II, ibidem 27 (1966), 289 - 323.

—— ['69] A construction of basis in $C^{(1)}(I^2)$, ibidem 33 (1969), 243 - 247.

—— ['74] Bases and approximation by splines, Proc.Internat. Congress Math.Vancouver 1974, 47 - 51.

—— ['75] The C(I) norms and orthogonal projections onto the subspace of polygonals, Trudy Mat.Inst.Steklov 134(1975), 415 - 418.

Z.Ciesielski and J.Domsta ['71] Construction of an orthonormal basis in $C^m(I^d)$ and $W_p^m(I^d)$, Studia Math. 41(1972),221 - 224.

Z.Ciesielski and T.Figiel['82] Spline approximation and Besov spaces on compact manifolds, Studia Math., to appear

—— ['83] Bases in Besov and Sobolev spaces on manifolds, in preparation.

I.K.Daugavet ['77] Introduction to the theory of approximation of functions (Russian), Izd.Leningr.Univ., Leningrad 1977.

M.M.Day ['73] Normed linear spaces, 3rd edition, Springer - Verlag, Berlin - Heidelberg - New York, 1973.

N.Dunford and J.T.Schwartz ['58], Linear operators, vol.I, Interscience, New York 1958.

H.W.Ellis and D.G.Kuehner ['60] On Schauder bases for spaces of continuous functions, Canad.Math.Bull.3 (1960), 173 - 184.

G.Faber ['10] Über die Orthogonalfunktionen des Herrn Haar, Jahresber.Deutsch.Math.Verein. 19(1910), 104 - 112.

—— ['14] Über die interpolatorische Darstellung stetiger Funktionen, ibidem 23(1914), 192 - 210.

C.Foiaş and I.Singer ['65], On bases in C([0,1]) and $L^1([0,1])$, Rev.Roumaine Math.Pures Appl. 10(1965), 913 - 960.

P.H.Franklin ['28] A set of continuous orthogonal functions, Math. Annalen 100(1928), 522 - 529.

B.Gelbaum and J.Gil de Lamadrid ['61] Bases of tensor products of Banach spaces, Pacific J.Math. 11(1961), 1281 - 1286.

B.G.Golubov ['71] Series with respect to the Haar system (Russian), Itogi Nauki, Matem. Analiz 1970, Moscow 1971, 109 - 146.

B.Grünbaum ['67] Convex polytopes, Interscience Publishers, London - York - Sydney, 1967.

V.I.Gurariĭ ['63] On bases in spaces of continuous functions (Russian), Doklady AN SSSR 148 (1963), 493 - 495

—— ['66a] Bases in spaces of continuous functions on compacta and some geometrical questions (Russian), Izvestia AN SSSR, Seria mat. 30(1966), 289 - 306.

—— ['66b] Subspaces and bases in spaces of continuous functions (Russian), Doklady AN SSSR 167(1966), 917 - 973.

L.A.Gurevič ['54] On a basis in the space of continuous functions defined on a closed bounded subset of the n - dimensional space (Russian), Trudy Voronež.Gos.Univ. 27(1954), 84 - 87.

A.Haar ['10] Zur Theorie der orthogonalen Funktionensysteme, Math. Annalen 69(1910), 331 - 371.

K.Hoffman ['62] Banach spaces of analytic functions, Prentice - Hall, Englewood Cliffs, N.J., 1962.

J.R.Holub ['70], Concerning biorthogonal sequences in C(S), Comment, Math.(Prace Mat.) 13(1970), 275 - 279.

S.Kaczmarz ['29] Über ein Orthogonalsystem, Comptes Rendus du I Congrès Math. du Pays Slaves, Warszawa 1929, 189 - 192.

S.Kaczmarz and H.Steinhaus ['35] Theorie der Orthogonalreihen, Monogr. Mat., Warszawa - Lwów 1935.

M.I.Kadec ['74] The relative projection constants, and a certain theorem of Z.A.Čanturija (Russian),Sakharth.SSR Mecn.Akad. Moambe 74(1974), 541 - 543 (MR 53,#11345).

M.I.Kadec and A.Pełczyński ['62] On bases, lacunary sequences and complemented subspaces in the spaces L^p, Studia Math. 21(1962), 161 - 176.

L.V.Kantorovič, B.Z.Vulih, A.G.Pinsker ['50] Functional analysis in partially ordered spaces (Russian), Gosud.Izd.Tehn - Teor. Lit.Moscow 1950.

J.L.Kelley ['55] General topology, Van Nostrand,Princeton 1955.

S.Kwapień and S.J.Szarek ['79] An estimation of the Lebesgue functions of biorthogonal system with an application to the non - existence of some bases in C and L^1, Studia Math. 66(1979), 185 - 200.

H.E.Lacey ['74] The isometric theory of classical Banach spaces, Springer Verlag, Berlin - Heidelberg - New York 1974

J.Lindenstrauss and A.Pełczyński ['68] Absolutely summing operators in $\mathcal{L}_p$ spaces and their applications, Studia Math. 29(1968), 275 - 326

—— ['71] Contributions to the theory of the classical Banach spaces, J.Functional Analysis 8(1971), 225 - 249.

J.Lindenstrauss and L.Tzafriri ['77] Classical Banach spaces I, Springer - Verlag, Berlin - Heidelberg - New York 1977.

—— ['79] Classical Banach spaces II, ibidem 1979.

J.Marcinkiewicz ['37] Quelques théorèmes sur les séries orthogonales, Ann.Soc.Polon.Math. 16(1937), 85 - 96.

V.A.Matveev ['67] Series in a Schauder system (Russian), Mat.Zametki 2(1967), 267 - 278.

M.A.Meletidi ['71] Bases in the space C and L^p(Russian),Mat. Zametki 10(1971), 635 - 640 ; Math.Notes 10(1971), 812 - 815.

E.Michael and A.Pełczyński ['66] Separable Banach spaces which admit l_n^∞ approximations,Israel J.Math. 4(1966), 189 - 198.

—— ['67] Peaked partition subspaces of C(X), Illinois J.Math. 11(1967), 555 - 562.

A.A.Miliutin ['66] Isomorphism of spaces of continuous functions on compacta of the power of the continuum (Russian), Teor.Funkcii Funkcional.Anal. i Priložen. 2(1966), 150 - 156.

I.P.Natanson ['49] Constructive theory of functions (Russian), Gos.Izd.Tehn.Teor.Lit.,Moscow 1949.

A.M.Olevskiǐ ['66a] Fourier series of continuous functions with respect to bounded orthonormal systems (Russian),Izv.AN SSSR Ser.Mat., 30(1966), 387 - 432.

—— ['66b] A certain orthonormal system and its applications, Mat. Sb. 71(113)(1966), 297 - 336.

—— ['70] The stability of the Schmidt orthogonalization operator (Russian), Izv. Akad.Nauk SSSR Ser.Math. 34(1970), 803 - 826.

—— ['75] Fourier series with respect to general orthogonal system systems, Ergebn.Mat., Springer Verlag,Berlin - Heidelberg - New York 1975.

W.Orlicz ['55] On perfectly convergent series in some function spaces, (Polish), Prace Mat. 1(1955), 393 - 414.

K.I.Oskolkov ['79] The upper bound of the norms of orthogonal projections onto subspaces of polygonals ,in: Approximation Theory, Banach Center Publ. 4(1979), 177 - 183.

P.Oswald ['77] On the C norm of orthogonal projections onto the subspace of polygonals, Mat. Zametki (Math.Notes) 21(1977), 495 - 502.

S.Paszkowski ['55] Sur l'approximation uniforme avec noeuds, Ann.Polonici Math. 2(1955), 129 - 146.

—— ['56] On the Weierstrass approximation theorem, Colloq. Math. 4(1956),206 - 210.

—— ['57] On approximation with nodes, Rozprawy Mat. 14(1957), 1 - 63.

A.Pełczyński ['58] On the isomorphism of the spaces M and m, Bull.Acad.Polon.Sci. 6(1958),695 - 696.

—— ['60] Projections in certain Banach spaces, Studia Math.19(1960), 209 - 228.

—— ['64] On simultaneous extension of continuous functions, Studia Math. 24(1964), 285 - 304.

—— ['79] Some aspects of the present theory of Banach spaces, in : S.Banach,Oeuvres, vol.2,PWN, Warszawa 1979, pp. 223 - 302.

A.Pełczyński and W.Szlenk ['65] An example of a non - shrinking basis, Rev. Roumaine Math.Pures Appl. 10(1965),961 - 966.

J.Radecki ['70] Orthogonal basis in the space $C_1[0,1]$, Studia Math. 35(1970), 123 - 163.

J.Ryll ['73] Schauder bases for the space of continuous functions on an n - dimensional cube, Commentationes Math.(Prace Mat.) 17 (1973), 201 - 213.

—— ['78] Interpolating bases for spaces of differentiable functions, Studia Math.63(1978), 125 - 144.

—— ['80] Hermite interpolating splines of several variables, in: Constructive function theory '77, Sofia 1980, 475 - 478.

K.M.Šaĭdukov ['57] The existence of an orthonormal basis in the class of polynomials (Russian),Naučn.Trudy Kazanskogo Inst. Inž.Stroit.Neft.Prom. 5(1957), 119 - 151.

R.Schatten ['50] A theory of cross - spaces, Annals of Math. Studies no 26, Princeton 1950.

J.P.Schauder ['27] Zur Theorie stetiger Abbildungen in Funktionalräumen, Math.Zeitschr. 26(1927), 47 - 65.

S.Schoenfeld ['69] Schauder bases in spaces of differentiable functions, Bull.Amer.Math.Soc. 75(1969), 586 - 590.

—— ['72] Schauder bases in the space $C^k(T^q)$, Trans. Amer.Math. Soc. 165 (1972), 309 - 318.

L.Schwartz ['43] Etude de sommes d'exponentielles réelles, Paris 1943.

Z.Semadeni ['63a] Isomorphic properties of Banach spaces of continuous functions, Studia Math. (seria specjalna), 1(1963), 93 - 108.

—— ['63b] Product Schauder bases and approximation with nodes in spaces of continuous functions, Bull.Acad Polon.Sci. 11(1963), 387 - 391.

—— ['71] Banach spaces of continuous functions,vol.1, Monogr. Mat., PWN,Warszawa 1971.

W.Sierpiński ['10] L'influence de l'ordre des termes sur la convergence uniforme d'une série (Polish, French summary), Spraw. Tow.Nauk. Warsz. (Comptes Rendus des Séances de la Société Scientifique de Varsovie) 3(1910), 354 - 357.

I.Singer ['70] Bases in Banach spaces I, Springer - Verlag, Berlin - Heidelberg - New York 1970.

—— ['81] Bases in Banach spaces II, ibidem 1981.

S.J.Szarek ['80] Non existence of Besselian basis in C(S), J. Functional Analysis 37(1980), 56 - 67.

—— ['79] Bases and biorthogonal sequences in the spaces C and L^1, Arkiv för Mat. 17(1979), 255 - 271.

W.Szlenk ['66] Une remarque sur l'ortogonalisation des bases de Schauder dans l'espace C,Colloq.Math. 15(1966), 297 - 301.

G.E.Tkebučava ['77] On the order of the growth of powers of lacunary algebraic basis in the space $C^m(I^d)$, Bull.Acad.Polon.Sci, Ser.Math.Astronom.Phys. 25(1977),461 - 464.

P.L.Ul'janov ['70] Certain properties of series in Schauder system (Russian), Mat.Zametki (Math.Notes) 7(1970), 431 - 442.

F.S.Vaher ['55] On a basis in the space of continuous functions defined on a compactum (Russian), Doklady AN SSSR 101(1955) 589 - 592.

J.L.Walsh ['23] A closed set of normal orthogonal functions, Amer. J.Math. 55(1923), 5 - 24.

H.E.Warren ['71] A special basis for C[0,1],Proc.Amer.Math.Soc. 27(1971), 495 - 499.

P.Wojtaszczyk ['73] Existence of some special bases in Banach spaces, Studia Math. 47(1973), 83 - 93.

W.Wolibner ['51] Sur un polynôme d'interpolation, Colloq.Math. 2(1951), 136 - 137.

Z.Wronicz ['79] On the application of the orthonormal Franklin system to the approximation of analytic functions, in : Approximation Theory, Banach Center Publ. 4(1979), 305 - 316.

H.Yamabe ['50] On an extension of the Helly's theorem, Osaka Math.J. 2(1950), 15 - 17.

INDEX OF SYMBOLS

$\mathbf{N} = \{1,2,\ldots\}$, $\mathbf{Z}$ - the set of integers,

$\mathbf{R}$ = the set of real numbers, $\mathbf{I} = [0,1]$,

$\mathbf{C}$ = the set of complex numbers,

$D_n, D, \tilde{D}, \hat{D}$ - various sets of dyadic numbers (see 2.1.1),

$C(X)$, $C^{(p)}(x)$, $C_o(X)$ - the space of continuous [with continuous p-th derivatives, continuous vanishing at ∞] real-valued functions on X (see 1.2.1, 3.9.1),

$\| \; \|$ - a norm on a generic Banach space, the supremum norm in a function space, the variation norm in case of a space of measures

$M(X)$ - the space of Radon measures on X

$L^p(X,\mu) = L^p(X) = \{f : \| f \|_p = (\int_X |f|^p d\mu)^{1/p} < \infty\}$, $1 \leq p < \infty$,

$L^\infty = \{f : \| f \|_\infty = \text{ess sup } |f| < \infty\}$,

$l_n^p = L^p\{1,\ldots,n\}$, $l^p = L^p(\mathbf{N})$,

$\text{diam } X = $ diameter of X, $\text{dist}(x,Y) = \inf\{\rho(x,y) : y \in Y\}$,

$\text{span } A = $ the linear span of the set A ,

$\text{vert}(T) = $ the set of vertices of T ,

ω - the smallest infinite ordinal (and cardinal) number,

$\omega_f(\delta) = \sup\{|f(x) - f(y)| : \rho(x,y) \leq \delta\}$ - modulus of continuity,

δ_{nm} - Kronecker's delta

$d_n = \sup\{\text{diam } \Delta : \Delta \in T_n\}$

Vol. 759: R. L. Epstein, Degrees of Unsolvability: Structure and Theory. XIV, 216 pages. 1979.

Vol. 760: H.-O. Georgii, Canonical Gibbs Measures. VIII, 190 pages. 1979.

Vol. 761: K. Johannson, Homotopy Equivalences of 3-Manifolds with Boundaries. 2, 303 pages. 1979.

Vol. 762: D. H. Sattinger, Group Theoretic Methods in Bifurcation Theory. V, 241 pages. 1979.

Vol. 763: Algebraic Topology, Aarhus 1978. Proceedings, 1978. Edited by J. L. Dupont and H. Madsen. VI, 695 pages. 1979.

Vol. 764: B. Srinivasan, Representations of Finite Chevalley Groups. XI, 177 pages. 1979.

Vol. 765: Padé Approximation and its Applications. Proceedings, 1979. Edited by L. Wuytack. VI, 392 pages. 1979.

Vol. 766: T. tom Dieck, Transformation Groups and Representation Theory. VIII, 309 pages. 1979.

Vol. 767: M. Namba, Families of Meromorphic Functions on Compact Riemann Surfaces. XII, 284 pages. 1979.

Vol. 768: R. S. Doran and J. Wichmann, Approximate Identities and Factorization in Banach Modules. X, 305 pages. 1979.

Vol. 769: J. Flum, M. Ziegler, Topological Model Theory. X, 151 pages. 1980.

Vol. 770: Séminaire Bourbaki vol. 1978/79 Exposés 525–542. IV, 341 pages. 1980.

Vol. 771: Approximation Methods for Navier-Stokes Problems. Proceedings, 1979. Edited by R. Rautmann. XVI, 581 pages. 1980.

Vol. 772: J. P. Levine, Algebraic Structure of Knot Modules. XI, 104 pages. 1980.

Vol. 773: Numerical Analysis. Proceedings, 1979. Edited by G. A. Watson. X, 184 pages. 1980.

Vol. 774: R. Azencott, Y. Guivarc'h, R. F. Gundy, Ecole d'Eté de Probabilités de Saint-Flour VIII-1978. Edited by P. L. Hennequin. XIII, 334 pages. 1980.

Vol. 775: Geometric Methods in Mathematical Physics. Proceedings, 1979. Edited by G. Kaiser and J. E. Marsden. VII, 257 pages. 1980.

Vol. 776: B. Gross, Arithmetic on Elliptic Curves with Complex Multiplication. V, 95 pages. 1980.

Vol. 777: Séminaire sur les Singularités des Surfaces. Proceedings, 1976-1977. Edited by M. Demazure, H. Pinkham and B. Teissier. IX, 339 pages. 1980.

Vol. 778: SK1 von Schiefkörpern. Proceedings, 1976. Edited by P. Draxl and M. Kneser. II, 124 pages. 1980.

Vol. 779: Euclidean Harmonic Analysis. Proceedings, 1979. Edited by J. J. Benedetto. III, 177 pages. 1980.

Vol. 780: L. Schwartz, Semi-Martingales sur des Variétés, et Martingales Conformes sur des Variétés Analytiques Complexes. XV, 132 pages. 1980.

Vol. 781: Harmonic Analysis Iraklion 1978. Proceedings 1978. Edited by N. Petridis, S. K. Pichorides and N. Varopoulos. V, 213 pages. 1980.

Vol. 782: Bifurcation and Nonlinear Eigenvalue Problems. Proceedings, 1978. Edited by C. Bardos, J. M. Lasry and M. Schatzman. VIII, 296 pages. 1980.

Vol. 783: A. Dinghas, Wertverteilung meromorpher Funktionen in ein- und mehrfach zusammenhängenden Gebieten. Edited by R. Nevanlinna and C. Andreian Cazacu. XIII, 145 pages. 1980.

Vol. 784: Séminaire de Probabilités XIV. Proceedings, 1978/79. Edited by J. Azéma and M. Yor. VIII, 546 pages. 1980.

Vol. 785: W. M. Schmidt, Diophantine Approximation. X, 299 pages. 1980.

Vol. 786: I. J. Maddox, Infinite Matrices of Operators. V, 122 pages. 1980.

Vol. 787: Potential Theory, Copenhagen 1979. Proceedings, 1979. Edited by C. Berg, G. Forst and B. Fuglede. VIII, 319 pages. 1980.

Vol. 788: Topology Symposium, Siegen 1979. Proceedings, 1979. Edited by U. Koschorke and W. D. Neumann. VIII, 495 pages. 1980.

Vol. 789: J. E. Humphreys, Arithmetic Groups. VII, 158 pages. 1980.

Vol. 790: W. Dicks, Groups, Trees and Projective Modules. IX, 127 pages. 1980.

Vol. 791: K. W. Bauer and S. Ruscheweyh, Differential Operators for Partial Differential Equations and Function Theoretic Applications. V, 258 pages. 1980.

Vol. 792: Geometry and Differential Geometry. Proceedings, 1979. Edited by R. Artzy and I. Vaisman. VI, 443 pages. 1980.

Vol. 793: J. Renault, A Groupoid Approach to C*-Algebras. III, 160 pages. 1980.

Vol. 794: Measure Theory, Oberwolfach 1979. Proceedings 1979. Edited by D. Kölzow. XV, 573 pages. 1980.

Vol. 795: Séminaire d'Algèbre Paul Dubreil et Marie-Paule Malliavin. Proceedings 1979. Edited by M. P. Malliavin. V, 433 pages. 1980.

Vol. 796: C. Constantinescu, Duality in Measure Theory. IV, 197 pages. 1980.

Vol. 797: S. Mäki, The Determination of Units in Real Cyclic Sextic Fields. III, 198 pages. 1980.

Vol. 798: Analytic Functions, Kozubnik 1979. Proceedings. Edited by J. Ławrynowicz. X, 476 pages. 1980.

Vol. 799: Functional Differential Equations and Bifurcation. Proceedings 1979. Edited by A. F. Izé. XXII, 409 pages. 1980.

Vol. 800: M.-F. Vignéras, Arithmétique des Algèbres de Quaternions. VII, 169 pages. 1980.

Vol. 801: K. Floret, Weakly Compact Sets. VII, 123 pages. 1980.

Vol. 802: J. Bair, R. Fourneau, Etude Géometrique des Espaces Vectoriels II. VII, 283 pages. 1980.

Vol. 803: F.-Y. Maeda, Dirichlet Integrals on Harmonic Spaces. X, 180 pages. 1980.

Vol. 804: M. Matsuda, First Order Algebraic Differential Equations. VII, 111 pages. 1980.

Vol. 805: O. Kowalski, Generalized Symmetric Spaces. XII, 187 pages. 1980.

Vol. 806: Burnside Groups. Proceedings, 1977. Edited by J. L. Mennicke. V, 274 pages. 1980.

Vol. 807: Fonctions de Plusieurs Variables Complexes IV. Proceedings, 1979. Edited by F. Norguet. IX, 198 pages. 1980.

Vol. 808: G. Maury et J. Raynaud, Ordres Maximaux au Sens de K. Asano. VIII, 192 pages. 1980.

Vol. 809: I. Gumowski and Ch. Mira, Recurences and Discrete Dynamic Systems. VI, 272 pages. 1980.

Vol. 810: Geometrical Approaches to Differential Equations. Proceedings 1979. Edited by R. Martini. VII, 339 pages. 1980.

Vol. 811: D. Normann, Recursion on the Countable Functionals. VIII, 191 pages. 1980.

Vol. 812: Y. Namikawa, Toroidal Compactification of Siegel Spaces. VIII, 162 pages. 1980.

Vol. 813: A. Campillo, Algebroid Curves in Positive Characteristic. V, 168 pages. 1980.

Vol. 814: Séminaire de Théorie du Potentiel, Paris, No. 5. Proceedings. Edited by F. Hirsch et G. Mokobodzki. IV, 239 pages. 1980.

Vol. 815: P. J. Slodowy, Simple Singularities and Simple Algebraic Groups. XI, 175 pages. 1980.

Vol. 816: L. Stoica, Local Operators and Markov Processes. VIII, 104 pages. 1980.